ZOOBIQUITY

ZOOBIQUITY

What Animals Can Teach Us

About Health and the Science of Healing

BARBARA NATTERSON-HOROWITZ, M.D.
AND KATHRYN BOWERS

ALFRED A. KNOPF · NEW YORK · 2012

THIS IS A BORZOI BOOK
PUBLISHED BY ALFRED A. KNOPF

www.aaknopf.com

Knopf, Borzoi Books, and the colophon are
registered trademarks of Random House, Inc.

Library of Congress Cataloging-in-Publication Data
Natterson-Horowitz, Barbara.
Zoobiquity / Barbara Natterson-Horowitz and Kathryn Bowers.—1st ed.
p. cm.
Includes bibliographical references and index.
ISBN 978-0-307-59348-1
I. Bowers, Kathryn.   II. Title.
[DNLM: 1. Disease Models, Animal.   2. Pathology.   3. Pathology, Veterinary.
4. Physiology, Comparative.   5. Psychology, Comparative.   QZ 33]
636.089607—dc23   2012005051

Jacket images: (top) © Jill Greenberg, courtesy of ClampArt Gallery,
NYC; (bottom) © David Noton Photography / Alamy

Jacket design by Chip Kidd

Manufactured in the United States of America

First Edition

For Zach, Jenn, and Charlie—BNH

For Andy and Emma—KSB

Contents

Authors' Note   ix

ONE   Dr. House, Meet Doctor Dolittle   3
Redefining the Boundaries of Medicine

TWO   The Feint of Heart   19
Why We Pass Out

THREE   Jews, Jaguars, and Jurassic Cancer   31
New Hope for an Ancient Diagnosis

FOUR   Roar-gasm   55
An Animal Guide to Human Sexuality

FIVE   Zoophoria   87
Getting High and Getting Clean

SIX   Scared to Death   109
Heart Attacks in the Wild

SEVEN   Fat Planet   132
Why Animals Get Fat and How They Get Thin

EIGHT **Grooming Gone Wild** 159
Pain, Pleasure, and the Origins of Self-Injury

NINE **Fear of Feeding** 176
Eating Disorders in the Animal Kingdom

TEN **The Koala and the Clap** 194
The Hidden Power of Infection

ELEVEN **Leaving the Nest** 212
Animal Adolescence and the Risky Business of Growing Up

TWELVE **Zoobiquity** 234

Acknowledgments 245

Notes 249

Index 293

Authors' Note

Although this work is a journalistic collaboration between two authors, we chose for stylistic reasons to write the book from Dr. Natterson-Horowitz's point of view. We felt her journey from focusing solely on human medicine to a broader, species-spanning approach demanded a first-person narrative structure. Most interviews in the book were conducted by both authors, although in a few cases only one author did the questioning. The final book is the result of a true partnership not just between Dr. Natterson-Horowitz and Ms. Bowers but among the many physicians, veterinarians, biologists, researchers, other dedicated professionals, and patients (whose names we've changed where necessary) who so generously shared their time, scholarship, and experiences with us.

ZOOBIQUITY

Dr. House, Meet Doctor Dolittle

Redefining the Boundaries of Medicine

In the spring of 2005, the chief veterinarian of the Los Angeles Zoo called me, an urgent edge to his voice.

"Uh, listen, Barbara? We've got an emperor tamarin in heart failure. Any chance you could come out today?"

I reached for my car keys. For thirteen years I'd been a cardiologist treating members of my own species at the UCLA Medical Center. From time to time, however, the zoo veterinarians asked me to weigh in on some of their more difficult animal cases. Because UCLA is a leading heart-transplant hospital, I'd had a front-row view of every type of human heart failure. But heart failure in a tamarin—a tiny, nonhuman primate? That I'd never seen. I threw my bag in the car and headed for the lush, 113-acre zoo nestled along the eastern edge of Griffith Park.

Into the tiled exam room the veterinary assistant carried a small bundle wrapped in a pink blanket.

"This is Spitzbuben," she said, lowering the animal gently into a Plexiglas-fronted examination box. My own heart did a little flip. Emperor tamarins are, in a word, adorable. About the size of kittens, these monkeys evolved in the treetops of the Central and South American rain forests. Their wispy, white Fu Manchu–style mustaches droop below enormous brown eyes. Swaddled in the pink blanket, staring up at me with that liquid gaze, Spitzbuben was pushing every maternal button I had.

When I'm with a human patient who seems anxious, especially a child, I crouch close and open my eyes wide. Over the years I've seen how this can establish a trust bond and put a nervous patient at ease. I did this with Spitzbuben. I wanted this defenseless little animal to understand how much I *felt* her vulnerability, how hard I would work to help her. I moved my face up to the box and stared deep in her eyes—animal to animal. It was working. She sat very still, her eyes locked on mine through the scratched plastic. I pursed my lips and cooed.

"Sooo brave, little Spitzbuben . . ."

Suddenly I felt a strong hand on my shoulder.

"Please stop making eye contact with her." I turned. The veterinarian smiled stiffly at me. "You'll give her capture myopathy."

A little surprised, I did as instructed and got out of the way. Human-animal bonding would have to wait, apparently. But I was puzzled. Capture myopathy? I'd been practicing medicine for almost twenty years and had never heard of that diagnosis. Myopathy, sure—that simply means a disease that affects a muscle. In my specialty, I see it most often as "cardiomyopathy," a degradation of the heart muscle. But what did that have to do with capture?

Just then, Spitzbuben's anesthesia took effect. "Time to intubate," the attending veterinarian instructed, focusing every person in the room on this critical and sometimes difficult procedure. I pushed capture myopathy out of my mind to be fully attentive to our animal patient.

But as soon as we were finished and Spitzbuben was safely back in her enclosure with the other tamarins, I looked up "capture myopathy." And there it was—in veterinary textbooks and journals going back decades. There was even an article about it in *Nature,* from 1974. Animals caught by predators may experience a catastrophic surge of adrenaline in their bloodstreams, which can "poison" their muscles. In the case of the heart, the overflow of stress hormones can injure the pumping chambers, making them weak and inefficient. It can kill, especially in the case of cautious and high-strung prey animals like deer, rodents, birds, and small primates. And there was more: locking eyes can contribute to capture myopathy. To Spitzbuben, my compassionate gaze wasn't communicating, "You're so cute; don't be afraid; I'm here to help you." It said: "I'm starving; you look delicious; I'm going to eat you."

Though this was my first encounter with the diagnosis, parts of it were startlingly familiar. Cardiology in the early 2000s was abuzz with a newly described syndrome called takotsubo cardiomyopathy. This distinctive condition presents with severe, crushing chest pain and a markedly abnormal EKG, much like a classic heart attack. We rush these patients to an operating suite for an angiogram, expecting to find a dangerous blood clot. But in takotsubo cases, the treating cardiologist finds perfectly healthy, "clean" coronary arteries. No clot. No blockage. No heart attack.

On closer inspection, doctors notice a strange, lightbulb-shaped bulge in the left ventricle. As the pumping engines for the circulatory system, ventricles must have a particular ovoid, lemonlike shape for strong, swift ejection of blood. If the end of the left ventricle balloons out, as it does in takotsubo hearts, the firm, healthy contractions are reduced to inefficient spasms—floppy and unpredictable.

But what's remarkable about takotsubo is what *causes* the bulge. Seeing a loved one die. Being left at the altar or losing your life savings with a bad roll of the dice. Intense, painful emotions in the brain can set off alarming, life-threatening physical changes in the heart. This new diagnosis was proof of the powerful connection between heart and mind. Takotsubo cardiomyopathy confirmed a relationship many doctors had considered more metaphoric than diagnostic.

As a clinical cardiologist, I needed to know how to recognize and treat takotsubo cardiomyopathy. But years before pursuing cardiology, I had completed a residency in psychiatry at the UCLA Neuropsychiatric Institute. Having also trained as a psychiatrist, I was captivated by this syndrome, which lay at the intersection of my two professional passions.

That background put me in a unique position that day at the zoo. I reflexively placed the human phenomenon side by side with the animal one. *Emotional trigger . . . surge of stress hormones . . . failing heart muscle . . . possible death.* An unexpected "aha!" suddenly hit me. Takotsubo in humans and the heart effects of capture myopathy in animals were almost certainly related—*perhaps even the same syndrome with different names.*

But a second, even stronger insight quickly followed this "aha." The key point wasn't the overlap of the two conditions. It was the gulf between them. For nearly four decades (and probably longer) veterinarians had known this could happen to animals—that extreme fear could damage muscles in general and heart muscles in particular. In fact, even the most basic veterinary training includes specific protocols for making sure animals being netted and examined don't die in the process. Yet here were the human doctors in early 2000 trumpeting the finding, savoring the fancy foreign name, and making academic careers out of a "discovery" that every vet student learned in the first year of school. These animal doctors knew something we human doctors had no clue existed. And if that was true . . . what else did the vets know that we didn't? What other "human" diseases were found in animals?

So I designed a challenge for myself. As an attending physician at UCLA I see a wide variety of maladies. By day on my rounds, I began making careful notes of the conditions I came across. At night, I combed veterinary databases and journals for their correlates, asking myself a simple question: "Do Animals Get [*fill in the disease*]?"

I started with the big killers. Do animals get breast cancer? Stress-induced heart attacks? Leukemia? How about melanoma? Fainting spells? Chlamydia? And night after night, condition after condition, the answer kept coming back "yes." The similarities clicked into place.

Jaguars get breast cancer and may carry the BRCA1 genetic mutation that predisposes many Jews of Ashkenazi descent and others to the disease. Rhinos in zoos get leukemia. Melanoma has been diagnosed in the bodies of animals from penguins to buffaloes. Western lowland gorillas die from a terrifying condition in which the body's biggest and most critical artery, the aorta, ruptures. Torn aortas also killed Lucille Ball, Albert Einstein, and the actor John Ritter, and strike thousands of less famous human beings every year.

I learned that koalas in Australia are in the middle of a rampant epidemic of chlamydia. Yes, *that* kind—sexually transmitted. Veterinarians there are racing to produce a koala chlamydia vaccine. That gave me an idea: doctors around the United States are seeing human chlamydia infection rates spike. Could the *koala* research inform *human* public health strategies? Since unprotected sex is the only kind koalas have (my searches for condom use by animals came up short), what might those koala experts know about the spread of sexually transmitted diseases in a population that practices nothing but "unsafe" sex?

I wondered about obesity and diabetes—two of the most pressing health concerns of our time. I burned midnight pixels investigating questions like: Do wild animals get medically obese? Do animals over-eat or binge-eat? Do they hoard food and eat in secret at night? I learned that yes, they do. Comparing animal grazers, gorgers, and regurgitators to human snackers, diners, and dieters transformed my views on conventional human nutritional advice—and on the obesity epidemic itself.

Very quickly, I found myself in a world of surprising and unfamiliar new ideas, the kinds I'd never been encouraged to entertain in all my years of medical training and practice. It was, frankly, humbling, and I started to see my role as a physician in a whole new way. I wondered: Shouldn't human and veterinary doctors be partnering, along with wildlife biologists, in the field, the lab, and the clinic? Maybe such collaborations would inspire a version of my takotsubo moment, but for breast cancer, obesity, infectious disease, or other health concerns. Perhaps they would even lead to cures.

The more I learned, the more a tantalizing question started creeping into my thoughts: Why *don't* we human doctors routinely cooperate with animal experts?

And as I searched for that answer, I learned something surprising. We used to. In fact, a century or two ago, in many communities, animals and humans were cared for by the same practitioner—the town doctor, as he set broken bones and delivered babies, was not deterred by the species barrier. A leading physician of that era named Rudolf Virchow, still renowned today as the father of modern pathology, put it this way: "Between animal and human medicine there is no dividing line—nor should there be. The object is different but the experience obtained constitutes the basis of all medicine." *

However, animal and human medicine began a decisive split around the turn of the twentieth century. Increasing urbanization meant fewer people relied on animals to make a living. Motorized vehicles began pushing work animals out of daily life. With them went a primary revenue stream for many veterinarians. And in the United States, federal legislation called the Morrill Land-Grant Acts of the late 1800s relegated veterinary schools to rural communities while academic medical centers rapidly rose to prominence in wealthier cities.

---

*One of Virchow's most illustrious students was the Canadian doctor William Osler, revered by American medical students as a father of modern medicine. What's less well known to physicians is that veterinarians also consider Osler a father of their profession. He was a key advocate for the comparative method and influential in shaping what became McGill University's School of Veterinary Medicine in Montreal.

As the golden age of modern medicine dawned, there was simply more money, prestige, and academic reward to be had in pursuing human patients. For physicians, this era all but erased their tarnished image as the leech purveyors and potion makers of times past. But veterinarians enjoyed little to none of this skyrocketing social status and its accompanying wealth. The two fields moved through the twentieth century for the most part on divided, yet parallel, paths.

Until 2007. That's when a veterinarian named Roger Mahr and a physician, Ron Davis, arranged a meeting in East Lansing, Michigan. They compared notes on similar problems they encountered in their animal and human patients: cancer, diabetes, the adverse effects of secondhand smoke, and the explosion of "zoonoses" (diseases that spread from animals to humans, like West Nile virus and avian flu). They called for physicians and veterinarians to stop segregating themselves based on the species of their patients and start learning from one another.

Because Davis was president of the American Medical Association (AMA) and Mahr headed the American Veterinary Medical Association (AVMA), their meeting carried more weight than the handful of previous attempts to reunify the fields.*

But the Davis-Mahr announcement received little notice in the popular media, or even among medical professionals, especially physicians. True, One Health (the favored term for this movement) has gotten notice from the World Health Organization, the United Nations, and the Centers for Disease Control and Prevention.† The Institute of Medicine, which is the health arm of the National Academy of Sciences, hosted a One Health summit in Washington, D.C., in 2009. And veterinary schools, including those at the University of Pennsylvania, Cornell, Tufts, UC Davis, Colorado State, and the University of Florida, have embarked on One Health collaborations in education, research, and clinical care.

Yet, the truth is that most physicians will go through their entire careers never interacting with veterinarians, at least not professionally.

---

*One of the first modern efforts at unification came in the 1960s from the eminent veterinary-epidemiologist Calvin Schwabe, who is regarded as a pioneer of this field.

†The movement has gone by several different names over the years, including comparative medicine and One Medicine.

Until I started consulting at the zoo, the only time I even thought about animal doctors was when I brought my own dogs in for an exam or vaccination. My veterinary colleagues tell me they regularly read human medical journals to keep up on the latest research and techniques. But most physicians I know—including myself, until recently—would never dream of consulting an animal-focused monthly, even one as highly respected as the *Journal of Veterinary Internal Medicine.*

I think I know why. Most physicians see animals and their illnesses as somehow "different." We humans have our diseases. Animals have theirs. And I suspect there's another reason. The human medical establishment has an undeniable, though unspoken, bias against veterinary medicine. While most physicians have many laudable attributes—tireless work ethics, the desire to help others, a sense of duty to the community, scientific rigor—we have some dirty laundry I must reluctantly air. Doctors, it may or may not surprise you to learn, can be snobs. Ask your (non-M.D.) podiatrist, optometrist, or orthodontist if he's ever felt condescension from someone with those two hallowed initials after her name, and you'll likely hear some juicy tidbits about physician arrogance or that special brand of M.D. noblesse oblige.

By the way, we do it even to each other. You won't find a group of cocky neurosurgical residents sharing coffee and muffins with the cheerful family practice team or the empathetic psych interns. There is an unwritten hierarchy. The more competitive, lucrative, procedure-driven, and "elite" specialties sit at the top of the physician self-importance pyramid. Given how readily physicians rank themselves based on which body part they minister to, just imagine the disdain they might work up for mere "animal docs." I'm sure it would shock some of my colleagues to learn that vet school is now harder to get into than med school.

When some vets tell me about this historical antipathy between our fields, many bristle about not being taken seriously as "real" doctors. But while it rankles when M.D.'s condescend, most vets simply take a resigned approach to their glitzier counterparts on the human side. Several have even confided to me a veterinarians' inside joke: *What do you call a physician? A veterinarian who can treat only one species.*

Still, among physicians, welcoming animal doctors as peers just "isn't done." As Darwin shrewdly observed, "we do not like to consider [animals] our equals." And yet, all of biology, the foundation of medicine itself, relies on the fact that we *are* animals. Indeed, we share the vast majority of our genetic code with other creatures.

And, of course, on some level we accept this vast biological overlap: almost every medicine we take—and prescribe—has been tested on animals. Indeed, if you asked most physicians what animals can teach us about human health, there is one place they would automatically point: the lab. But that is precisely not what I am talking about.

This book isn't about animal testing. Nor is it about the complex and important ethical issues of lab animal investigation. Instead, it introduces a new approach that could improve the health of both human *and* animal patients. This approach is based on a simple reality: animals in jungles, oceans, forests, and our homes sometimes get sick—just as we do. Veterinarians see and treat these illnesses among a wide variety of species. And yet physicians largely ignore this. That's a major blind spot, because we could improve the health of all species by learning how animals live, die, get sick, and heal in their *natural* settings.

As I started to focus on sameness, instead of being distracted by difference, it changed how I viewed my patients, their diseases, and even what it means to be a doctor. The line between "human" and "animal" started to blur. It was unsettling at first. Every echocardiogram I performed—on humans at UCLA and animals at the L.A. Zoo—suddenly exploded with familiarity and new meaning. Every mitral valve, every left ventricular apex, carried the echoes of our shared evolution and health challenges.

The cardiologist in me was thrilled with this new perspective, the myriad overlaps. But as a psychiatrist, I wasn't so sure. Physical similarities were one thing. Blood, bones, and beating hearts animate not just primates and other mammals but also birds, reptiles, and even fish. Still, I assumed, our uniquely developed human brains meant the similarities ended with our bodies. Certainly the overlap couldn't extend to our minds and emotions. So I came at the question from a psychiatric perspective.

Do animals get . . . obsessive-compulsive disorder (OCD)? Clinical depression? Substance addiction and abuse? Anxiety disorders? Do animals ever take their own lives? And again I sat back, a little astounded, while my research yielded a series of fascinating and surprising answers.

Octopuses and stallions sometimes self-mutilate, in ways that echo the self-injuring patients we call "cutters." Chimpanzees in the wild experience depression and sometimes die of it. The compulsions psychiatrists treat in their patients with OCD resemble behaviors veterinarians see in animal patients and call "stereotypies."

Suddenly, the benefits for human *mental* health seemed enormous. Perhaps a human patient compulsively burning himself with cigarettes could improve if his therapist talked shop with a bird specialist who had treated dozens of parrots with feather-picking disorder. Maybe Princess Diana or Angelina Jolie (who both publicly admitted cutting themselves with blades) could have found solace in discussing their urges with an equestrian expert who treats horses that compulsively bite themselves.

Significantly for addicts and their therapists, species from birds to elephants are known to seek out psychotropic berries and plants for the presumed purpose of changing their sensory states—a.k.a. getting high. Bighorn sheep, water buffaloes, jaguars, and primates of many kinds consume—and then show the effects of—narcotics, hallucinogens, and other intoxicants. Naturalists have been noting these behaviors in the field for decades. Is a treatment—or at least a new perspective—for alcoholism or addiction lying dormant in all that animal research?

I also searched for veterinary examples of depression and suicide. It seemed unlikely that animals would experience the same psychiatric urges to kill themselves that humans do. While the similar nature of their emotions has been persuasively described by behaviorists and veterinarians, I doubted that other animals share our foresight of death or knowledge of its power. Still I asked, "Do animals commit suicide?"

Well, they don't tie nooses around their necks or shoot themselves with revolvers, and they don't leave notes explaining why they did it. But examples of what appears to be grief-related and life-threatening "self-neglect" (refusing food and water) crop up throughout the scientific literature and in accounts that veterinarians and pet owners tell. And insect suicide, driven by parasitic infection, has been well documented by entomologists.

Which raises an interesting issue. Our physical body structures evolved over hundreds of millions of years. Perhaps modern human emotions too have evolved over millennia. Has natural selection played a role in what we feel, from anxiety, grief, and shame to pride, joy, and even schadenfreude?

Although Darwin himself studied and wrote extensively about natural selection's influence on human and animal emotions, none of my psychiatric training even touched on the possibility that human feelings could have evolutionary roots. In fact, it was almost the opposite. My education included stern warnings against the tantalizing pull to anthropomorphize. In those days, noticing pain or sadness on the face of an animal was criticized as projection, fantasy, or sloppy sentimentality. But scientific advancements of the past two decades suggest that we should adopt an updated perspective. Seeing too much of ourselves in other animals might not be the problem we think it is. Underappreciating our own animal natures may be the greater limitation.

As a psychiatrist, I was officially convinced. Remaining ignorant of the mental and physical disorders of animals, I began to feel, was as narrow-minded as refusing to seek out important human research simply because it was reported in a foreign language.

Still, the skeptic in me looked for any reason to explain away the similarities. Perhaps it was simply our shared environment. And after all, we humans have commandeered the food chain, imposing our dominant diets, weapons, and diseases on everything below us.

So I began to look anew at conditions I'd long assumed to be uniquely human and modern. And with that I came across some remarkable findings: dinosaurs with gout, arthritis, stress fractures . . . even cancer. Not so long ago, paleontologists uncovered a mass in the fossilized skull of a *Gorgosaurus,* a close relative of *Tyrannosaurus rex.* A brain tumor, they said, had brought down one of the Earth's most notorious carnivores, connecting a late-Mesozoic cancer patient to human brain cancer victims, including the composer George Gershwin, reggae artist Bob Marley, and U.S. Senator Ted Kennedy.

Having spent a career taking care of human patients in the here and now, I was suddenly confronted by a shifted boundary. Cancer has struck and killed its victims for at least seventy million years. I wondered how this knowledge might redefine how patients and physicians view the disease . . . or even how oncologists might search for ways to cure it.

Around this time I started working with Kathryn Bowers, a science journalist. A nondoctor with a background in social science and literature, she saw wider implications in these medical similarities. She urged me to view my overlapping experiences at the zoo and the hospital in a broader context. Together we began to research and write this book, bringing together medicine, evolution, anthropology, and zoology.

We started with a survey of how philosophers and scientists through the centuries have positioned our species among our fellow creatures. Clearly, for as long as humans have been able to ponder it, we've been of two minds about the apparent fact that we *are* animals. Judging by the written record going back at least as far as Plato, our ancestors acknowledged the obvious similarities between us and the so-called lesser creatures. Plato mused, "Man is the plumeless genus of bipeds; birds are the plumed." At the same time, people have long wanted to preserve a definition of humanity that kept us on a higher plane.

With *The Origin of Species,* Charles Darwin gave us a new (and, to many, unnerving) way to conceive of ourselves in relation to animals—positing that man and beast exist as different branches of the same tree rather than on different sides of a schism. Scholars of all stripes weighed in on whether and how humans were related to apes and other species.

In the mid-twentieth century, this debate was reignited by *The Naked Ape.* With studied objectivity, Desmond Morris, a zoologist and former curator of mammals at the London Zoo, described human feeding, sleeping, fighting, and parenting the way a biologist would document animal behavior in the field.

At about the time Morris was pointing out how similar we are to apes, two pioneering primatologists were documenting the many ways *apes* act like *us.* Jane Goodall was among the first to observe wild chimpanzees using tools and engaging in a type of organized warfare. For nearly twenty years, Dian Fossey lived near a group of gorillas in Rwanda, studying their vocalizations and social organization. Fossey's and Goodall's authoritative writings and memorable media appearances about the apes' distinct personalities and extended family relationships fed a growing public interest in human-ape crossover even as the two women advanced serious scientific knowledge.

Subsequently, many scholars attempted to demystify contemporary human life by studying animals and evolutionary biology. Two clashing powerhouses were the Harvard-based polymaths Edward O. Wilson and the late Stephen Jay Gould.

Wilson rocked academia and the wider public discourse in 1975 with the publication of *Sociobiology.* Inspired by his extensive research on ants, Wilson connected social behavior in animals to evolutionary forces, including natural selection. When extended to human societies, this suggested that our genes outline many aspects of our nature and behavior. But Wilson's theories were introduced in a particularly inhospitable climate. A mere three decades after eugenic theories were used to justify genocide, the world was not ready to hear that any aspects of human nature might be genetically predetermined. And as the civil rights and feminist movements were gearing up to dismantle centuries of racial, gender, and economic discrimination, public opinion would simply not tolerate theories with even a faint suggestion that "biology is destiny." Furthermore, with the scientific revolutions of molecular biology and genome mapping a decade and a half in the future, Wilson didn't yet have access to the high-tech tools that would ultimately back up many of his theories.

Wilson was harshly branded by some of his academic colleagues as a racist, sexist "determinist." One of his main detractors was Gould, a prominent paleontologist, geologist, and historian of science (who also happened to be one of my advisers on the undergraduate thesis I wrote about Darwin's influence on public perceptions of physical deformity). In books such as *The Panda's Thumb,* Gould argued that the subtleties of the human condition cannot be understood solely through natural selection. He cautioned readers that an overly genetic explanation of human behavior could reinforce regressive social agendas. His views matched the academic climate of the 1970s and '80s—the same era in which New Historicists were reinterpreting literature and deconstructionists dismantling Western civilization courses.

It was during this fertile period that Richard Dawkins published such provocative books as *The Selfish Gene* and *The Blind Watchmaker.* Dawkins characterized evolution as an unsentimental process, a self-interested and unceasing race among rival genes. Criticized, like Wilson, for having overstated the dominance of genetics over culture, Dawkins, an Oxford professor, nonetheless continues to probe the biological basis of human behavior, including its role in religion and belief in God. In a later work, *The Ancestor's Tale,* Dawkins explored the concept of a unified biology, identifying the shared ancestry across species—among them hippos, jellyfish, and single-celled organisms.

In 2005, *Nature* published a study that redefined the conversation: the human genome is 98.6 percent similar to that of chimpanzees. That single statistic inspired many people, and not only scientists, to reconsider what defines us as humans. Now, instead of trying to prove the *existence* of a connection between animals and humans, the race is on to explore the depth and breadth of this enormous overlap.

The challenge has led scientists to explore far beyond great apes. Biologists are rapidly uncovering ancient genetic similarities that link diverse species—mammals, reptiles, birds, and even insects. The discovery is astonishing: nearly identical clusters of genes have been passed down for billions of years, from cell to cell and organism to organism. These remarkably unchanged gene groups code for similar structures and even similar reflexes across species. In other words, a common genetic "blueprint" instructed the embryos of Shamu, Secretariat, and Kate Middleton to grow different, yet homologous, limbs: steering flippers, thundering hooves, and regal, waving arms. *Deep homology* is the term coined by biologists Sean B. Carroll, Neil Shubin, and Cliff Tabin to describe these genetic kernels we share with nearly all creatures. Deep homology explains how genes taken from a sighted mouse and placed into a blind fruit fly cause the insect to grow structurally accurate fly eyes. And it is a deep homology that genetically connects keen, light-responsive vision in a hawk to photosensitivity in green algae. Deep homology traces our molecular lineage to our most ancient common ancestors. It proves that all living organisms, including plants, are long-lost relatives.

Today, the specific nature/nurture controversy that so dominated the academic scene in the 1980s is something of a historical footnote. Advances in molecular biology, genetics, and neuroscience have shifted the debate away from *whether* there's a genetic basis for behavior and toward a more nuanced conversation about how genes, culture, and environment *interact*. This has given rise to a burgeoning new field called "epigenetics." Among other things, epigenetics considers how infection, toxins, food, other organisms, and even cultural practices can turn genes on and off to alter an animal's development.

Think about what that means. Evolution doesn't just happen over huge numbers of generations or millions of years. It can happen to you or me, or any animal, within our own lifetimes. Amazingly, epigenetic changes to our DNA mean that the genes we pass on to our children can differ from the ones we inherited. Epigenetics and deep homology are two sides of the evolutionary coin. Epigenetics helps explain rapid evolutionary changes and highlights the role environments can play in genetic health. Deep homology reminds us of our ancient origins and the glacial pace at which much evolutionary change occurs.

This stunning new perspective has started to change many fields, including biology, medicine, and psychology. When it was published in 2008, *Your Inner Fish*—Neil Shubin's illuminating journey through our shared anatomy with ancient life forms—ignited excitement about the power of comparative biology to inspire new ideas in modern medicine. Shubin, a paleontologist and biologist at the University of Chicago, joins Randolph Nesse, George Williams, Peter Gluckman, and Stephen Stearns in advancing a new field of evolutionary medicine in their books *Why We Get Sick, The Principles of Evolutionary Medicine,* and *Evolution in Health and Disease.* Other influential scientists who've blazed trails through the shared terrain of human and animal biology include Sean B. Carroll (*Endless Forms Most Beautiful*), Jared Diamond (*The Third Chimpanzee*), Steven Pinker (*The Blank Slate*), Frans de Waal (*Our Inner Ape*), Robert Sapolsky (*A Primate's Memoir*), and Jerry Coyne (*Why Evolution Is True*), to name just a few.

Interest in the mental life of animals, dismissed for many years as too speculative and an exercise in anthropomorphizing, has gained greater acceptance, too. Books by Temple Grandin (*Animals Make Us Human* and *Animals in Translation*), Jeffrey Moussaieff Masson (*When Elephants Weep*), Marc Bekoff (*The Emotional Lives of Animals*), and Alexandra Horowitz (*Inside of a Dog*) have demonstrated animal cognition and behavior that resemble what we might call foresight, regret, shame, guilt, revenge, and love.

Yet, while inspiring and illuminating, their books left me wanting a concrete way I could use their insights to improve my work as a physician. I wanted to break down the wall between physicians, veterinarians, and evolutionary biologists because together we are uniquely situated to explore the animal-human overlap where it matters most urgently—in the effort to heal our patients.

What had captivated me as a physician, what launched me on a journey that reshaped my entire approach to medicine, was a simple idea: to distill these decades of evolutionary research together with the collective wisdom of animal caregivers into a form both my patients and I could use within the four walls of my examining room.

Kathryn and I had found, practically without exception, an animal correlate to every human disease we could think of—from "Jurassic cancer" to "diseases of civilization." What we lacked was a name for this new fusion of veterinary, human, and evolutionary medicine.

Finding nothing in the literature, we decided to come up with our own: "zoobiquity." From the Greek for "animal," *zo,* and the Latin for "everywhere," *ubique,* "zoobiquity" joins two cultures (Greek and Latin), just as we are joining the "cultures" of human and animal medicine.

Zoobiquity looks to animals, and the doctors who care for them, for answers to humankind's pressing concerns. It peers back into our deep past—pausing but not stopping at great apes or even primates on the evolutionary timeline. It opens our minds to the common illnesses and shared vulnerabilities of the mammals, reptiles, birds, fish, insects, and even the bacteria with whom we evolved and share Earth.

Engineers already seek inspiration from the natural world, a field called biomimetics. Wings and fins inspire designers to create vehicles that float and fly more efficiently. Cockroaches helped solve the pressing problem of how to keep a robot stable as it climbs over uneven terrain, after researchers copied the insect's double-tripod legs and produced a machine that rarely tips over and can right itself when it does. Termites, mosquitoes, toucans, glowworms, and moths are just a few of the animals with superpower-like adaptations that scientists are trying to bring to a human market.

Now it's medicine's turn. I was in the right place at the right time to put takotsubo together with capture myopathy. (You'll find more on this finding in Chapter 6, "Scared to Death.") Zoobiquity encourages similar interdisciplinary experiences for other physicians. And this field-merging approach could have other important benefits. If studies funded by the National Institutes of Health expanded the boundaries of their inquiry by adding the simple question "Do animals get ____?" the benefits of scientific investigation could be vastly amplified.

A comparative approach could extend far beyond the walls of a human or veterinary hospital. It could help aspiring businessmen or middle school girls navigate complex hierarchies—by exposing similar challenges within a school of salmon or a herd of bighorn sheep. It points out the overlaps in the ways animals protect and defend their territories—and how and why we humans create borders, castes, kingdoms, and prisons. It dangles the possibility that human parenting could be informed by a greater knowledge of how our animal cousins solve issues of child care, sibling rivalry, and infertility.

Of course, human beings are unique as a species. Contained in our mere 1.4 percent genetic difference from chimpanzees are the physical, cognitive, and emotional features responsible for Mozart, the Mars rover, and the study of molecular biology itself. But the magnificent glare of this crucial but tiny percentage blinds us to our 98.6 percent sameness. Zoobiquity encourages us to look away, for a moment, from the obvious yet narrow range of differences and embrace the many enormous similarities.

Sadly, Spitzbuben the tamarin later died—not, I hasten to add, because of my attempt to befriend her. After her necropsy (the term for an animal autopsy), I took a slide of some of her heart cells to one of the most respected cardiac pathologists in the country, a colleague of mine at UCLA, Michael Fishbein.

As we peered through Fishbein's microscope, I noted how the damaged heart muscle cells seemed ensnared and strangled by the surrounding tissue. I felt a jolt of dreadful recognition as I spotted familiar-looking pink and purple shapes illuminated in the glaring white circle of the microscope's frame. Although the abnormal cardiac cells belonged to a furry, tailed tree dweller, they were essentially identical to human heart cells with the disease.

But this was more than a cellular display of our common ancestry with animals. The patterns illustrated a simple fact well known to veterinarians but unknown or ignored by modern physicians. Animals and humans share a vulnerability to the same infections, illnesses, and injuries.

As he had done so many times before with human heart specimens, Fishbein studied the slide carefully before he spoke. "Cardiomyopathy," I recall him observing. "Could be viral—looks just like a human's."

His phrase contained the essence of zoobiquity. Undistracted by fur and a tail, we saw, under that microscope, not "heart disease in a tamarin" but, rather, "heart disease in a primate"—gorilla, gibbon, chimpanzee, tamarin . . . or human.

As I heard Fishbein's words, my single-species focus officially died. Emerging in its place was zoobiquity, a connecting, *species-spanning* approach to the diagnostic challenges and therapeutic puzzles of clinical medicine. I would never look at another heart, human or animal, the same way again.

TWO

The Feint of Heart

Why We Pass Out

An urban hospital's emergency room only occasionally resembles its television doppelgängers on shows like *Grey's Anatomy* and *House, M.D.* Yes, we do see those whirlwinds of frantic activity around gunshot wounds, heart attacks, and drug overdoses. But in between come the calmer, less grim interludes. They arrive in the form of familiar characters: the hypochondriac, the overly vigilant parent, and, of course, the fainter.

As trivial as it might seem, fainting—what doctors call syncope—is so prevalent that it accounts for 3 percent of ER visits and 6 percent of hospitalizations in the United States. In UCLA's emergency department, we care for plenty of TV drama–worthy cases, including the victims of earthquakes, multicar crashes, and gang wars. But we also have fainters coming in almost every night—in fact, emergency rooms handle more fainting episodes than they do firearms injuries, suicide attempts, and third-degree burns combined.

About a third of all adults have fully fainted at least once in their lives. Nearly all of us have experienced that woozy, prefaint feeling, where all you can do is grope for a nearby chair and hang your head over your knees. And it's nothing to laugh at: syncope can be a symptom of serious heart ailments and can also cause severe injury—for instance, if you crack your head on your trip to the floor.

A cardiologist routinely cares for fainting patients. Although it may seem like an ailment of the brain, syncope is actually a complex interplay between the brain and the heart. At UCLA's medical school, where I lecture on fainting, I explain that a loss of consciousness often occurs when the brain is abruptly deprived of blood and oxygen. The specific causes vary, but more often than not, the heart is a prime suspect.

We all know that we can get dizzy when we stand up too fast. That kind of fainting comes from the basic physics challenge of moving liquid blood around the body against gravity. And fainting caused by a serious heart condition—where the heart is unable to pump a steady supply of blood to the brain—is relatively easy to diagnose.

But for the more storied form of syncope—the emotionally triggered faint that's been employed as a plot point by writers from Shakespeare and Austen to J. K. Rowling and Stephen King—its basic cause remains a mystery.

Yet this kind of fainting, called vasovagal syncope (VVS), is so common that casualty assistance officers delivering the news of a soldier's death to family members are trained to treat it. Nurses deal with fainting so often during blood draws that they keep ammonia inhalants (modern-day smelling salts) within arm's reach. And every obstetrician knows that some of the biggest fainters around are the husbands of women in labor. At the point of highest emotion (the baby's head crowning or popping out of the uterus during a C-section) the *thunk* of the father's head hitting the floor occasionally precedes the cry of the newborn.

Yet all my intellectual knowledge and hands-on experience with fainting didn't prepare me for what I would encounter when I took my twelve-year-old daughter to get her ears pierced. Instead of entrusting her pure, unsullied lobes to the high school kid at the mall jewelry store, I had, in my maternal wisdom, chosen the cleanest, safest venue I could think of: the starched and sterilized medical office of a family friend who's a plastic surgeon. On the happy day, my excited daughter settled into a comfortable, overstuffed chair designed for recipients of Botox injections. She gave me a brave smile. The doctor marked her ears with a green pen. He held up a hand mirror so my daughter could approve the placement. Then he drew out the silver piercing gun . . . I watched my daughter's smile fade . . . The gun moved closer and closer to her head . . . and had almost reached her left ear when—crash! Before I could even say, "You're doing great, honey," she had keeled over.

Believe me, my daughter wasn't in that office under duress. For years she had been begging me for pierced ears; she *wanted* to be there. And we could not have chosen a less threatening environment. Yet some instinct in her body or mind had insisted that she'd be better off unconscious than "present" in that moment. And clearly her brain and heart had followed orders and triggered a fainting response.

Later, as I mulled it over, I found myself focusing on the convoluted logic of fainting. If that piercing gun had been a real weapon, wouldn't she have been better off making an escape or putting up a fight, rather than falling helpless at the feet of the attacker? How has this odd response remained in the gene pool? Why didn't evolution take out the fainters long ago, in favor of fighters and fliers?*

For clues to puzzles about human bodies and behaviors, we can look to creatures whose daily realities are less detached from their evolutionary roots than are the lives of modern Western urbanites. Vasovagal syncope is the perfect starting point for a zoobiquitous expedition. I realized that I had never, despite years of treating human fainters, thought to ask one basic question: Do animals faint?

A survey of any veterinarian's patients quickly confirms that yes, they sometimes do. In dog breeds from rottweilers to Chihuahuas, syncope can follow everyday activities like barking and jumping, frolicking, grooming, and bathing. Some canines faint when they're roused to sudden activity after being at rest. Some vasovagal fainting in dogs and cats happens when they're physically restrained against their will, an especially terrifying situation for many pets. Remarkably, as is the case with many humans, some pet patients have been reported to faint in response

---

*There are some theories. One, the "clot-production" hypothesis, posits that a slow heartbeat or full-on faint helps animals avoid bleeding to death after an attack, since slow-moving blood under low pressure clots better. The less plausible "human violent conflict" hypothesis suggests a Paleolithic-era origin, speculating that fainting evolved in women and children as a way of taking them (but not men) out of harm's way during tribal warfare.

to needles: a Yorkie after a blood draw . . . a kitten after having urine drawn from its bladder with a syringe . . . a Cavalier King Charles after a vaccination.

What about wild animals? This is a harder question to get at, but zoo veterinarians have seen chimpanzees faint, especially when the animals are stressed or dehydrated. Wildlife veterinarians have seen screech owls and juncos fall into a torporlike state when they're handled during blood draws. And Charles Darwin reported catching a robin that "fainted so completely that for a time I thought it dead." He also saw a terrified canary "not only tremble and turn white about the base of the bill, but faint."

Fainting episodes often begin in the same way and in the same situations as the well-known fight-or-flight response. When animals, including human animals, sense a possibly mortal threat, adrenaline and other hormones (called catecholamines) flood into our bloodstreams. Our hearts race. Our blood pressures soar. We breathe faster. Crucially, we get a burst of energy, allowing us to either escape from the threat or battle it off.

But as you'll soon see, the old duality of "fight or flight" needs an update. Many animals have at their disposal an additional trick to boost their odds of living through a dangerous encounter. It's not just fight or flight. It's fight, flight, or *faint*.

Remarkably, fainting begins the same way as the other two fear responses—with a high-emotion stressor and a surge of adrenaline. But from there fainting follows a different route. Instead of the heart beating *faster* (tachycardia), it *plummets* (bradycardia). Instead of blood pressure *surging*, it *plunges*. Detecting low-pressure, slow-moving blood, sensors throughout the body signal to the brain that something is terribly wrong: a failing heart or a catastrophic loss of blood. In a protective response, the brain shuts the system down by fainting.

For anyone who's had a racing pulse after being scared, this slowing of the heart seems counterintuitive. But you've felt it. Imagine that wave of intense nausea you had when you lost your passport in Beijing or discovered a partner was cheating on you. Reflect on that "I think I'm going to vomit" feeling that washed over you following a career-jeopardizing mistake or a near miss between your carful of kids and a sixteen-wheeler. It's also the woozy feeling you may get before stepping in front of an audience, as you anticipate hundreds or thousands of eyes being trained on you. (For more on the heart's sometimes deadly response to eye gaze, see Chapter 6, "Scared to Death.")

That extreme, sick feeling is the vagal response. It's caused by the part of the nervous system responsible for "digesting and resting": the *parasympathetic* system. For a few crucial seconds the *sympathetic* system (which controls "fight or flight") withdraws and the parasympathetic system takes over. A pulse check during those awful moments of vagal nausea would reveal a slowed heart rate. In some cases, but not all, it slows enough to cause a loss of consciousness, what most of us call a faint.

Although losing a passport won't induce dread in a chipmunk, other stressful situations will. Alarm-triggered slowing of the heart has been documented across the animal kingdom. Woodchucks, rabbits, fawns, and monkeys have all shown marked slowing of the heart (and a decrease in blood pressure) in response to fear. Willow grouse, caimans, cats, squirrels, mice, alligators, many species of fish, and, yes, even chipmunks display this cardiac trick as well. And while it isn't always followed by a faint (it isn't always in people, either), this switch to a vagal state and the slowing of the heart in response to stress is as commonplace as it is curious. It's exactly what happened to my daughter in the ear-piercing chair. For years, I've known this by the human medical term "fear-induced, vagally mediated bradycardia." Once I started looking into it, I came across a different term, one used by veterinarians: "alarm bradycardia." It sounded enticingly similar to our term—not to mention more succinct. And sure enough, the two terms describe exactly the same condition.

One noticeable difference between animal and human fainting is that, while animals frequently get alarm bradycardia, they seem to fully faint less often than humans. Then again, for every actual faint we see in the ER, we know there are many more cases where people feel the swoon, the nausea, the light-headedness of bradycardia while never completely blacking out. It isn't unreasonable to call this syndrome, both in humans and animals, "near-fainting while conscious." And since so many species do it, it brings us back to a fundamental question: Do animals whose hearts go into super slo-mo at times of high stress have a survival advantage?

There are a few possible answers, the first of which you've probably already guessed. Alarm bradycardia can help an animal feign death and thus possibly fool a predator into passing it by.

One study demonstrated that inexperienced foxes could be fooled by ducks whose slowed nervous responses made them seem dead. Older foxes, though, having been tricked out of a meal or two in the past, had wised up. These savvy hunters knew to kill the duck on the spot or possibly bite its legs off, to ensure that it didn't miraculously "rise from the dead."

This trick of the heart and mind has saved people from imminent, actual harm. In 1941, twenty-one-year-old Nina Morecki was fleeing a concentration camp and her Nazi pursuers in the Polish woods when she fainted. After regaining consciousness, she found herself surrounded by the dead bodies of her less fortunate comrades. Other grim analogues might include those survivors of mass killings who play dead until they can escape. This strategy has been described in survivors' accounts from the Babi Yar massacre during World War II, the Rwandan genocide in 1994, and shooting rampages like the one at Virginia Tech in 2007.

Another common side effect of near fainting while conscious is both disgusting and tactically brilliant. A vagal state can make an animal lose control of its bodily functions. Some animals urinate or defecate under extreme emotion or fear. Many predators find urine or feces repugnant and will leave. Dogs are known to retreat at the smell of skunk; frightened shrews produce such foul odors from their anal pockets that even ravenous badgers keep their distance. Vomiting by the would-be prey can have the same conveniently repellant effect on the predator.

This potentially embarrassing loss of bodily control in response to fear is one vestige we humans probably wish we'd evolved out of. But in fact it may occasionally serve a protective function for us as well. Rape-prevention educators sometimes instruct women to urinate or vomit if rape is imminent. In some cases, the attacker will be repulsed and withdraw. A more common phenomenon is seen in women who successfully avoid sexual assault by fainting or by entering a "near fainting while conscious" state. Psychologists have studied cases of this and compared them to immobility reactions in animals. They suggest that when fighting back isn't an option, *not* struggling may defuse the situation and reduce the likelihood of rape.* While far from foolproof, fainting succeeds enough of the time to warrant a serious consideration of its evolutionary roots.

It's a grim irony that the one group of people who may best understand the vital role fainting plays in giving the body a needed respite is a group dedicated to inflicting pain: torturers. Many narratives taken from torture victims contain a nauseatingly familiar refrain. Under the terror and physical violation, many victims pass out. But, horrifically, when they come to, the torturer is cued to resume his assault. You could say that by overriding the body's protective response—the faint—the torturer adds yet another level of affliction, the way sleep deprivation keeps a body from having a restorative break.†

A slowed heart offers another key survival advantage. It helps a vulnerable animal keep still. Canadian scientists studying white-tailed deer tracked what happened when they played recorded wolf howls to fawns. The baby deer responded with "very predictable" alarm bradycardia, slowing their hearts and quieting their bodies. Think of the survival edge this physiological trick gives to fawns, who often get left alone for long periods while their mothers go off to forage. A slowed heart rate keeps them from rustling around when danger is nearby. In other words, it helps them hide. Is this physiology present in young humans?

This is the kind of experiment we would never do on infants; terrifying them on purpose to test their heart rates would certainly get the researcher excoriated, if not arrested. And yet, remarkably, an accident of geopolitical fate has given us a small window into how the very youngest members of our own species respond to primal terrors.

---

*Female robberflies sometimes employ a similar tactic to thwart unwanted sexual advances. Writes entomologist Göran Arnqvist, "If grasped by a male, they exhibit thanatosis (playing dead). Once the female ceases to move the male apparently no longer recognizes the lifeless female as a potential partner, loses interest and so releases the female." Whether or how this insect rape-prevention strategy has implication for human beings is unclear, but Arnqvist posits that since feigning death is so widespread in the insect world, females may have adapted this strategy to protect themselves from unwanted copulations.

†Some experts believe that crucifixion is death by recurrent vasovagal syncope. During this horrible form of torture (from which the word *excruciating* is derived), you're restrained from collapsing into a restorative horizontal position. You faint and recover without respite, and eventually succumb to low blood pressure and oxygen deprivation.

On the night of January 18, 1991, during the Gulf War, Scud missiles launched by Iraqi troops began exploding in Israeli communities. Citizens were alerted by howling air-raid sirens that blared from outdoor speakers and on the radio and TV. Since there was a terrifying possibility that the bombs were carrying chemical payloads in addition to their explosive power, the frightened populations had been instructed to don gas masks and seek shelter when they heard the wail of a siren.

In the maternity ward of a Tel Aviv–area hospital that night, three women were in labor. As is standard practice, they had been fitted with fetal heart monitors that strapped around their bellies to keep track of their babies' heartbeats. At three a.m., a sudden, terrifying shriek of a Scud alert siren penetrated the walls of the maternity ward—and, apparently, the wombs of the expectant mothers. As hospital staff scrambled to put gas masks on themselves and their patients, the nurses noticed something highly unusual on the fetal monitors. The heart rates of all three of the about-to-be-born infants suddenly and unexpectedly.... *plummeted.* From a healthy and brisk 100 to 120 beats per minute they slowed by half, to a frightening 40 to 60. The tiny hearts "lay low" like this for two minutes and then returned to normal.

All three babies, who hadn't yet even heard their parents' voices outside the womb, responded physiologically, with bradycardia, to the sound of danger. Some of the slowing may have resulted from the sounds of the siren itself and some from maternal stress hormones entering the fetus's body in response to the siren. Either way, these obstetrical observations strongly suggest that even prior to birth itself, we're equipped with unconscious anti-predator defenses, including a potent alarm bradycardia response. All three babies were ultimately born healthy, as well as apparently armed with the full complement of survival instincts we all possess but rarely think about.

Hiding in the face of danger—what scientists call crypsis—is one of nature's most common and effective strategies for staying out of a predator's stomach. Some animals depend on body shapes and camouflage to help them hide. And some conceal themselves by performing instinctive or learned behaviors like freezing, hiding, or crouching. Many animals do all of these. The stillness of a slowed heartbeat is just one of many tools prey animals employ to help them "disappear," at least as far as a predator is concerned.

Freezing, hiding, and crouching—with the help of slowed hearts—connect our nervous systems to the vast range of species with whom we share common ancestry. Examining fainting through the lens of veterinary medicine allowed me to reframe this common but puzzling cardiac event as a possible anti-predation strategy. And this hypothesis, in turn, helped me understand the powerful feedback loop between heart and brain that results, for some of us, in lost consciousness or passing out. Exploring why took me into the watery habitat of our ancient ancestors.

Astronotus ocellatus, commonly known as an oscar, is a freshwater fish related to tilapia. Energetic and affectionate, oscars' reputation as "puppies of the aquarium" comes from the enthusiastic greetings they give their owners, complete with tail wagging, acrobatic flips, and finger nibbling. But when oscars get stressed out—for example, when you're cleaning their tank—they can seemingly go lifeless. Lying on their sides, completely still, they lose color and breathe more slowly. Their fins stop moving. They sometimes stay this way even when nudged.

If I were able to place an aquatic version of my stethoscope over the heart of that very still—yet alive—fish at the bottom of the tank, I would hear another clue as to why fainting may have survived so many grueling rounds of natural selection. Or, rather, the clue would be in what I *wouldn't* hear: a robustly beating heart. Instead, I'd notice the familiar, super-slow rhythm of a bradycardic heart, dominated by lengthy pauses between beats.*

To understand the significance of this decelerated, less noticeable heart rhythm, consider the physiology of a master predator: the shark. Along with certain other underwater predators, like rays and catfish, sharks come equipped with heartbeat *detectors.* Called ampullary organs, these specialized sensory cells pick up the weak electrical pulses put out by the beating hearts of other fish. The hunters' internal ears may also scan for fish heartbeats, picking out the *lub-dubs* like doctors do through stethoscopes. Predators can lock onto the telling signals and home in on them

---

*The fish heart has two cardiac chambers separated by a rudimentary valve; the mammalian heart has four chambers and four cardiac valves. When the heart's valves close, they create clicks we call heart sounds. In humans, the shutting of the heart's valves generates the iconic *lub-dub* sound.

with lethal accuracy, even when their target is some distance away or hiding under sand. Which means: underwater, a beating heart can be a deadly giveaway.*

Every one of us has this "tell." Whether you're a human being, a salamander, or a canary, your telltale heart starts beating in the early days after conception and keeps going until the day you die.

But if a fish underwater could silence its beacon, it would become acoustically invisible. It might even be able to evade a predator. Anyone who's seen a submarine movie will recognize this principle. The commander of a submarine being tracked by enemy sonar will invariably order his crew to "run silent"—which involves everything from shutting off radios to cutting the engine to suppress the heartbeat of the submarine. Once the threat has passed, they fire up the engines again and the sub speeds away to safety.

Knowing this, we can see why natural selection might have favored fish that had the good fortune to faint their way out of becoming dinner. Having a heart that radically slows in response to real or perceived threats might have been a major advantage, offering protection before an attack even got under way. Fainting and "near fainting while conscious" may have evolved as a lifesaving "third option," offering a protective alternative to the more storied "fight or flight."

As we know, the heart-slowing reflex triggered by states of high arousal, such as fear, pain, or distress, is a core feature of vasovagal fainting in human beings. Alarm bradycardia has protected animals across all classes of vertebrates, and persists in us today precisely because its protective power is so deeply embedded into the autonomic nervous system, which has been passed down from our ancient water-dwelling ancestors. This hypothesis connects the acutely slowing heart of a hunted fish in the water to a human fainter in the ER.

In some ways, it's hard to think of ourselves as prey. Human beings today are so dominant on our planet that we can (and do) wipe out whole species, sometimes without even knowing it. Most of us in developed countries will make it through our entire lives without ever facing a realistic threat from a nonhuman animal predator. An evolutionary

---

*The Volvo car company once offered a heartbeat detector as an option on some of its models. Volvo claimed the machine could alert you to the presence of an intruder in the backseat of your car *before* you got behind the wheel.

vestige like fainting seems as ill-fitted to our modern times as a chariot repair shop. But a zoobiquitous approach lets us understand reflexes and behaviors in ourselves that mirror anti-predation strategies in other animals.

Picture the myriad defenses nature has bestowed on many adult animals: quills, antlers, talons, noxious smells, and deadly poisons. While all can be useful during an attack, they also serve as "don't mess with me" warnings that can prevent an attack in the first place. The same goes for a peculiar jumping maneuver among deer and gazelles called "stotting." The stotting animal springs up, lands stiffly on all four legs, then springs up again and again, moving away from a predator as if on a pogo stick. Scientists argue about how this behavior helps an animal escape. It seems like a colossal waste of energy—energy that could be spent on running away. But the whole point seems to be to show off superb stamina. Stotting tells a predator that the animal has energy to spare and even *thinking* about giving chase is a waste of time.

Wildlife biologists call these physical traits and behaviors "signals of unprofitability." They send a clear message to a predator: Move along, find an easier target.

We humans use signals of unprofitability for protection, too. Picture a bodyguard flexing his bulging biceps. Think about how you might instinctively pull yourself up to your full height and walk with an exaggerated swagger on an unnervingly quiet street at night. Imagine the sign on your lawn advertising that you have a burglar alarm inside, or the teams of lawyers employed by big companies to fight lawsuits. The message in each case is the same: Find another victim. This one's too much trouble.

Indeed, maintaining and advertising a strong defense is a fundamental drive across species. In a conversation I had with the late Harvard evolutionary biologist Karel Liem, he explained that nearly every animal behavior has elements of self-protection, or anti-predation, at its core.

And the physiology of fainting is no different. Simply being still can confer survival advantages. Of course, it doesn't always work, but it does enough of the time to make it a respectable option of last resort.

Yet respect is rarely the reaction that greets fainters. Alarm bradycardia, vagal nausea, freezing in place, feigning death, and full-on fainting are almost always taken as signs of weakness or cowardice, portrayed in literature and film as shorthand for the lily-livered. Franklin Pierce's episode of battlefield fainting, for example, landed him the moniker "the Fainting General," which dogged him even after he became president of the United States in 1853. Few would characterize George H. W. Bush, Margaret Thatcher, David Petraeus, Fidel Castro, or Janet Reno as weak-willed, yet all suffered fainting spells while in office. To an observer, passing out might seem helpless, a physiologic act of surrender, even defeat. But, given fainting's protective power, perhaps it's time to revise this derogatory and uninformed view of syncope.

Fight, flight, or faint. Fainting is the body's way of flipping a circuit breaker. It halts the action and perhaps even a pursuer. It can defuse conflict. It can enable escape. Fainting and its related spectrum of "slowing down" behaviors remain with us because over hundreds of millions of years they have helped animals evade death. Embedded in fainting's ancient physiology is an important lesson for how we respond to the things that scare us. Fighting or fleeing your enemy may work some of the time. But when fighting is futile and fleeing not an option, just being still may offer an even more powerful form of protection.

Teens at the ear-piercing salon, fawns hidden in leaves, blood donors, and fish escaping predators have all inherited fainting's death-evading neurocircuitry. Their conversing hearts and minds have bestowed upon them a respite—a momentary, deceitful reprieve that for eons has sometimes meant a way out.

THREE

Jews, Jaguars, and Jurassic Cancer

New Hope for an Ancient Diagnosis

As veterans streamed home from Asia and Europe after World War II, doctors in the United States were battling a deadly threat on the home front. Five times the number of Americans who died at Iwo Jima and Omaha Beach were dying every year of heart disease. In response, the National Heart Institute launched what became the gold standard in long-term medical investigations: the Framingham Heart Study. Every two years, starting in 1948 and continuing today, thousands of men and women from that Massachusetts city go to special doctors. They give blood and other lab samples, have comprehensive physicals, and answer question after question about their eating, exercise, work, and leisure habits.

As the data piled up over decades, researchers began to discern patterns. High blood pressure and smoking led to heart disease. Age and gender influenced risk. It's hard to believe that this information we now accept as routine was ever unknown. Even today, Framingham's half century (and counting) of statistics is paying dividends as researchers mine the data for long-term trends in stroke and dementia, osteoporosis and arthritis. The iconic study is now in its third generation, having enrolled many of the children and *grandchildren* of the original participants.

Longitudinal medical studies—those with large populations and elongated time frames—are hard to pull off. What makes them so valuable is exactly what can make them so frustrating. Even when lots of people sign up, many drop out. Participants lose interest. They forget to go to their physicals. They move away and don't leave a forwarding address. They blow off the third or thirteenth or thirty-third questionnaire.

But the challenge hasn't daunted Dr. Michael Guy. In 2012 he began enrolling three thousand participants in what is perhaps the most ambitious new longitudinal study in more than a decade. Its focus is cancer in a population that has a staggering 60 percent risk of dying from the disease.

And his research team knows for sure that their test subjects aren't likely to cheat or fib or flee. They won't fudge answers on their surveys or tell a researcher only what she wants to hear. They'll be loyal and enthusiastic and obedient. The researchers know this because they chose their participants deliberately and wisely. They are all golden retrievers.

Before you start picturing a floppy-eared puppy in a sterile wire lab cage, let me explain. The dogs enrolled in the Canine Lifetime Health Project—a long-term cancer study Guy sometimes calls "Framingham for Dogs"—are beloved pets. Recruited from normal homes all over the United States, they live in yards and bedrooms, romp with children and other dogs, eat the food their owners carefully select and prepare for them. They walk neighborhood sidewalks and play fetch in local parks.

Like the human participants in the Framingham study, each dog in the Canine Lifetime Health Project will be followed for the rest of its life. As the data roll in, epidemiologists, oncologists, and statisticians will scrutinize the dogs' diets to see if nutrients or portion sizes contribute to developing cancer. They will pore over environmental exposures—from secondhand smoke to household cleansers. They will measure how far the dogs live from power lines and freeways to determine whether any cancers cluster in significant ways. The researchers will analyze the genetic code of each dog, comparing it to the others and to the complete canine genome (completed in 2005 on the DNA of a female boxer named Tasha).

This unprecedented study, undertaken by the nonprofit Morris Animal Foundation, could radically shift our approach to cancer in dogs. And the effort may yield knowledge that will benefit not only future generations of pets but also the animals at the other end of the leash. Dog cancer has many stories to tell about human cancer: where it comes from, why it migrates, and, possibly, how to stop it in its tracks. A multispecies take on cancer research means our special relationship with man's best friend is about to get even closer.

Except for some grizzling on her muzzle, Tessa's fur was glossy black—a striking contrast with her streetlight-yellow vest. The bright garment, as snug as a Partridge Family costume, was covered with embroidered patches. A few advertised dog food companies. One identified Tessa as a "Dock Dog," an elite animal athlete whose jumping and fetching prowess makes the average pet look like a Little Leaguer going up against Derek Jeter. But the most noticeable feature of Tessa's vest were the two words stitched in black thread across her midriff: "Cancer Survivor."

Tessa is a black Labrador retriever I met in the spring of 2010 at a gathering of pet patients who had battled illnesses and won. Although a brown lesion on the gum behind her lower left fang was still visible, her mouth cancer had been in remission for two years. As I patted Tessa's furry, wedge-shaped head, her owner, Linda Hettich, explained how she had discovered her dog had the disease. They were playing fetch and Tessa brought back a bloody tennis ball. A trip to the vet confirmed a cancer diagnosis, and Tessa went into treatment. Although Hettich's distinct alto voice (she's the noon anchor for a Los Angeles news radio station) conveyed gratitude that Tessa's cancer had not returned, her face betrayed a certain grim anxiety. Tessa was not her first dog to have cancer. A few years earlier, her beloved mutt, Kadin, had died of it. Hettich admitted in a whisper that she sometimes wonders why two of her dogs have fought the disease.

"With Kadin, there was a tremendous amount of guilt," she told me. Now that Tessa has had cancer, she said, there are moments when she wonders, "I'm two for two—what did I do?"

That didn't surprise me a bit. I'd heard "What did I do?" before; that question frequently plagues many human cancer patients, too.

One of my roles at UCLA involves caring for people who've developed heart problems as a side effect of their cancer treatments. Sometimes they share with me their personal theories for why they drew the cancer card. Often, it's something they did: *My cell phone. My deodorant. My char-grilled salmon. My microwave. My lipstick. My plastic Evian bottle. My years as a flight attendant.* Or something they didn't do: *Missing church. Not exercising. Skipping mammograms.* Something that was done to them: *My father's nicotine addiction. The fluoride in my water. The new carpet at my office.* Or general stress: *A lingering lawsuit. A mountainous credit card balance. Caring for an aging parent.*

I understand that these narratives allow patients to feel a modicum of control in the face of a terrifying diagnosis. Because that in itself can be healing, I usually just listen quietly as I measure their blood pressure, check their pulses, and place my stethoscope over their heart. But some seem to be seeking medical absolution, so I gently remind them of something they've surely heard before: cancer has many causes. Within the DNA we inherited from our parents, from our great-great-great-grandparents, and from ancient animal ancestors lie the blueprints and machinery that instruct cells to create and maintain our body parts. But when this machinery contains errors and then malfunctions, the out-of-control growth we call cancer can develop.

Here's what I mean. Living, growing organisms must constantly replace old and dying cells with fresh, new ones. Making a new cell requires copying every single one of the almost three billion building blocks (called nucleotides) in the cell's DNA. This provides the daughter cell with the exact same information as its parent. When all goes well (and, astonishingly, it usually does), the DNA is copied exactly. But occasionally, about once every ten thousand nucleotides, a mistake is made. Chemical codes can be left out, duplicated, or put in the wrong place.

Much of the time, these slipups—called mutations—are caught by the cell's chemical "proofreaders" and fixed before they wind up in a new cell. Often, a "typo" sneaks through but it's not significant and the cell can continue along normally, even with the misprint. Sometimes these mistakes occur in critical regions of the DNA and actually enhance cell function. These minor changes, over time, can produce new traits, new behaviors, and even new species. For example, alterations or mutations are responsible for size differences in dog breeds. Slight variations in the genes that direct skeletal growth create the most obvious difference between a Chihuahua and a Great Dane.

However, some mutations harm the cell's function. For example, normal cells carry "suicide codes" in their DNA. When a cell gets old or is damaged beyond repair, these codes spring into action and cause the cell to self-destruct in a process called apoptosis. But cells can develop mutations in the very genes that direct the destruction. When the destruction instructions go awry or malfunction, the damaged cells will stay alive. They then can replicate—mistakes and all. When that happens, the new defective cell, like its parent, lacks normal cell death instructions. Now there are two cells, each with DNA mistakes and missing the appropriate controls. When these faulty cells replicate, they become four, then eight, then sixteen. Soon an entire population of immortal cells has grown without restraint. This is cancer: initially normal cells, grown out of control, now with different DNA instructions.

When the out-of-control, mutation-containing cells cluster together, they form a tumor. Sometimes the mutated cells find their way into the bloodstream or lymph system, which are essentially superhighways with mass access to the rest of the body. When the cells travel far from their origins and then replicate in the new location, that's metastasis. Some cancers, like melanoma, metastasize readily. Others, like chordomas found on the base of the skull, are less ambitious and grow primarily in one region. (By the way, this is the most basic difference between the cancers we call "benign" and those we say are "malignant." Any mass of abnormal cells is a tumor, but benign growths tend to remain in the same location and refrain from invading nearby tissues.)

But whether a cancer is sluggish or fleet, a homebody or an adventurer, tumor-forming or what we call "liquid," what underlies its enormous burden of suffering and death is nothing but errors in the genetic code. Many behaviors and factors in the environment promote these errors and lead to cancer. Smoking, sun exposure, excess alcohol consumption, and obesity have all been linked to DNA damage and to various cancers.

There's also a catalog of known, toxic substances that, given adequate exposure, can almost certainly trigger cancer: naturally occurring radon (and other radioactive substances), asbestos, chromium-6, formaldehyde, benzene, and others. The National Institutes of Health (NIH) flags fifty-four documented carcinogens implicated in human cancers. More research will surely add to this list.

With so many toxins in our environment and so many cancer diagnoses in our communities, it's easy to point to our polluted surroundings and connect the cancer-causing dots to our neighbors' suffering. Cancer, many people believe, is unnatural—a disease of our own making. In fact, cancer prevention has become a marketing tool. The simple act of choosing milk or deodorant or tuna fish can feel like a high-stakes exercise in cancer avoidance. Sorting out what's Madison Avenue from what's medically accurate has become a challenge for patients and a responsibility for their doctors.

But cancer can also develop in people who didn't smoke, drink, or tan and who avoided microwaving food in plastic and cooking on Teflon. It strikes yoga practitioners, breast-feeders, and organic gardeners; infants, five-year-olds, fifteen-year-olds, fifty-five-year-olds, and eighty-five-year-olds. And, pointedly, it's not uncommon to see elderly patients who have done everything "wrong" . . . but show no trace of the disease.

The impulse to blame ourselves or our cultures for our diseases is not unique to modern society or to cancer. As the medical historian Charles Rosenberg has pointed out, "The desire to explain sickness and death in terms of volition—of acts done or left undone—is ancient and powerful."

What insights can a species-spanning approach bring? Even the briefest survey of cancer in other animals sheds light on a critical but overlooked truth: where cells divide, where DNA replicates, and where growth occurs, there will be cancer. Cancer is as natural a part of the animal kingdom as birth, reproduction, and death. And, as we'll see, it's as old as the dinosaurs. Literally.

Tessa was just one of the million or so dogs who get diagnosed with cancer each year. Intriguingly, many canine cancers behave very similarly to human cancers. Lethal prostate cancer runs a similar clinical course in men and male dogs. Breast cancer may seek bone tissue in female dogs, just as it can preferentially metastasize to the skeleton in women. Osteosarcoma, which tends to hit human teenagers during their growth spurts, strikes with similar ferocity in many large and giant-breed dogs.

Sadly, many outcomes are similar, too. As in people, many cancers in dogs become resistant to therapy. And in both species they can recur, even after a patient has been given the all clear.

Dogs aren't the only animals in our lives who get cancer. When a cat presents with fever and jaundice, the vet must consider leukemia or lymphoma, leading feline killers in the United States. And when a cat's owner discovers a lump in her pet's breast, it may turn out to be a highly aggressive form of breast cancer also diagnosed in many women. For some cats with breast cancer, lumpectomy may suffice. For others, radical mastectomy of the entire chain of all eight mammary glands must be performed.

Rabbit hysterectomies are commonly recommended due to the high risk of uterine cancer as these pets age. Parakeets are prone to developing tumors on their kidneys, ovaries, or testes. And cancer patients can also be reptiles. Zoo veterinarians have reported on leukemia in pythons and boa constrictors, lymphoma in death adders and hognose snakes, and mesothelioma in rattlesnakes.

Pediatricians of fair-skinned children aren't the only doctors who worry about skin cancer in their patients. Equine sunburn is thought to cause skin cancer in light-colored horses, although this "gray horse melanoma" may connect more to a genetic issue in the breed than to too many hours spent basking in the sun. Still, because as many as 80 percent of gray horses will get skin cancer of some kind, their concerned owners, along with those of horses with white "socks" on their legs or blazes on their noses, sometimes apply zinc oxide sunscreen to exposed skin areas. Others, like the parents of towheaded toddlers, insist that their horses wear a hood when out of the stable.

If your dermatologist reminds you to remove your nail polish before coming in for your yearly mole scan, that's because she wants to check not only for melanoma but also for squamous cell carcinoma, a common form of skin cancer and the same kind Tessa had. Tessa's was in her mouth, but it can also start under a toenail. That's similar to what happened to a zoo rhinoceros I once examined. Her cancer grew under her horn—which is made from keratin, exactly the same protein that makes up our finger- and toenails. Cattle also develop squamous cell carcinomas in the pale skin encircling their eyes. Some Herefords have been intentionally bred for darker pigmentation around the eyes, which gives them a little more protection from the sun and seems to reduce the incidence of cancer.

Strike-branding livestock with sheet-metal strips heated to 300° to 600°F can cause tumors to grow around these permanent markings. Like branded cattle, humans who modify their bodies with branding are at increased risk of cancer at the sites of these injuries. Even tattooing may be associated with a rare form of skin cancer.

Cancer strikes across ecosystems and throughout the animal kingdom. Osteosarcoma, the cancer that forced Ted Kennedy's son, Ted Junior, to undergo an amputation in the early 1970s, attacks the bones of wolves, grizzly bears, camels, and polar bears. Paul Allen, the cofounder of Microsoft, successfully battled Hodgkin's lymphoma. Sadly, a killer whale from Iceland succumbed to this cancer of the immune system after months of fever, vomiting, and weight loss. And the neuroendocrine cancer that claimed the life of Apple cofounder Steve Jobs, while rare in humans, is a fairly common tumor of the domestic ferret and has been diagnosed in German shepherds, Cocker spaniels, Irish setters, and other dog breeds.

Wild sea turtles around the world are dying in large numbers from cancerous tumors possibly triggered by a herpes virus. Genital cancers have become rampant in marine mammals, from North American sea lions to South American dolphins to open-ocean sperm whales. Many of these cancers are brought on by rampaging strains of the papilloma virus, which in humans can cause cervical cancers and genital warts.

So severely is the disease assailing some animal groups that three wild species are facing extinction because of cancer. Tasmanian devils, found only on their namesake island off mainland Australia, are in the midst of an epidemic of devil facial tumor disease, a cancer that spreads when they fight. Deaths from cancer are hindering conservation of endangered Attwater's prairie chickens, which used to thrive across Texas, and Western barred bandicoots, an Australian marsupial.

Cancer can grow in insects, including fruit flies and cockroaches. The disease can even be destructive in the plant world, although plant tumors, sometimes called "galls," cannot metastasize and so, for plants, cancer is a chronic condition, not a leading killer. Although cancer rarely kills the plant, it does decrease its vigor.

One thing is clear: cancer is not unique to humans. And neither is it a product of our modern times. More than 3,500 years ago, before soup cans were lined with bisphenol A–laced plastic, before hormones were pumped into meat, and before methylparabens were added to shampoos, Egyptian physicians described human breasts with "bulging tumors."

Ancient Greek doctors, including Hippocrates, explicated cancer in their medical texts (and coined the term *karcinos,* which means "crab"). The disease appears in ancient Indian Ayurvedic and Persian medical books and in Chinese folklore. Galen, the renowned second-century Greek physician who practiced in Rome, said breast cancer was the most common of the many cancers he saw. In fact, as James S. Olson writes in *Bathsheba's Breast,* "Among ancients, breast cancer *was* cancer," primarily because it was the one they could easily see.

In the last few decades, paleopathologists have used X-rays and other methods to survey Egyptian mummies. They've examined Bronze Age skeletons from Britain and preserved corpses from Papua New Guinea and the Andes. While their data is admittedly limited—no soft tissue, DNA degradation—the researchers widely agree that cancer did indeed exist in human antiquity. But it's even older than that.

In 1997, amateur fossil hunters happened upon the fossilized remains of a female meat eater known as *Gorgosaurus,* a lanky cousin of *T. rex.* The paleontologists from the Black Hills Institute of Geological Research who examined her became intrigued by a puzzling finding. In spite of her fearsome, five-inch-long serrated dagger teeth and impressive twenty-five-foot height, this *Gorgosaurus* was riddled with injuries: a lower-leg fracture, fused vertebrae in her tail, a shattered shoulder, broken ribs, and a raging, pus-filled jaw infection. Examining the fossils with electron microscopes and plain radiographs revealed a possible explanation for these multiple injuries. The scans showed evidence of a mass in the dinosaur's skull. While paleontologists have argued over the nature of this mass, some experts believe it to be the fossilized remains of a brain tumor.

A tumor positioned in the ancient animal's skull would have pressed on her cerebellum and brainstem. These areas are critical regulators of motor activity, balance, memory, and autonomic functions like heart rate. What this meant for the dinosaur is written into her injured skeleton. Researchers suggest that the burgeoning tumor likely affected her daily life.

"As the tumor grew, the dinosaur—a female perhaps three years old—would have forgotten where she left her last kill, and then she would have forgotten to go to the bathroom," said one. A tumor in that position meant she wouldn't have been able to move quickly or make rapid predatory decisions. Like many humans with brain tumors, this ancient creature might have had pain—excruciating headaches upon waking and when bearing down for a bowel movement or any time she bent her head lower than her heart, perhaps to drink or feed or mate.

Other paleo-oncologists have found tumors in hadrosaurs, the duck-billed prey favored by *T. Rex*. At the University of Pittsburgh, medical students learn about cancer by studying a 150-million-year-old diseased dinosaur bone on loan from the Carnegie Museum of Natural History. And evidence of probable metastatic cancer has been found in the bone of a Jurassic dinosaur that lived some 200 million years ago.

Because dinosaur DNA would have been subject to transcription errors similar to those humans face, it's not surprising that tumors formed in prehistoric creatures. On the other hand, environmental factors may have played a role as well. For most of us, "carcinogens" is synonymous with "man-made toxins." In fact, however, many mutation triggers are as natural as flowers, plants, and sunshine.

At times, even the most pristine, "natural" corners of our planet can become as polluted as a Superfund site. A couple of million years ago, for example, you wouldn't have wanted to be living in what is now Yellowstone National Park's unspoiled Hayden Valley. That's when the region's supervolcano spewed ash over an area that would now cover sixteen states. About sixty-five million years ago, in an area of west-central India called the Deccan Traps, a monster volcano belched more than a quarter of a million cubic miles of lava over the landscape and filled the air with toxic gases like sulfur dioxide. Ionizing radiation, toxic volcanic spew, or even Mesozoic food sources may have wreaked havoc with the DNA of the living creatures inhabiting the Earth in these areas and during these periods. In fact, cycads and conifers, the oldest living seed-bearing plants, and staples of dinosaurs' diets, contain potent carcinogens. This means that we are not the first (or only) species on Earth whose diet or environment has been infiltrated with carcinogenic substances.

"Jurassic cancer" demonstrates that while we humans may have coined the term "cancer," we certainly didn't create the condition. In fact, the sheer ubiquity of cancer makes it an intrinsic part of life. Yes, toxic exposures created by humans have amplified the risk, in some cases greatly. Several examples of cancer in animals I named earlier have been linked to environmental poisons (more on that in a moment). But the *potential* to get cancer is simply part of being a living creature on Earth, an organism with cells containing replicating DNA.

The vulnerability of DNA to mutation means that cancer "becomes a statistical inevitability in nature—a matter of chance and necessity," as Mel Greaves wrote in *Cancer: The Evolutionary Legacy.*

While nothing is likely to dull the devastation a patient feels upon hearing the dreaded words "You have cancer," perhaps there's a small measure of solace to be found in the knowledge that the disease is at least as old as the dinosaurs and as universal among today's animals as hearts and blood and bones. But a zoobiquitous approach to cancer research could promise more than psychological balm. It could lead to breakthroughs in treatments, therapies, and our understanding of the risks. In fact, it's already starting to do just that.

Imagine two animals: a tiny bumblebee bat (weight: .07 of an ounce, the size of a penny) and an enormous blue whale (weight: 420,000 pounds, the size of twenty-five elephants). The huge whale has vastly more cells in its body than the tiny bat and trillions more cell divisions over its longer life. Which animal would you predict would be more likely to get cancer? Because we know that cancer stems from a single cell's faulty replication, you might think that animals with more cells, more replications, and more mutations would have more cancer.

Genomics researchers at the University of Pennsylvania tested this hypothesis by calculating the number of cells in the human colon and comparing it to the number of cells in the colon of a giant blue whale. They concluded that if cell division and "proofreading" were identical across species, all whales ought to have colorectal cancer by the time they hit their eightieth birthday.

But as far as we know, they don't. In fact, larger species, overall, seem to get cancer less often than smaller species. This fascinating observation is called Peto's paradox, after the British cancer epidemiologist Sir Richard Peto, who recognized and first described this biologically surprising reality.

To be clear, Peto wasn't talking about the size differences between large and small members of the *same* species—say, seven-foot, six-inch basketball player Yao Ming and four-foot, eight-inch gymnast Kerri

Strug. Rather, the paradox describes cancer rates *between* species—like bats and whales. In fact, within species, larger individuals may actually have a greater susceptibility to some tumors. Osteosarcoma, for example, a malignant bone cancer seen in adolescence, occurs more commonly in tall teens. Similarly, osteosarcoma in dogs is seen most frequently in larger, long-limbed breeds like Great Danes, Dobermans, and Saint Bernards.

The implication of Peto's paradox is that there's something special about DNA replication in large animals—something that may protect them from cancer. Large-animal DNA might be more effective at repairing itself. Perhaps the cells of megafauna divide with greater fidelity to the original and so are less susceptible to cancer-causing mutations. Or maybe they contain better DNA proofreaders and lower mutation rates. Larger animals might have better tumor-suppression genes. More efficient immune systems. Or maybe their cells are just better at programmed cell suicide—apoptosis.

If nothing else, Peto's paradox shows that unexpected hypotheses can emerge from a comparative approach. But human cancer specialists don't read the *Journal of Cetacean Research and Management.* And marine biologists don't regularly attend the American Society of Clinical Oncology's annual meeting. Important clues about the nature and behavior of cancer across species remain unconnected.

Making matters more difficult, obtaining truly accurate statistical information about cancer rates in wild species is challenging at best. Doing a necropsy on every dead animal in the wild is a practical impossibility. And human-style cancer screening for wild animals is equally implausible. If regular colonoscopies were possible in wild whales, we might find clues to their cancer-protection mechanisms.

Bringing together experts in wildlife biology, human oncology, and veterinary medicine could expand our understanding of cancer itself. Academics are increasingly recognizing the benefits of an interdisciplinary approach. The National Cancer Institute's Comparative Oncology Program and enterprises like the Center for Evolution and Cancer at the University of California, San Francisco, are expanding research on cancer. The next major cancer breakthrough might not come from a basic scientist working on a genetically engineered mouse in a sterile lab, but from a veterinary oncologist thinking about bumblebee bats, blue whales, and Saint Bernards.

Perhaps one of the most promising places to look for cancer clues among our fellow species is the disease that ranks among the top killers of female humans: breast cancer. Breast cancer strikes mammals from cougars, kangaroos, and llamas to sea lions, beluga whales, and black-footed ferrets. Some breast cancer in women (and the occasional man) is connected to a mutation of a gene called BRCA1. All humans have a BRCA1 gene. It's on our seventeenth chromosome. But some of us (about 1 in 800) are born with a mutated version. For Jewish women of Ashkenazi descent, it's as high as 1 in 50.

BRCA1 appears to be an especially skilled molecular copy editor. When it's working right, it catches mistakes in the DNA every time a cell divides. It corrects typos and restores deletions. Like a great editor, BRCA1 keeps the DNA elegant, supple, concise, and true to its original intentions. But when BRCA1 is or becomes mutated, the DNA codes can get garbled and confused. Over generations of division this can lead to cancerous cell replication.

Many organisms have vulnerable BRCA1 genes. And in some animals, its malfunction seems to result in breast cancer, just as it does in humans. In one Swedish study, the presence of a BRCA1 mutation made English springer spaniels four times more likely to develop breast cancer. Jaguars in zoos in the United States that were on progestin-based birth control showed patterns of breast cancer that were very similar to those of women with BRCA1 mutations.* Zoo veterinarians report high incidence in other big cats, too, including tigers, lions, and leopards.

And yet, having a BRCA1 mutation doesn't automatically lead to breast cancer. BRCA1-related breast cancer results when genetic predisposition meets something that activates it—including hormonal and environmental exposures. Researchers call these triggers "second hits." And studying a variety of animals could help pinpoint which combination of genes and triggers leads to cancer.

This leads to the counterintuitive possibility that when it comes to breast cancer, a jaguar originating in South America and an English springer spaniel living in Sweden might be medically more relevant to an Ashkenazi Jewish woman than her next-door neighbor is. In medical jargon, we call these spontaneously occurring cancers "natural animal

---

*Sadly, Linda Munson, the U.C. Davis veterinary pathologist leading the research, died before the jaguar genome was fully sequenced and scanned for clues to BRCA1 relevance, although her early research pointed to a connection.

models," and they're prized by scientists for their power to expose the true biology of disease.

Unlike the jaguars and spaniels who carry an increased risk of breast cancer, some groups of mammals, intriguingly, may be protected from it.* The latte you sipped this morning contained milk from an animal sorority that very rarely gets breast cancer. Professional lactators—the dairy cows and goats that make milk for a living—have rates of mammary cancer that are so low as to be statistically insignificant. That animals who lactate early and long seem to have some protection against breast cancer is not only fascinating; it parallels human epidemiologic data that tie breast-feeding to reduced mammary cancer risk.

The species-spanning protective power of breast-feeding—or the hormonal states associated with it—could hint at a new form of prevention. For example, if induction of lactation, a few times a year, could be shown to dramatically reduce a woman's lifetime risk of developing the leading lethal cancer among females, it could transform preventive medicine. This may sound odd, but it's just one step stranger than many other health-maintenance routines we take for granted. Women have their cervixes swabbed and their breasts X-rayed. For many, taking hormone-altering daily birth control pills has become standard, quelling endometriosis, suppressing acne, or even guaranteeing a period-free honeymoon. We have our colons scoped, our skin scanned for moles. You might not wrinkle your nose at the idea of scheduling a preventative, induced lactation if you knew that your risk of breast cancer might plummet to the level enjoyed by professional animal lactators.

It may also be that lactation is protective because it decreases the exposure to estrogen a breast receives with every menstrual cycle. An approach suggested to me by Chris Bonar, the chief veterinarian of the Dallas World Aquarium, could help clarify exactly what it is about breast-feeding that seems to be protective. He noted that female mammals have different numbers of reproductive cycles per year. Some wild bats, for example, have vaginal menstrual bleeding every thirty-three days, a monthly cycle similar to that of some primates, includ-

---

*Here I should pause to dispel the myth—often repeated—that sharks "don't get cancer." Tumors of many kinds, some metastatic, have been found in numerous species of sharks. Rumors to the contrary are likely promulgated by those hawking alternative remedies at the expense of wild species.

ing humans. In contrast, sheep and pigs are polyestrus and ovulate only several times a year. Female ring-tailed lemurs, bears, foxes, and wolves usually cycle just once a year. But breast-feeding disrupts reproductive cycles in mothers. So by comparing breast cancer rates in female animals with different cycle frequency—and different hormone exposures—comparative oncologists could home in on an important distinction: how much breast-feeding's power to protect comes from lactation itself and how much from disrupting the hormones that accompany reproductive cycles.

Another thing we can learn from animal cancer is the extent to which it's caused by outside invaders: viruses. Veterinary oncologists see this all the time. Lymphomas and leukemias among cattle and cats are quite frequently viral. Many of the cancers sweeping sea creatures from turtles to dolphins are rooted in papilloma and herpes viruses.

As we know, cancer starts as a cell with mutated DNA. Few things in nature rival viruses for their skill at tinkering with DNA. But human physicians tasked with fending off and treating so-called lifestyle cancers, such as those caused by smoking, drinking, or overeating, tend to think "infectious trigger" only when it comes to a narrow range of malignancies. Every oncologist and many patients, for example, know that viruses are responsible for Kaposi's sarcoma, some leukemias and lymphomas, and some liver cancers. "Cancer à deux" (cervical and penile cancers shared by sexual partners) is spread by the human papilloma virus.

In fact, worldwide, about 20 percent of human cancers are viral. In Asia, the leading cause of liver cancers is viruses: hepatitis B and C. Across Africa's "lymphoma belt,"* the Epstein-Barr virus is a known driver of Burkitt's lymphoma. Human papilloma virus and hepatitis B and C are on the NIH's list of known carcinogens. The idea that cancers spread virally has led some epidemiologists to call for treating cancer as an infectious disease. This is something veterinarians already do.

Peto's paradox, Jews and jaguars, professional lactators, viral triggers—in these cases, a zoobiquitous approach can help us generate new hypotheses about the causes of cancer. But animals may be able to help us in a

*According to the WHO, "a region that extends from West to East Africa between the 10th degree north and 10th degree south of the equator and continues south down the Eastern coast of Africa."

more urgent or timely way. They may be able to warn us of impending disease—before it actually strikes us.

In 1982, dead beluga whales began drifting ashore along the St. Lawrence Estuary in northeastern Canada. The leading cause of death was a grim list of cancers. Intestinal. Skin. Stomach. Breast. Uterine. Ovarian. Neuroendocrine. Bladder.

The St. Lawrence belugas, it turned out, were saturated with heavy metals, as well as other industrial and agricultural contaminants, including dichlorodiphenyltrichloroethane (DDT), polychlorinated biphenyls (PCBs), and polycyclic aromatic hydrocarbons (PAHs). Researchers from the University of Montreal didn't have to look far for the source of these artificial intruders. Lining the coast were aluminum smelters. Every year for decades, these factories pumped tons of PAHs into the water and released other contaminants into the air. Day after day, these compounds had drifted through the water and accumulated on the ocean floor. Mussels and other sea-dwelling organisms absorbed them. When the belugas scooped into the sediment to feed, they received a double dose of toxins—in the sand and silt as well as in their food.

The contaminants were linked to the whale cancers and the die-off. Significantly, another group of animals living around the St. Lawrence Estuary at the same time shared the whales' unusual cancer patterns: humans.

When animals die in clusters, we're wise to pay attention. Emerging infectious diseases like SARS and avian influenza often show up first in animals. Endocrine-disrupting chemicals may manifest in animals before they affect human fertility. Animals even forewarn us of biological attacks or chemical leaks; when anthrax escaped from a Soviet military facility in 1979, for example, the first to die were nearby livestock.

Sometimes the animal warnings point to cancer. Although PCB production and DDT use have been banned in the United States for more than thirty years, researchers now suspect that the toxins may be contributing to a significant increase in the outbreak of sea lion cancers off California. For thirty years, starting in the 1940s, manufacturing companies used this part of the Pacific to dump millions of pounds of those chemicals. Although the Environmental Protection Agency instituted a clean-up in 2000, there's one reservoir that's hard to get at: the animals' own bodies. Through gestation and lactation, up to 90 percent of a mother's contaminant load can be "dumped" into her first-born pup. Veterinary oncologists believe that either repeated "hits" of the toxins are causing the animals' cells to mutate or they're suppressing the animals' immune systems so much that the herpes virus that causes the cancer has a better chance to replicate. If that's true, these animal cancers could be warnings to humans living in areas polluted by similar chemicals that the dangers of toxins might go beyond direct exposure. They may be passed down through generations (meaning their presence can be felt long after a toxic site is cleaned up) and/or have secondary effects on the immune system.

Industrial pollutants are causing animals to suffer and die. Responsibility for these animal illnesses lies squarely with our species. Indeed, if animals could lawyer up, we humans would probably find ourselves as the defendants in any number of class-action lawsuits. Beluga whales shouldn't have to die terrible deaths from cancer because we allow our industries to foul the waters where they eat and breed.

So with the proviso that in an ideal world we wouldn't give animals cancer for the convenience and greed of certain industries (many of which we all partake in, whether petroleum or plastic or pesticides), the sad fact that animals get cancer can be helpful to humans if we think of them as sentinels. And one way to honor their sacrifice is to do something about it. Not pretend it doesn't affect us. As governments, as societies, as a species, we need to act when we see disease emerging in clusters of animals—to save them and to save ourselves.

As humans, we don't live in the waters of the St. Lawrence Estuary or the Pacific kelp beds off California. We live in condo complexes and single-family homes, studio apartments, farmhouses, and trailers. And in all of those places, who lives with us? Dogs.

Hundreds of millions of dogs around the world coexist with us as pets. At the simplest, most expedient level, this means they can serve as in-house sentinels to warn about or confirm cancer risks. One study of nose and sinus cancers in dogs, for example, found a strong correlation with the use of indoor coal or kerosene heaters. The longer the dog's nose, the greater its chance of getting this cancer, possibly because of a greater exposed nasal surface area. Bladder cancer and lymphoma, both linked to pesticides, have been reported in pet dogs, with a higher risk for bladder cancer found in female dogs that were obese. And military dogs who served in Vietnam had higher-than-usual rates of testicular cancer, possibly because of their exposure to a variety of chemicals, infections, and medications during their tours of duty.*

Where we don't overlap might be telling, too. Dogs rarely get colon cancer. Lung cancer is also atypical, although short- and medium-nosed dogs living in homes with smokers are susceptible. Canine breast cancer is rare in countries that promote spaying but quite common where most female dogs remain reproductively intact. In humans, too, oophorectomy (removal of the ovaries) and premature ovarian failure dramatically reduce breast cancer risk.

But beyond their possible use as canine canaries in our domestic coal mines, dogs may be ideal proxies for studying the actual biology of how cancer works in our own bodies. Currently, the vast majority of cancer studies are done on mice. So-called humanized mice have been specially bred to mimic our gene patterns. Often their immune systems have been altered to allow the cancer to grow. Most lab mice are "given" cancer; usually it doesn't arise spontaneously in their bodies. For decades, this "artificial" cancer has yielded useful insights into tumor biology—how cells divide, how tumors form, how they metastasize and spread to other parts of the body. But the genesis of the disease, its complexity, how it becomes resistant to therapy, and how it recurs are really not answerable in mouse models. Even side by side, under a microscope, mouse tumors are very different from human tumors.

Our human tumors, it turns out, are remarkably similar to those of the animals with whom we live: our pet dogs. Dog cancer cells and human cancer cells are nearly indistinguishable. Dogs live longer than mice, so researchers can observe the cancer—and the treatments—over the long term. And, unlike most lab mice, pet dogs' immune systems are intact, allowing oncologists to study how a cancer acts when chal-

---

*Cats have served as sentinels, too: one study linked oral cancer in cats to environmental tobacco smoke.

lenged by natural defenses. Dogs are also simply much bigger than mice. This has implications, both practical—the tumors are physically easier to see—and philosophical (think of Peto's paradox).

Here I must pause to make absolutely clear that I'm *not* talking about experimenting on dogs in labs. I'm talking about the opposite. Observing cancer while caring for companion animals—pets—who develop cancer spontaneously and receive treatment from veterinarians.

This novel approach is known as comparative oncology. Recognizing that studying naturally occurring cancers in the animals that share our homes might unlock some of cancer's mysteries, the National Cancer Institute launched the Comparative Oncology Program (COP) in 2004. One of the COP's early innovations was to pool the brain trusts of twenty top-tier veterinary teaching hospitals in the United States and Canada. This network, called the Comparative Oncology Trials Consortium, conducts clinical trials in pet dogs, searching for new anticancer drugs and treatments for human patients. (The trials are sponsored by pharmaceutical companies hoping to bring new human therapies to market.) But while pet health may not be the intended goal of the program, some of the advances that benefit humans will come back full circle to enhance the health of animals, too.

Comparative oncology has already improved the health of many animals, including us. It's not a stretch to say that new cures have come out of cross-species cancer comparisons (although physicians and veterinarians alike tend to manage expectations with more clinical terms like "novel therapeutic strategies" and "positive survival rates"). For example, the limb-sparing technique that human doctors use today to save teenagers with osteosarcoma from amputations was pioneered in dogs by a veterinary oncologist, Stephen Withrow, and his team at Colorado State University, working jointly with physicians. And a potential cure for malignant lymphoma using transplanted stem cells was first successful in twelve pet dogs under treatment at Fred Hutchinson Cancer Research Center in Seattle, paving the way for this technique to be used in human beings.

Veterinary gene chasers are currently looking at DNA for molecular clues to canine lymphoma, bladder cancer, and brain cancer. And here's why genes are relevant: picture a Great Dane looming over a Chihuahua or a Saint Bernard sniffing a pug. Members of *Canis lupus familiaris*, although belonging to the same species, can look and act extremely different from one another. But those desirable differences—traits honed over centuries of selective breeding and codified in the American Kennel Club Blue Book—carry an ironic and sometimes tragic Trojan horse. As Kerstin Lindblad-Toh, the MIT molecular biologist who led the canine genome-mapping project, explained to me, breeding for desirable traits inadvertently selects for and transmits other mutations, some of which can cause cancer.

The way German families from the Black Forest region are susceptible to kidney and retinal cancers, or Ashkenazi Jews to breast, ovarian, and colon cancers, certain dog breeds are prone to certain cancers. German shepherds, for example, can develop a kind of heritable kidney tumor. As the veterinary oncologists Melissa Paoloni and Chand Khanna explain in a review published in *Nature Reviews Cancer*, the genetic mutation that causes the dog cancer is similar to the one that leads to Birt-Hogg-Dubé syndrome in people—which makes them vulnerable to kidney cancer, too. Salukis, descended from the royal dogs of ancient Egypt, are among the oldest breeds. Their chromosomal legacy codes for their lean, regal elegance but also for a one-in-three chance of developing hemangiosarcoma, a highly aggressive tumor of the heart, liver, and spleen occasionally seen by human cardiologists, hepatologists, and oncologists.*

Paoloni and Khanna note that chow chows have higher-than-usual rates of gastric carcinoma and melanoma. Boxers lead the list for developing mast-cell cancer as well as brain tumors. Bladder cancer disproportionately strikes Scottish terriers. Histiocytic sarcoma (an extremely

---

*When certain populations show the same mutation, it's usually a result of the "founder effect." That's when a long line of descendants arises from a very few progenitors and for some reason—geographic, cultural—remains isolated. The founder effect has been noticed in populations from microbes and plants to animals, including humans. The mutation that causes cystic fibrosis, for example, can be traced to one person. The founding individual who first carried the BRCA1 mutation in Ashkenazi Jewish families is thought to have lived more than two thousand years ago.

Geneticists frequently see founder effects in bottleneck populations. These are groups where certain factors mean that many descendants come from few ancestors. Cheetahs are a natural bottleneck population. As their numbers dwindle, they rely on the genes of fewer and fewer breeding members to keep their species alive. This is an issue for many endangered species. With domesticated dogs, we humans create bottleneck populations on purpose: by breeding all future descendants from a set number of progenitors, we limit the genes in that pool, including the mutated ones.

complicated cancer that hides out in locations like the spleen) favors flat-coated retrievers and Bernese mountain dogs.

But noticing where cancer *isn't* can be as instructive as noticing where it *is*. As Paoloni and Khanna point out, remarkably (although so far still inexplicably), two breeds of dogs seem to get cancer less often than the others: beagles and dachsunds. Like the professional lactators who rarely get breast cancer, these extra-healthy dog breeds may point to behaviors or physiology that offer cancer protection.

Despite all the possibilities that lie in comparative oncology, only a fraction of human doctors ever think beyond the mouse. As a UCLA oncologist colleague confirmed to me, even the smartest human cancer researchers *never* talk about naturally occurring animal cancers.

And while initiatives like the COP are slowly changing that, zoobiquitous collaborations between physicians and veterinarians are, at present, all too rare. If we could change this, the world of cancer care and cancer research might look quite different. I learned this for myself when I heard the story of a fortuitous meeting of two oncologists, one a physician, the other a veterinarian, that resulted in a radical new treatment for melanoma.

In many ways, the dinner crowd at New York's Princeton Club that autumn evening in 1999 was like any other. Blue blazers and regimental ties. Silvering temples. Smart skirts and pearls and pumps. Conversation probably tumbled around the Y2K bug, an exciting new HBO series called *The Sopranos,* and gas prices that were climbing to a steep $1.40 per gallon after hovering below the dollar mark for most of the preceding summer. Silently surveying it all, as it had for decades, was the cold metal eye of a bronze tiger on the wall.

But at one table, the banter was anything but ordinary. Around the starched white tablecloth, ice clinking in the water glasses, sat a dozen or so scientists intently strategizing about lymphoma. With one exception, they were all human cancer experts.

Listening quietly at first was the sole outlier, Philip Bergman. Bergman, who is tall, with thick, wavy dark hair and a groomed Van Dyke beard, is a veterinarian. He has the calm, measured voice and lack of extraneous movement that mark nearly every animal doc I've met. That night, though, he was feeling a little out of his element. As he told me a few years later, he kept thinking: "This is the *Princeton* Club. I'm a veterinarian. I don't really belong here." (Never mind that he spent years training at the M. D. Anderson Cancer Center and holds multiple degrees, including a Ph.D. in human cancer biology.)

Near Bergman sat Jedd Wolchok, an M.D. and Ph.D. with board certifications in human internal medicine and oncology. Wolchok was a rising star at Memorial Sloan-Kettering, one of the leading cancer research hospitals in the world. Suddenly Wolchok turned to Bergman. And out of his mouth came a most zoobiquitous question.

"Do dogs," he asked, "get melanoma?"

It was the right question, the right person, the right moment. Bergman happened to be one of the world's few experts in how this difficult, aggressive form of cancer attacks dogs. And he was looking for his next big project.

Bergman and Wolchok started comparing human and canine melanoma. They quickly learned, as Bergman put it, that "the diseases are essentially one and the same." In humans as in dogs, malignant melanomas often show up in the mouth, on foot pads, and under finger- and toenails. In both species, it metastasizes to the same "weird spots," favoring the adrenal glands, heart, liver, brain membranes, and lungs. In humans, melanoma resists chemotherapy. Surgery and radiation often don't keep it from spreading. It has a nasty trait of recurring, even after treatment. Same thing in dogs. Sadly, both humans and dogs have a very low survival rate with this cancer. Once diagnosed with advanced canine malignant melanoma, dogs can have as little as four and a half months to live. Human patients with metastatic melanoma often live less than one year. Wolchok and Bergman both knew that, for the sake of patients of both species, new approaches to malignant melanoma were "desperately needed."

Wolchok confided to Bergman that he was on the trail of a novel therapy, one that would trick a patient's immune system into attacking its own cancer.* His team at Sloan-Kettering had had some early suc-

---

*It's called xenogeneic plasmid DNA vaccination. Essentially, it "hides" proteins from a foreign species within the cells of a patient with cancer. When these foreign proteins circulate through the blood and lymph, the immune system senses the alien proteins. It thinks there's an invader afoot, and mounts an attack on its own cells. Getting the immune system to attack itself is called "breaking tolerance"—and it's so hard, said Bergman, that it's the "holy grail of cancer immunotherapy."

cess with mice. But they needed to know how the remedy might fare in animals with spontaneously occurring tumors, intact immune systems, and longer life spans. Bergman realized instantly that dogs could be that animal.

In three short months, Bergman had a trial up and running. He recruited nine pet dogs: a Siberian husky, a Lhasa apso, a bichon frise, and a German shepherd, as well as two cocker spaniels and three mixed breeds. All had been diagnosed with various stages of melanoma. For most of these pets, the experimental treatment was their last chance—and it was eagerly embraced by their grateful owners.

The therapy—which involved injecting *human* DNA into the dogs' thigh muscles*—worked even better than Bergman and Wolchok expected. Overall, the dogs' tumors shrank. Their survival rates soared. When the news of the success got out, Bergman started getting calls and e-mails from desperate dog owners all over the world. One client flew to New York from Napa Valley every two weeks so his dog could receive the injections. Another moved from Hong Kong with her pet and took up residence near Bergman's New York office. Before long, Bergman had more volunteers for the new therapy than he could handle. With financial support from the drug company Merial and help from Sloan-Kettering to produce the drug, Bergman launched another round of trials. And even when the spots were filled, the owners of canine cancer patients kept calling.

The treatment was ultimately tested in more than 350 pet dogs—and prolonged life so well that more than half the animals who got the injections exceeded their cancer-shortened life expectancies. In 2009, Merial released the vaccine to veterinary oncologists, under the name Oncept, making the treatment available to thousands of family pets stricken with cancer.

---

*From a *human* melanoma cell donated by an anonymous patient, the gene jocks at Sloan-Kettering extracted human tyrosinase cDNA. They shaped each strand into a ring and cloned it millions of times. Then Bergman injected these tiny doughnuts of DNA, called plasmids, into the dogs' thigh muscles, using a high-pressure, needle-free delivery system sort of like a high-tech air gun.

Deep inside the dogs' muscle and white blood cells, the plasmids started making *human* tyrosinase. Then the cells released the human proteins into the dogs' blood and lymph . . . which is where they encountered the immune system's fighter cells, called T cells. Not recognizing the human tyrosinase, the dogs' T cells attacked it. This immune response sparked the T cells to go after canine tyrosinase in the dogs' tumor cells as well.

Wolchok's four zoobiquitous words—"Do dogs get melanoma?"—sparked an intense collaboration, one that may have permanently changed the way veterinarians treat the disease in canine patients. And the translational potential is enormous. Bergman and Wolchok's success is inspiring work on a similar vaccine for melanoma in humans.*

Yet Bergman knows that, even with the success of Oncept, human medicine may still take a while to wake up to the possibilities of interspecies collaboration.

"Almost without fail, when I tell this story to groups of human doctors," he said to me—adding politely yet pointedly, "no offense to your colleagues"—"someone will come up to me afterward and ask, 'How did you convince those dog owners to let you give their pets cancer?' " Bergman chuckled. "I have to explain. These are not lab dogs. We didn't 'give' them cancer."

What he actually gave them was another shot at life.

---

*For the time being mice, not dogs, are providing the foreign tyrosinase.

FOUR

Roar-gasm

An Animal Guide to Human Sexuality

Lancelot* was having a rough morning. He kicked puffs of dirt up from the barn floor and snorted. A handful of students hovered around him warily, gauging his movements. He froze for a moment, standing tall on his dark brown legs and shifting his muscular haunches.

"Urine!" shouted Joel Viloria, the barn supervisor. In an instant, a student appeared with a plastic pouch of "liquid gold," urine that had been collected from a mare in heat and then frozen. Joel wafted the urinary ice cube under Lancelot's velvety nose. Clearly stimulated by the aroma, the stallion flared his nostrils, and his head reared back.

"Give him another look at the mare," Viloria commanded tensely. The thousand-pound stallion was led past a stall at the side of the barn. There, bathed in streaks of February sunlight, stood a pale young horse, her tail raised obligingly and receptively in a classic equine come-hither stance. Lancelot beelined for her.

"Okay, go!" urged Viloria, his voice firm but calm. Quickly, the stallion was steered away from the mare. But one of Lancelot's big eyes remained

---

*Not his real name.

on her, rolling sideways in the socket, as he was led away. "That's right, good," encouraged Viloria when Lancelot mounted one end of the padded metal breeding apparatus horsemen call a phantom mare.

The stallion struggled, his sleek forelegs gripping the sides of the metallic mare as though he were copulating with a real horse. But he slipped off. A student gently guided him to remount. Distracted, he tried. But again he slid back. This time, when the student tried to return his attention to the phantom, Lancelot pulled away and refused to get back on.

"All right, that's three—he's not in the mood. Take him back," said Viloria. The stallion was led away to his corral, his dark chocolate tail swishing across his flanks.

As Viloria later explained to me, the University of California, Davis, horse barn he oversees abides by a strict three-mount rule. When producing semen for breeding, each stallion is given three chances to accomplish what in nature seems like a very straightforward task. But this is not nature. First, the horse must become aroused and erect. He must next mount the metal-and-vinyl phantom. Then he must insert his penis into a lubricated, warmed metal tube underneath the phantom, thrust a few times, and ejaculate into the one-gallon plastic condom lining the tube. If on the third attempt he still hasn't produced a specimen, he's declared done for the day and led back to his corral for a restful (though possibly frustrating) afternoon and night.

Experienced horse breeders like Viloria know that even very practiced stallions sometimes can't perform sexually. As one website counsels, "Most people think of stallions as being big and tough, but they actually are quite sensitive. Circumstances have to be to their liking for them to feel comfortable with breeding."

Even when copulating with a live mare instead of producing a semen sample for artificial insemination, stallions can suffer from stage fright, intimidation, distraction, and inexperience. Male horses who are punished for sexual behaviors by rough handlers or mean mares in their youth may develop inhibitions around mounting and copulation as adults.* Some stallions show sexual interest in mares but will not mount. Some will mount but will not penetrate. Others get through the first

---

*At the other end of the spectrum, stallion experts know that "too much serious sexual experience too early" is detrimental to normal libido. Stallions "overused" as youngsters often develop low libido or even impotence as adults.

two stages but cannot ejaculate. And there are stallions who will mount mares only when another particular horse is present or watching. Among some social animals, including horses, the highest-ranking males dominate mating. The runners-up are deprived of most sexual opportunities. Their lower status and forced celibacy put them at risk for what veterinarians call "psychological castration," the eventual inability to have sex at all.

As Jessica Jahiel, an author and expert in the fields of horses and horsemanship, writes, "Pain, fear, and confusion can all lead to vastly decreased libido and sometimes an inability to breed."

Like veterinarians, human physicians also encounter patients in whom fear, pain, and confusion (and many other factors) interfere with the ability to have erections. As medical students we're taught to ask every patient about their sexual function and satisfaction, and we know we should because sexual performance is a useful measure of cardiovascular fitness. But the truth is that many doctors find it easier to ask Mr. Green whether he can walk up two flights of stairs without symptoms than whether he has chest pain during intercourse. Unless a patient brings up a specific sexual problem during a visit, physicians aren't likely to inquire about the quality and frequency of a patient's erections, ejaculations, and orgasms.

Cultural barriers, time constraints, and even prudishness get in the way of in-depth discussions about sex between physicians and patients. So, while a patient's sex life contains key information about his or her overall health, most physicians will address only those sexual problems that a patient feels need fixing.

Veterinarians, on the other hand, see and deal with sex much more as a normal part of their patients' lives. The first time I attended morning rounds at the Los Angeles Zoo, I was surprised by the careful attention the vets and keepers paid to the sexual activity of the animals in their care. How much, how often, and with whom—it was all valuable information relating to the physical and mental well-being of their patients. And the discussions proceeded without the uncomfortable silences and flushed faces I've seen in human exam rooms.

Spend any time around animals and you'll notice that sex comes in many forms. Some species commit to monogamous lifelong partnerships. Others are outrageously promiscuous and spread sexually transmitted diseases (STDs). There are species that engage in heterosexual behavior at one phase of their life and homosexuality during another. There are animals who rape, animals who trick partners into sex, animals who force themselves on their young. There are also animals who engage in what appears to be lengthy foreplay. Animals who fellate their partners. Animals who secure a form of consent before engaging in intercourse.

Careful scientific scrutiny of the shared biology and behavior of animal sex sheds light on the evolutionary background of human sexuality. A zoobiquitous survey of animal erections, copulations, ejaculations, and even orgasms could advance the treatment of human sexual dysfunction. And it might even uncover ways to enhance our sexual pleasures.

In this chapter we'll journey through a world of insect foreplay, comparative clitorology, and the shared pleasure of orgasm. But there's no better place to start this tour of the sex lives of animals, both human and nonhuman, than with the extraordinary feat of biomechanical engineering called the male penile erection.

Not surprisingly, when physicians study penises, we tend to focus on the human variety. But our world is abristle with phalluses and has been for at least half a billion years. Today and every day since at least the early Paleozoic era, in meadows, oceans, streams, and the air, many trillions of erections preceded trillions of copulations, which preceded trillions of ejaculations. Some erections sprouted readily and penetrated easily. Others flickered to life and abruptly terminated. Some were measurable in yards. Others were microscopic. Some were stiffened by blood; others by a similar fluid called hemolymph; others by skeletal supports made of cartilage or bone. Some erections culminated in mere seconds; others lasted hours.

It wasn't always this way. The earliest single-celled organisms on Earth simply cloned themselves. Some of their descendants still do. But as complex multicelled organisms evolved and eventually "discovered" the ability to mix their gametes, they gained a giant genetic advantage. (For more on this, see Chapter 10, "The Koala and the Clap.") Since these ancient creatures lived in the sea, the earliest sex was a straightforward process of spraying sperm and eggs into the water. The lucky few connected.

In that massive free-for-all, the fittest sperm reached the eggs and were rewarded with the prize of bringing their DNA into natural selection's next round. Sometimes the fittest sperm were the strongest swimmers. Sometimes they were the ones deposited nearest the eggs. Others developed ways to follow molecular scent pathways to find the eggs. Or they bundled together in teams to improve their timing and accuracy. As sperm perfected ingenious rudders, tails, chemical markers, and swimming strategies, the genital hardware that ejected them was evolving, too.

One innovation was internal fertilization, which allowed males to place their sperm not only near females but right inside them, next to the eggs. This allowed both males and females a measure of control over their offspring's DNA. Females could audition males before allowing them to mate. Sperm was less likely to be spilled on barren ground. And one invention that effectively accomplished this combination of choice and precision was the penis.*

The oldest penis on record goes back 425 million years. It belonged to a crustacean found preserved under ancient volcanic ash at the bottom of a sea that used to cover Herefordshire, in England. The paleontologists who found the shrimplike creature named it *Colymbosathon ecplecticos,* from the Greek for "astounding swimmer with a large penis." Before it was found, the oldest known phallus was 400 million years old. It belonged to a fossilized daddy longlegs from Scotland.

When dinosaurs roamed the ur-continent of Pangaea about 200 million years later, their penises roamed with them. Paleontologists have speculated about dinosaurs' mating apparatuses and behaviors, using what they know of crocodilians and birds, today's relatives of those prehistoric creatures. The erect penis of a male titanosaur, for example, may have been twelve feet in length. Experts speculate that the male sauropod, with a body the length of a school bus, approached the massive, receptive female from behind. Like his crocodilian and avian descen-

---

*Not all internal fertilization requires a penis. As the behavioral ecologist Tim Birkhead has noted, male cockroaches, scorpions, and newts produce a sperm packet called a spermatophore that they attach or place near the female's reproductive opening. Most squid, octopuses, and cuttlefish use a specially modified limb to transfer spermatophores into the female. Many birds simply touch their genital regions together when they have sex.

dants, he likely inserted his penis from this dorsal position and, at climax, ejaculated sperm through a vessel running along the outside of his organ.

Nowadays, Earth's penises exist in multivaried splendor. Spiny anteaters sport four-headed varieties that rotate between copulations. Although most birds don't have penises,* the phalluses of Argentine lake ducks are nearly eight inches long (almost as long as an ostrich's), corkscrew-shaped, and festooned with dense, brushy spines that sharpen to hard spikes at the base. Despite a thirty-three-inch member and a penis-to-body ratio of seven to one, *Limax redii,* a Swiss slug, doesn't have the most impressive proportions in nature. That title goes to *Balanus glandula,* which wows the tide pool with its prodigious barnacle penis.† Permanently cemented to a tidal rock, the barnacle sports a penis forty times the size of its body. Barnacle penises, as long as they are, vary in their girth. Barnacles living in rougher waters sport thicker, stronger, and sturdier members. But those in calmer surroundings extend their longer filamentous penises in search of distant barnacle "vaginas."

Fleas and some worms also have hugely proportioned penises. And some animals have more than one. Several species of marine flatworms have dozens of penises. Some snakes and lizards are doubly endowed; switching between their two hemipenes during multiple copulations increases their sperm count by a factor of five. As for insects, so exuberantly inventive are their male genitalia that entomologists scrutinize them to classify entire species.

If you haven't thought much about the procreative thrustings of other animals, especially those you can't see, you're not alone. Many animals are nocturnal, extremely small, shy, or just very careful to mate where other animals (including curious biologists) can't see them. Inaccessibility to these covert proceedings has been a barrier to the comparative study of sexuality.‡ But the challenges of achieving up-close analyses of

---

*Birkhead notes, "It is generally assumed that most birds lost their penises over evolutionary time—probably as a weight-saving adaptation to flight, for their reptilian ancestors possess one (or, in some cases, two)."

†Although barnacles are generally hermaphrodites (they possess both male and female genitalia), they prefer to have sex with other barnacles as opposed to with themselves.

‡Nonetheless, interest in comparative aspects of male genitalia has had a long history, starting with Paleolithic cave paintings and continuing to the Icelandic Phallological

these animals in flagrante delicto have meant gaps in knowledge and frank misinformation.

The sexcapades of krill, for example, have been seriously underestimated. These tiny shrimplike creatures make up the bulk of the diets of important aquatic megafauna, including whales. It had long been assumed that krill reproduce by mixing their eggs and sperm near the surface of the water. In 2011, however, the journal *Plankton Research* reported the surprising discovery that Atlantic krill—all 500 million *tons* of them—mate at depth. In these deep, dark, underwater orgies, krill use internal fertilization techniques that involve penetrative sex.

Since arising more than 200 million years ago, all male mammals have had penises, each achieving erection in one of three ways. An actual penis bone, called a baculum, offers a stiffening assist to many male bats, rodents, carnivores, and most nonhuman primates. A rope of thick tissue running down the center of the shaft partially stiffens the fibro-elastic penises of pigs, cattle, and whales. (The popular chew toy sold in pet stores called a bully stick is made by drying out this bull penis structure.)

But humans, along with armadillos and horses (not to mention several nonmammals like turtles, snakes, lizards, and some birds), have what's called an inflatable penis. These organs thicken and harden using only hydraulics and internal compartments of spongy tissue that fill up with blood or other body fluids.

From a biomechanical perspective, these inflatable kinds of penises are really quite extraordinary. As Diane A. Kelly, a biologist and penis expert at the University of Massachusetts, Amherst, explained to me, creating a structure adequately stiff for penetration that is also strong enough to withstand intravaginal thrusting is a tricky mechanical challenge. The steps that go into building a hard penis have an elegant flow that would please any professor of engineering.

It starts with the deceptively inert, flaccid penis. A penis in repose,

---

Museum. Devoted exclusively to phallology—the study and collection of penises—it houses embalmed or dried penises from most of the mammalian species in Iceland. On display at the museum, for visitors' scrutiny, are the embalmed members of a narwhal, polar bear, Arctic fox, reindeer, and many species of whale. Most of the specimens are housed in jars of formaldahyde, but an impressive (though flaccid) elephant penis hangs from a wall.

although it seems floppy and relaxed, is actually in a state of constant, moderate contraction. The tube of smooth muscle that runs down its center is mildly tensed. So are the linings of the thousands of tiny blood vessels that crisscross the organ. Further contraction of this muscle and the arteries is what accounts for shrinkage in cold weather or water. So although a penis in the process of erecting can seem like it's springing into action, it's actually submitting to a crucial, and opposite, process. First it must relax.

The command to relax comes from the pudendal nerves. When the smooth muscle lets go, arteries deep in the penis dilate. The channels suddenly open up. Blood rushes in, straightening the vessels and filling millions of tiny pockets in the two tubes of spongy tissue (called corpus cavernosum) that run the length of the penis shaft.

Next comes a key chemical reaction. When arteries dilate anywhere in your body—whether in your cheeks when you blush, your gut when you eat, or your genitals when aroused—they release nitric oxide.* In the penis, this very special molecule (not to be confused with nitrous oxide, your dentist's laughing gas, or nitrogen dioxide, the air pollutant) signals the smooth muscles to relax even further. More blood rushes in. By this point, the penis is crowded with liquid, and the increased volume compresses nearby veins, blocking their blood from flowing back out again. The chamber becomes tenser and tenser with trapped liquid, assisted by other structures that tighten and constrict. Pressure soars inside the fleshy tube. Most erections reach an internal pressure of one hundred millimeters of mercury—comparable to that which a boa constrictor might use to suffocate its prey.

To protect the organ from rupturing under this intense force, a complex net of collagen fibers surrounds the outside of the penis, under the skin. As Kelly describes, the collagen strands are arranged in deeply folded, alternating perpendicular layers along the length of the penis. This allows them to pleat open efficiently when the erection is under way. Not only does this collagen "skeleton" strengthen the erection, it gives the structure a resistance to bending that engineers call "flexural stiff-

---

*In the 1990s, scientists figured out that nitric oxide could be delivered in pill form and thus was born Viagra and other erection-enhancing drugs. This discovery restored sexual function to millions of men and garnered the 1998 Nobel Prize in Medicine for my UCLA colleague Lou Ignarro, as well as Robert Furchgott, and Ferid Murad.

ness." (Kelly says it's a trick shared by pufferfish, whose expandable skin also contains highly crimped, alternating strands of collagen.) When the penis is not being used for copulating or mating displays, the erective construction has the added benefit of folding away for neat storage. Being able to stow your penis provides more than simple convenience. A study on certain fish that cannot retract their reproductive organs—because they're modified, permanently stiffened anal fins—showed that males with longer ones suffered higher rates of predation than those with organs that were less obviously on display.

When the erection is complete, and stimulation has reached what doctors poetically if vaguely describe as the "point of no return," a spinal cord reflex causes a sudden burst of muscle contractions throughout the genital area, starting with the neck of the bladder. In rippling chains of contractions fueled by massive outflow from the sympathetic nervous system, the muscles around the testes and scrotum tense, followed by those of the epididymis, vas deferens, seminal vesicles, prostate gland, urethra, penis, and anal sphincter. The rapid clamping and unclamping of these muscles, at intervals less than a second apart, spurts semen out of the urethra. A few slower spasms may follow that initial explosion of muscle activity. This sequence has been preserved across a wide spectrum of mammalian species.

The comparative study of ejaculation has focused mostly on primates and rodents. But all male mammals descend from shared ancestral ejaculators. The penises of mammals from narwhals to marmosets to kangaroos propel semen in nearly identical ways. And the ejaculation of a male human today even shares basic physiology with reptiles, amphibians, and sharks and rays. Ejaculation isn't new. In fact, the human seminal propulsion system has ancient origins. This makes it not only intriguing but plausible that the human male's experience of ejaculation may be shared by other animals. With the mechanics being so similar, the question is, do other animals experience the intense pleasure that drives so many men to such good and bad behavior?

The experience of orgasm is not only legendary but also measurable. Electroencephalograms show brain-wave shifts, including an increase in slow-frequency theta waves, which are associated with deep relaxation. Many men describe a feeling of euphoria intriguingly similar to what heroin users describe experiencing when they plunge a needle into a blood vessel and discharge the drug into their system. The brains of ejaculating male rats are known to release powerful chemicals, including heroin-related opioids, oxytocin, and vasopressin. Taken together, the muscle contractions, brain changes, chemical rewards, and relaxed feelings add up to create the male orgasm.

After ejaculation and orgasm, a process called detumescence, or deflating, begins. Neurohormonally, this sequence is essentially simply the reverse of erection. The smooth muscle of the penis shaft contracts. So do the penile arteries. Blood flow to the penis decreases. With less pressure pushing them shut, the veins open up and normal drainage resumes. Chemicals associated with the sympathetic nervous system begin to take over. And before you know it, the penis is back in its resting state of slight contraction.

Clearly, a lot has to happen for this amazing structure to build on cue. But with that many dependent steps, a lot can go wrong. To complicate matters, human erections can be achieved in essentially two ways: through thought or through touch.

As most men can attest, the penis is perfectly capable of achieving an erection purely from direct stimulation. It's called a reflexogenic erection and is regulated by nerves in the lower spine. Reflexogenic erections are well known to prepubescent boys, men in deep states of REM sleep, and men with spinal cord injuries (in whom the nerves connecting brain to penis have been severed). Reflexogenic erections are as unconsciously controlled as digesting and breathing; they can spring up when a man least suspects or wants them.*

Reflexogenic, early-model proto-erections, in species such as barnacles and mollusks, evolved long before reptilian or mammalian penile stiffening. While effective at penetration and sperm delivery, these erection 1.0s lacked what more evolved erections offer: opportunistic engorgement and strategic deflation.

An important advancement in the evolution of erection was the addition of input from the brain. This allowed the brain to send signals to the

---

*If you are an ER doc in São Paulo, you are most likely aware that erections can arise from another surprising source: the venomous bite of the Brazilian spider *Phoneutria nigriventer*. While potentially toxic and possibly fatal, the venom can also induce an erection lasting many hours. Not surprisingly the venom has been marketed to males for whom more conventional pharmaceuticals have not provided success.

penis through the spinal cord. From an evolutionary perspective, these psychogenic or "cerebrally elicited" erections are a savvy improvement on the reflexogenic type. Involving the brain in a process as intricate and crucial as an erection expands the animal's reproductive opportunities and physical safety. It allows him to judge and respond to his environment before firing up or shutting down an erection. It enables sensory inputs like seeing, smelling, touching, or even thinking (fantasizing) about someone or something sexy to trigger the erection cascade. And it facilitates nearly instantaneous shutdown when a predator—or, more likely, a competitor—reveals himself.

And this is true whether the male is a moose, a mole, or a man.

My tour of the U.C. Davis horse barn included a visit to a small white room about the size of a galley kitchen in a New York City apartment. Where a Viking range might have been, a high-tech, semen-spinning machine stood instead. Nearby was a refrigerator-freezer, for storing ejaculate and frozen urine. The urine, as I'd seen in the breeding shed, plays an essential role in the sensory stimulation leading to psychogenic erection.

When a randy stallion walks past a mare who's in heat, she will often instantly, reflexively let loose a steaming stream of urine. This serves a strategic purpose. Urine contains telltale molecules that indicate a female's ovulatory status. When women buy a box of plastic ovulation-predictor sticks, they are purchasing fertility-detection technology that a stallion's nostrils can provide for free.

Male horses (along with many other animals, including camel, deer, rodents, cats, and even elephants) can detect these compounds by sniffing and tasting urine. Their sense of smell is enhanced by a characteristic grimace they make called a "flehmen." This one-sided lift of the upper lip resembles an animal version of Elvis Presley's famous sexy sneer. As the animal raises his lip, he inhales, wafting the odor molecules into contact with his vomeronasal organ, a sensitive scent detector located near the roof of the mouth. Humans perform a similar chemoreception when they slurp and swirl wine around the roof of the mouth to bring the aromatic molecules in closer contact with the sensitive receptors in their gums and nostrils. Whether or not we once had a vomeronasal organ is up for debate. Some biologists believe humans have lost it; others question whether we ever had one.

What we *do* continue to share with animals that possess a vomeronasal organ—and perform the flehmen grimace—is the seventh cranial nerve. Also called the facial nerve, this brain-body communication line runs between the face and the emotional centers of the brain. This same nerve, originating in essentially the same place in the brainstems of many animals and all humans, transforms anger into a dog's snarl, surprise into a macaque's widened eyes, and joy into a child's smile.*

If you picture a human doing a flehmen, you'll notice something unmistakable. The single-sided, upturned lip is also an instantly recognizable expression of disgust. If you try it yourself, you might even feel a little ripple of repulsion. Yet a pantheon of sexy male rock stars—from a strutting Mick Jagger to a sneering Billy Idol—has exploited this ancient multitasking neural circuit to flash-flehmen their female audiences, with swoon-inducing effects. Elvis's ancient lip curl, perhaps more so than his swinging pelvis, sent adolescent girls into shuddering seizures of excitement and ecstasy. Having seen what a stallion's flehmen indicates to a receptive mare, I can understand why Elvis's blatant sexuality may have threatened a whole generation of fathers in the 1950s.

That the flehmen signals both lust and disgust is thanks to the intertwining anatomic connections with the brainstem. And it may help explain why so much about our genital and urinary functions both fascinates and repels us. Male urine can communicate chemically to females as well as the other way around. Male porcupines court females by showering them with urine during precopulatory courtship. Male goats spray urine on their faces and trademark billy-goat beards as an olfactory indication of their sexual readiness. Elk bucks similarly wallow in urine during rutting seasons.

Nonmammals communicate using urine, too. Courting female crayfish release a stream that attracts interested males. The urine of male swordtail fish is full of pheromones. Males swim upstream and then uri-

---

*Compared to other animals, humans have evolved facial muscles that are complex and numerous. The reason a dog's or cat's face may not seem as expressive as a human's is not that they lack an interior experience and even emotional input to the facial nerve. Rather, they have fewer facial muscles and fewer branches off the facial nerve to control them.

nate, so that the sex-communicating liquid flows downstream, where it can be "read" by available females.*

When Lancelot was led past the fertile mare, he was allowed to look at her but not touch. Surely part of her effect was olfactory. In addition, the sight of her raised tail would have been a visual invitation. Sight cues are another very powerful stimulant of psychogenic erection. Visual cues—we might call them nature's pornography—excite many animals.

For example, the reddening and swelling of the female genital area (called perineal "elaboration") shown by many monkeys and apes signal that they are ready to mate. Males of these various species respond to the size of the swellings and are most visually excited by the largest ones.

Blindfolded bulls in a test situation were much less likely to mate with an unfamiliar cow than bulls that could see. Restricted vision impaired the bulls' performance.

Visual stimulation has an interesting effect on females, too. Male courtship displays are the iconic stock images of nature documentaries, from mighty rams butting heads to gentle bowerbirds presenting their paramours with nests intricately inlaid with flowers, shells, stones, and berries. The reason for all this visual excitement is to entice a female into copulation—by communicating not just sexual readiness but superior genetic fitness. However, these displays might also have an invisible effect that extends beyond the act of copulation and actually improves survival of any offspring.

Researchers in Morocco were desperately trying to improve the reproduction rates of an endangered bird called the houbara bustard. These natives of northern Africa have been hunted into near oblivion for their supposedly aphrodisiac meat. After the artificial breeding program had produced disappointing hatch rates, investigators realized that some of the females they were inseminating by hand had never actually seen a mature male bustard. So they decided to try an experiment. Instead of simply placing a sample of sperm into the females, they first gave them a look at a sexy male bustard—one who was strutting around in the

---

*Psychiatrists have historically considered sexual interest in urine to be pathological. They have viewed "water sports," "golden showers," and bathing in or consuming urine enjoyed by urophiliacs as abnormal acts by disturbed patients. It's interesting to note the broad range of species for whom urine plays an important role in attraction and arousal.

characteristic houbara pre-mating ritual, with his white head and neck feathers puffed out like a rock star wearing a boa. Females who had been primed with this sight—no matter whose sperm they were eventually fertilized with—were more likely to lay viable eggs. Their chicks were more likely to hatch and were stronger when they did. The reason: primed with the sexy sight, the females added more testosterone to their eggs. This made them grow faster and stronger. It spurred the chicks to create more testosterone on their own, giving them a hormonal head start in life. Of course, this was not a conscious choice the mother birds made; rather, it was a physiological response to a visual cue. Similarly, pig breeders have found that sows who were "courted" by boars before artificial insemination—or even just exposed to boar odor—had higher conception rates.

In the animal world, females are not just passive vessels receiving sperm but active participants who can influence the outcome of their breedings through sperm selection and egg enhancement. This is a new area of study that could have important implications for improving animal breeding programs around the world. And it could also help women struggling with infertility. Assisted reproduction has come a long way in the past decade, but although the male specimen-collection rooms of fertility clinics are well stocked with racy magazines, women are not regularly advised to be "visually motivated" during monthly egg development. Maybe surfing YouTube for clips of a drenched and brooding Colin Firth, riding crop in hand, would have an enhancing effect on egg recruitment and growth, whether a woman is going through in vitro fertilization (IVF) or trying naturally for conception.

Yet another brain-based erection enhancer enters male brains through the ears. Canoodling horses nicker and whinny; in-the-mood boars "chant." The biologist Bruce Bagemihl notes that "female Kob antelope whistle, male Gorillas pant, female Roufous Rat Kangaroos growl, male Blackbuck antelopes bark, female Koalas bellow, male Ocellated Antbirds carol, female Squirrel Monkeys purr and male Lions moan and hum." Any one of those sounds, produced for a willing recipient, could trigger a neural cascade that would result in or enhance an erection. One fascinating study revealed that female Barbary macaques timed their "loud and distinctive" copulatory utterances to coincide with—or perhaps influence the likelihood of—their mate's ejaculation. Bulls have been found to get erections when played a recording of the sounds of a cow in estrus.

But the brain's ability to translate sensory inputs into an erection has a deflating flip side. Sometimes, instead of encouraging an erection, the brain squelches it.

A mating animal is vulnerable. It's necessarily distracted from its environment. It's momentarily disconnected from other important survival activities, like food gathering and territory defense. A psychogenic component to erection means that if a male's brain detects danger, threat, competition, or diminishing returns, it can terminate the erection.

But this physiology sets the stage for the number one reason human males visit doctors with sexual complaints: erectile dysfunction, or ED.* That's when erections consistently don't achieve the hardness required for penetration or last as long as they once did. Although not life-threatening, as medical problems go, ED can profoundly affect quality of life and the social well-being of men and their partners. Worldwide, one in ten men suffers from ED, thirty million of them in the United States alone. Preoccupation with penile stiffness sustains a multibillion-dollar industry that peddles drugs, devices, dietary aids . . . and not a small amount of snake oil.

According to Arthur L. Burnett, an expert in neurourology at Johns Hopkins University, our understanding of erectile dysfunction has done "an about-face" in the past four decades. Doctors used to think ED developed out of inevitable yet vague factors like aging and hormonal imbalances—or that it was entirely psychological. In the heyday of psychoanalysis, a man's inability to achieve a firm erection was presumed to be a consequence of his unresolved internal conflicts.

Today, Burnett told me, ED is seen as a "truly physical problem." Because human erections are utterly dependent on blood flow, conditions like diabetes, hypertension, clogged arteries, vein disorders, and weak pulses—anything that impedes the proper surging of blood around the body—can induce or exacerbate ED. So can having one's nerves severed during prostate surgery.

Men are no longer told that ED is all in their heads or that they've got an emotional problem. On the other hand, although most ED is physical

---

*Some five hundred years ago, Leonardo da Vinci remarked, "The penis does not obey the order of its master, who tries to erect or shrink it at will . . . the penis must be said to have its own mind." A few centuries later, Leo Tolstoy grimly noted, "Man survives earthquakes, experiences the horrors of illness, and all of the tortures of the soul. But the most tormenting tragedy of all time is, and will be, the tragedy of the bedroom."

in origin, psychogenic ED—when a man feels willing but his medically capable penis won't cooperate—does indeed exist. And it can trigger confusion and distress for patients and couples.

As we've seen, animal penises also stiffen and soften in response to environmental and other triggers. It may be that what we call psychogenic ED in humans has roots in protective physiology that is shared by aroused males of many species.

Ring-tailed lemurs, the wide-eyed primates immortalized by the voice of comedian Sacha Baron Cohen in the animated film *Madagascar,* typically mate just once a year. During that discrete window every autumn, the females are fertile for a single, fleeting eight- to twenty-four-hour stretch, and the males' testosterone levels rise. This "wild Halloween party," in the words of Andrea Katz, curator of the Duke Lemur Center, creates a high-stakes frenzy of competition among the males and provocative teasing from the females. As Lisa Gould, an anthropologist at the University of Victoria and an expert on ring-tailed lemur behavior, told me, male-male competition is so frenetic during breeding season that she has seen males jump on other mating males and push them off females. Sometimes males become seriously injured during male-male combat over the chance to mount a female. One encounter she witnessed was especially interesting. A lower-ranked male was desperately trying to complete a copulation amid the fray of mating animals.

"He was incredibly nervous and kept looking around. He kept jumping on and off and looking around behind him. I don't think he ever completed his mating," she said. Gould explained that, in lemurs at least, these so-called failed copulations are probably the result of social stress and competition. The neurological input of vigilance or fear could affect copulation success. Gould also pointed out that each male is different. The lemur she observed nervously looking around while he tried to copulate was clearly very concerned about his surroundings and social challengers. Another male might thrive on that kind of competition. No two animals are exactly alike, and individual animals will show ranges of tolerance to various types of stressors.

What biologists call failed copulations physicians might call a lost erection or erectile dysfunction. And physiologically, they're similar. Fear and anxiety interfere with an erection's critical first step: relaxation. Remember, to build an erection the penis must first relax. If the brain senses danger, the surge of adrenaline and other hormones will terminate the relaxation sequence and squelch the nascent erection.* Any animal capable of having an erection can, and will, sometimes lose it.

And that's a good thing, because coitus interruptus can be a lifesaver. Imagine the fate of an animal who continued to mate despite looming danger. Sometimes the greatest threats come not from outside predators but from school-, flock-, and herdmates. And in male animals, we see that social intimidation can inhibit erections. The mere presence of a dominant ram shuts down sexual activity in subordinate male sheep, for example. Among deer and other ungulates living in hierarchical groups, often only the top-ranking males can mate. The control of mating by dominant males has been seen in birds, reptiles, and mammals. The celibate subordinate males, deprived of sex, may lose the ability to get an erection. This form of animal erectile dysfunction, or "psychological castration," may be temporary and reversible, but it also may last a lifetime.

But modern human sexual encounters are not typically interrupted by predators leaping from the bushes or sexual competitors grabbing mates away. So I asked the UCLA urologist and impotence expert Jacob Rajfer whether psychological stress could interfere with male erections. "Yes," he told me simply. "Some men under stress have difficulty with their erections." When I asked him to specify what kinds of stress, he just laughed. Getting older, problems on the job, or relationship troubles can strain the modern male. Wary dread can come in the form of a looming deadline, a lawsuit, or crushing credit card debt. Stress extinguishes erections—whether human or animal—through activation of the sympathetic nervous system. Failed copulations across species are connected through ancient neural feedback loops that protect mating males. This illustrates the remarkable power of the brain over the penis. While losing an erection in this way can be frustrating and perhaps embarrassing for a patient, the connection itself isn't pathological. And it's certainly not unique to individual men. To keep animals from getting eaten or beaten up in the act, threats *should* trip the circuit breaker of sexual activity.

During millions of years of matings in a dangerous world, some

---

*Fear can in some cases enhance arousal. "Mile-High Club" members and others who are stimulated by having sex in dangerously public places will attest to this. The neurocircuitry of desire and fear converge in the brain's amygdala.

males have evolved the ability to ejaculate as quickly as possible. Males that could speedily transfer sperm before a jealous competitor or hungry predator pounced would live to mate another day, and their sperm would stand a better chance of fertilizing. Moreover, speedy ejaculation might also have given a reproductive advantage to males attempting to inseminate many females in a short time frame.

But, like the coitus interruptus of erectile dysfunction, coitus accelerando has been pathologized. We human doctors call it premature ejaculation (PE). According to Arthur Burnett, the Johns Hopkins urologist, ejaculation problems are actually more common than erection issues, although fewer men seek medical attention for them. For other animals, however, it's not necessarily a problem. In fact, it may be an advantage.

In a 1984 paper, Lawrence Hong, a sociologist at California State University, Los Angeles, suggested that "an expeditious partner who mounts quickly, ejaculates immediately, and dismounts forthwith might be best for the female."

Indeed, many animals do transfer sperm promptly.* Human males take, on average, three to six minutes to go from vaginal penetration to ejaculation. Our closest genetic relatives, chimps and bonobos, do it in about thirty seconds. Stallions typically ejaculate after only six to eight thrusts. Some birds that lack penises transfer sperm in just a fraction of a second, by touching their genital openings to the females', a process that has been described as a "cloacal kiss." Small marine iguanas native to the Galápagos Islands have developed the ultimate in early seminal emissions: they can ejaculate *prior* to copulation. Ordinarily, marine iguanas need almost three minutes to ejaculate inside a female. This benefits larger, higher-ranking iguanas with the right and brawn to push a smaller male off a female. So smaller iguanas masturbate and cleverly store their semen in a special pouch. In the first few seconds after penetration, the sneaky little iguana slips the semen into his partner's genital opening, called a cloaca. By the time the larger iguana pushes him off, his swimmers may well be on their way to fertilizing the female's eggs.

---

*True, some animals take much longer. Rats may ejaculate rather quickly, but only after a long chase-and-mount pattern, which first involves eight to ten penile penetrations into the female's vagina. Some animals, including some cats and some insects, "lock" together after intercourse, using genital barbs and spines, inflatable body parts, and physical force. Sometimes the delay is used for inserting a copulatory plug made of mucus or gel. But for many couplings, it's an advantage to make it as speedy as possible.

There may be another benefit to speedy mating. Time-limited contact, particularly between moist mucus membranes, reduces the risk of transmitting pathological microorganisms. For many animals, parasitic infection poses a deadly threat. In these populations, fast copulation might be advantageous. (For more on animals' STDs, see Chapter 10, "The Koala and the Clap.")

Jacob Rajfer, the UCLA urologist, pointed out to me that PE affects a remarkably consistent 30 to 33 percent of men of all ages—from their early twenties through eighties. The incidence of ED, on the other hand, tends to increase with age. To Rajfer, this means PE is what doctors call a "normal variant"—and likely highly heritable. He summed PE up in a sentence: "I don't consider it pathological."

Whether the thrusting that precedes ejaculation lasts three hours or three seconds, the delivery of the payload to the female fulfills the reproductive function. PE as pathology is a new wrinkle in the long evolutionary success story called premature ejaculation. And this knowledge should console early climaxers. Because while it's sometimes embarrassing or perhaps unsatisfying today, instant ejaculation—and the neurocircuitry underlying it—has given hundreds of millions of ancestral ejaculators a head start in what biologists call the "sperm competition."

In my first year of med school, the highlight of the spring semester was Movie Night. Carrying tubs of popcorn, sodas, and giant bags of candy—some of us wearing pajamas and slippers—my classmates and I packed into the university auditorium and settled in. The lights dimmed, we sat back, and for the next four hours, we watched tape after tape after tape after tape . . . of the hardest-core pornography our professors had been able to procure.

The thinking was that, as future doctors, we needed to be familiar with the tremendous range of behaviors the human body, mind, and libido can get up to. We had to be able to conceal our shock (or arousal?) when a patient confessed some kinky predilection. We required background knowledge to be able to reassure concerned patients when they were normal. We needed to know what *was* normal and what even the sex industry considered out-of-the-box. And frankly, for many of us book-smart science nerds, we just needed to have our eyes opened.

Veterinarians require no such seminar. As hilariously—and exhaustively—cataloged by writers like Mary Roach, Marlene Zuk, Tim Birkhead, Olivia Judson, and Sarah Blaffer Hrdy, the sex lives of animals make for reading that is almost comically pornographic.

These writers are documenting a massive shift in what biologists know—or are willing to admit they've observed—about the sex lives of other animals. If you've ever watched, mortified, as your dog mounted a dinner guest's leg, you may have given some thought to animal masturbation. But until recently—even with much evidence to the contrary— genteel biologists held that animals don't masturbate. Their fig leaf was the argument that masturbation is not procreative; ergo, animals would have no evolutionary urge to pursue it. In fact, however, both sexes of many species are inventive self-pleasurers in the wild. Orangutans self-stimulate using dildos they make out of wood and bark. Deer find their antlers autoerotic. Birds masturbate by mounting and rubbing against clumps of mud and grass. Daddy longlegs spin two threads of silk upon which to rub and stimulate their genitals. Male elephants and horses rub their erections against their bellies. Lions, vampire bats, walruses, and baboons use paws, feet, flippers, and tails to stimulate their own genitals. Livestock farmers and large-animal veterinarians have long noticed masturbation in bulls, rams, boars, and billy goats, and have even calculated what time of day it is most likely to occur (5 a.m. appears to be a favorite hour for many bulls).

Mutual masturbation has been documented in many species. Bats and hedgehogs include oral sex as a common component of their sexual encounters. Blowhole sex in dolphins has been observed by marine biologists. Bighorn sheep and bison have frequent (homosexual) anal sex. Spinner dolphins, herons, and swallows engage in group sex. And bonobos . . . well, those well-publicized lascivious close human relatives seem to do it all.

Male-male and female-female mounting has long been observed in livestock (in fact, watching for females to mount each other is an old rancher's trick for discerning when cows are in estrus and ready to breed). But until the last decade, animal homosexual behavior was explained away, pathologized, or entirely ignored by academia and even by popular naturalists. That changed around the end of the 1990s, with the publication of several books, including Bruce Bagemihl's *Biological Exuberance,* Marlene Zuk's *Sexual Selections,* and Joan Roughgarden's

Evolution's Rainbow, which brim with examples of the hundreds of species that show homosexual, bisexual, and transgendered tendencies and behaviors. Bagemihl includes a many-hundred-page "Wondrous Bestiary" that catalogs sightings and descriptions of these behaviors in wild primates, marine mammals, hoofed mammals, carnivores, marsupials, rodents, and bats, as well as a huge range of birds and even butterflies, beetles, and frogs. Roughgarden details genital licking and anal sex in bighorn sheep; penis fencing in bonobos; lesbian mounting in Japanese macaques; and the all-male orgies of giraffes, orcas, manatees, and gray whales . . . among many other combinations, species, and activities. Marlene Zuk and Nathan W. Bailey have studied female same-sex parenting behavior in Laysan albatrosses and reported on the genetics of fruit fly homosexual behavior.

Clearly it's time to put to rest the idea that homosexuality is somehow unnatural, particularly if your definition of "unnatural" is something not found in nature. Indeed, as Bagemihl notes, "The capacity for behavioral plasticity—including homosexuality—may strengthen the ability of a species to respond 'creatively' to a highly changeable and 'unpredictable' world."

It should be noted that same-sex sexual behavior is not necessarily the same as same-sex *preference* or *orientation,* both of which are harder to prove and less well documented in the wild than individual same-sex activities. Nevertheless, many human and nonhuman animals alike regularly engage in same-sex behaviors including oral sex, anal sex, group sex, and mutual masturbation.

Even the patterns of heterosexual animal couplings have been radically updated in just a few short years. Females that were long thought to be monogamous have been unmasked as cheaters and home wreckers. Until recently, the conventional wisdom held that males of most species were the rovers, the spreaders of seed; females remained faithful to one male, if not for life, then at least for a whole mating season. But research using DNA to confirm the paternity of offspring suggests that female promiscuity is not only common, it's practically the norm. As the behavioral ecologist Tim Birkhead puts it in *Promiscuity,* his engaging book about sperm competition, animal paternity studies have spelled the "near elimination of the idea of male and female sexual monogamy." In "organisms as different as snails, honeybees, mites, spiders, fish, frogs, lizards, snakes, birds and mammals . . . multiple paternity is widespread." The reasons, according to Birkhead? Female "infidelity" can improve the genetic quality of offspring and sometimes results in females' securing fitness-enhancing resources for themselves and their children.

We've inherited a complex sexual heritage from our animal ancestors. The broad range of human sexual interests and practices attests to this. But we humans are also able to project the consequences of our actions. For better or worse, we live in cultures that have rules and taboos. Our sexual behavior cannot be divorced from that, and it's a mistake to look to "nature" for moral guidance. In the words of Marlene Zuk: "Using information about animal behavior to justify social or political ideology is wrong . . . .People need to be able to make decisions about their lives without worrying about keeping up with the bonobos."

Some sexual practices so repel us as human beings that we consider them immoral and have made them illegal: rape, pedophilia, incest, necrophilia, bestiality. But millions of times a day, millions of animals engage in them. Normal reproduction for species of insects, scorpions, ducks, and apes requires an act of rape (often called "coercive copulation" by biologists). New York City's bedbug epidemic has made it common knowledge that bedbugs (and their relatives) mate through a technique called "traumatic" or "hypodermic" insemination, in which the male mounts a female, stabs her with his scimitar-sharp male organ, and ejaculates directly into her bloodstream. An animal form of necrophilia can be found in creatures from frogs to mallard ducks, which have been observed copulating with dead members of their own species. Sex with relatives and immature members of the group occurs in primates and many other vertebrate and invertebrate species; some evolutionary biologists think that the parent-child clashes we associate with adolescence may have emerged to protect newly sexualized animals from early mating attempts by their own relatives.

And interspecies sex—known as bestiality when humans engage in it—has been occurring for an extremely long time, possibly as long as sex itself has been going on. Respectable scientific studies theorize that sex between different species actually serves the evolutionary purpose of creating new variations. Birkhead notes, "Breeding males are usually highly motivated and often indiscriminate. Ejaculation carries little cost and there has therefore probably been little selection against males copulating with the wrong species. Indeed, selection may have favoured a lack of discrimination among males, since he who hesitates is lost."

But while there are lines between what humans can and should do sexually, there is an important human takeaway from the study of manatee oral sex, or bighorn sheep anal intercourse, or bat cunnilingus. What every one of these sexual behaviors—from stallion masturbation to monkey fellatio to frog necrophilia—reminds us is that sex is not always linked to reproduction. In fact, it could be argued that the vast majority of sexual activity in animals does not have procreation as its goal.

Marlene Zuk agrees: "Even in nonhumans, sex can be about more than reproduction . . . Even sex is not always about sex, at least in the short term," she writes. Anders Ågmo, a behavioral neuroscientist, takes it a step further. He has called the production of offspring "an accidental physiological side effect of sex behavior."

For animals, sex provides benefits besides reproduction. And the same is true for us. In social mammals, sex promotes bonding between individuals and strengthens relationships. And the repetitive touching, stroking, or hugging that may accompany sex provides some of grooming's well-known soothing benefits.

Social bonding. Relationship building. Soothing. Are we missing anything? What about pleasure? The pursuit of pleasure may underlie many animals' interest in sex. But if pleasure is an important driver of sexual activity, then pity the 25 percent of women who claim to derive no enjoyment from it at all. The search for ways to help them brings us to another crossroads of human and veterinary medicine.

While Lancelot, the horse at U.C. Davis, was in the barn struggling with the phantom, about a dozen female horses were outside in a special corral called the mare hotel. This equine inn was less like the Four Seasons and more like Nevada's infamous Mustang Ranch. Here the mares were getting teased. If you're like me, a city girl who had little more than a goldfish growing up, mare teasing is a jaw-dropping sight.

A handler led a stallion toward the corral of mares. One by one, he paused the male horse in front of a female. Some mares responded instantly. Their tails shot up, revealing glistening, engorged labia. They released a hot gush of urine. They shoved their hindquarters toward the stallion. Some swayed their backs in a slightly crouched position, as though inviting the male to mount them. Others displayed what Cornell veterinarian and animal behavior expert Katherine Houpt calls "the mating face" —"in which the [mare's] ears are swiveled back and her lips hang loose."

Other females, however, glanced up at the stallion and went right back to munching their hay. Some took one look and lunged toward the male with flattened ears, showing their teeth and whinnying threateningly.

The different behaviors depended on whether the mares were about to ovulate. Those who responded to the sexually available stallion with precopulatory behaviors were either ovulating or just about to. They were, the handler told me, "receptive." Those who ignored or pushed the male away were "nonreceptive."

Human women don't, thank goodness, raise our tails and urinate when a man glances our way on or around day fourteen of our cycles. We're said to have what's called "concealed" ovulation, meaning we lack the obvious "advertisements" of ovulatory status. But evolutionary scholars, like UCLA's Martie Haselton, are starting to take a closer look at clues we do give off—some of which are less subtle than we might think. Women have been found to dress more provocatively and roam farther from home when they're ovulating. Men perceive ovulating women to be more attractive; strippers get higher tips during the most fertile parts of their cycles. College-aged women phoned their fathers significantly less often during ovulation than at other times of their cycles—a behavior hypothesized to be some ancient defense mechanism against intrafamily attraction. But even when human females aren't ovulating, they may seek the pleasures of sex and orgasm.

Physically, female orgasm is very similar to male orgasm. Parasympathetic buildup abruptly shifts into a sympathetic burst of muscle contractions that end with a flood of rewarding neurochemicals and brain-wave changes. The sensory and physical similarity of orgasm across genders arises from nearly identical networks of nerves and hormones. In developing fetuses, male and female genitals spring from the same primordial cells. Indeed, the embryos of many species—whether human or canine or crocodilian—begin with no specific gender. Influences such as hormones, temperature, and environmental effects lead to the development of a penis in males but suppress its growth in females. In other words, a wife's labia and her husband's scrotum were once embryonically identical, as were her clitoris and the glans and upper shaft of his penis.

A quick comparative survey of animal sexuality reveals that the clitoris is not a uniquely human organ. This "tender button" has been found in a huge variety of female creatures, including horses, small rodents, a wide variety of primates, raccoons, walruses, seals, bears, and pigs. Bonobo clitorises and labia can swell to the size of soccer balls. Thanks to high circulating testosterone, the African spotted hyena sports a clitoris so large it's called a pseudopenis. Clitoris licking in these fiercely matriarchal societies is a sign of submission. Oversized clitorises are shared by European moles, some lemurs, monkeys, and a Southeast Asian carnivore called a binturong.

Significantly, all these animal clitorises—just like animal penises—are densely packed with nerve endings. This means the suite of sensations that makes up an orgasm may be shared as well across genders and species.

And yet, even with the physical ability to feel orgasms, many women don't. An estimated 40 percent of all women worldwide have sexual complaints. These include dyspareunia (generalized pain during sexual activity) and vaginismus (a rare affliction in which the muscles of the vagina clamp painfully and uncontrollably shut, prohibiting penile entry).

But by far the most common female sexual dysfunctions are low desire, impaired arousal, sexual aversion, inhibition, and inorgasmia. These conditions—sometimes collectively called hypoactive sexual desire disorder (HSDD)—can be persistent and distressing. They affect as many as one-fourth of all women worldwide. In the United States, although estimates vary, some 20 percent of women are thought to have HSDD. This suggests that every year more women suffer from low desire and failure to orgasm than are diagnosed with breast cancer, heart attacks, osteoporosis, and kidney stones *combined*. Like male erectile and ejaculatory dysfunction, female hyposexuality is not, on its own, life threatening. But it can create severe quality-of-life challenges that, in turn, can carry serious health risks, such as depression.

Low desire and HSDD can be situational (directed toward one partner) or general (a lack of interest in all sex). The patient may complain of other symptoms, including depression, anxiety, conflict, fatigue, and stress. Her detachment can range from dull, "close your eyes and think of England" resignation to actively finding sex distasteful or repulsive. Phobic and panic responses can occur at this extreme end of the disorder. Some women feel a physical urge to push away their partner. Some may feel like kicking, biting, hitting, or verbally lashing out.

Doctors treat HSDD with psychotherapy and by prescribing testosterone supplements, which can boost sex drive not just in men but in women as well. These interventions are, however, usually only moderately successful. Testosterone lacks FDA approval for treating HSDD (it must be prescribed "off-label"), and studies suggest that the time a woman spends on her therapist's couch does little to improve the quality of the time she spends with her mate on a mattress. Patients may be instructed to stop taking certain medications, particularly the selective serotonin reuptake inhibitor (SSRI) antidepressants, such as Prozac, Paxil, and Zoloft, because they can dull the libido. Beyond these basic fixes, however, HSDD's treatment outlook is somewhat gloomy. One online medical encyclopedia warns, "Cases of dissatisfaction by both partners often do not respond to such therapy, and frequently culminate in separation, finding a new sexual partner, and divorce."

I asked Dr. Janet Roser what she would prescribe if she noticed a female avoiding male sexual attention, ignoring solicitations, even lashing out at an unwanted advance. She said, "Nothing—if the patient is not in heat." Roser is a neuroendocrinologist who treats the horses at the U.C. Davis barn. For her, hearing about waning sexual interest leads immediately to this assumption: the female is not in estrus. She is nonreceptive. And being nonreceptive is a perfectly normal—indeed, expected—state for a female animal when she is not nearing ovulation.

As I'd seen at the horse barn when the handlers teased the mares, nonreceptive female horses may whinny, bite, and lunge or kick at approaching stallions. Many other female animals have equally obvious ways of making it clear to advancing males that they are not interested in sex. Female rats scratch, bite, and vocalize. Female cats hiss or strike out with their claws. Female macaques gang up on approaching males. Female llamas spit at their pursuers and run away from them, while female vampire bats lunge threateningly with their infamous canine teeth exposed. Nonreceptive female butterflies twist their abdomens upward and away from incoming males. Female fruit flies display the same behavior; some may even kick a pursuing male. Some beetles have sliding plates made of chitin they can position across their vaginal openings to deflect unwanted penetration.

There are several scenarios in which nonhuman females have sex when they're not fertile or receptive. Entomologists Randy Thornhill and John Alcock have described a phenomenon they call "convenience polyandry" in which a female will accept (or endure) the advances of a particularly aggressive or persistent male just to get him to leave her alone. And it's interesting to observe differences in receptive behavior between natural settings and captivity. James Pfaus, a Concordia University psychologist who studies the neurobiology of sexual behavior, told me that a female macaque housed with one male will copulate daily with him. When she's in estrus, she accepts more frequent liaisons, sometimes two or three per day. However, when she's returned to a more natural macaque social group—in which breeding females band together and solicit sex from males only on the days they're fertile—she copulates only around ovulation. Coercive copulation or rape is another scenario in which females have sex during their nonreceptive phases, although it must be said that males of many species do respect females' nonreceptive signals. If a female tells him to back off, some males will seek sex elsewhere—usually from another willing female but, in some species at certain times of the year, from another male.

Unenthusiastic acceptance of sex. Avoidance of it if possible. Occasional outright hostility or violence toward the interested male partner. When we put animal nonreceptivity side by side with female hypoactive sexual desire disorder, it's possible to see some intriguing crossovers. I suspect that low desire as a diagnosis is so high simply because women expect to be receptive to sex 24/7, regardless of where we are in our cycles. Although sexual response can occur during days outside of our windows of ovulation, in fact, women are fertile for only three to five days during each month. This may make women less receptive at other times.

Receptivity in animals is guided by surges in female sex hormones. These hormones, working through complex neural pathways in the spinal cord and brain, can cause certain predictable mating behaviors and even body postures. One stance in particular is a dead giveaway that a female is receptive. Ranchers, biologists, breeders, and veterinarians will recognize it. It's called lordosis. Lordosis is a very specific, hormonally driven posture in which the female arches her lower spine into a swayback, with her buttocks tipped rearward. Her pelvis softens and stretches. If she has a tail, during lordosis she might raise it or hold it to one side, exposing her genitals. Horses, cats, and rats have an exaggerated lordotic response, but it's also seen in sows, guinea pigs, and some primates. According to Donald Pfaff, an expert on lordosis based at Rockefeller University, it's a widespread neurochemical response seen in all quadruped females. Basically, he writes, a nerve signal triggered by the touch of a mounting male "ascend[s] the spinal cord of the female to reach her hindbrain and then her midbrain. There nerve cells receive a sex-hormone-influenced signal from the ventromedial hypothalamus. If the female has received adequate doses of estrogens and progesterone, that signal from the hypothalamus says 'Go, Mate, Do Lordosis Behavior.' If not, the signal is 'Resist, Kick, Flee the Male.' "

Like some erections, lordosis is considered to be reflexogenic—a spontaneous, hormone-driven response stimulated by touch. Receptive female elephant seals, for example, have been seen spreading their flippers and raising their tail ends when their "harem master" places a foreflipper on their back before mating. However, intriguingly, fear and anxiety can interfere with lordosis, so it may be that, like psychogenic erection, the brain can play a role in enhancing or shutting down the response.

While some sex researchers insist that human females don't display a lordosis reflex, Pfaff has pointed out that a "large number of mechanisms for hormone action in the central nervous system [are] known to be conserved as we move from animal brain to human brain tissue." He thinks it's possible to apply "basic, reductionistic principles . . . to all mammals, human patients included." Indeed, as he colorfully puts it in *Man and Woman: An Inside Story,* "The most elementary functions of the hypothalamus, such as the female's ovulation or the male's erection and ejaculation work quite similarly . . . proved true from 'fish to philosopher,' from 'mouse to Madonna.' "

The swaying back and vaginal presentation of a lordotic animal occur in association with a cascade of hormones, neurotransmitters, and muscular contractions. And the components of this cascade are shared by human women. We may not be wired for the overt and reflexive lordosis displayed by rats or cats. But the lordosis posture is certainly something human men find alluring and women feel sexy engaging in.* Once you start looking for them, media images of human lordosis are all around us. One of the most famous pinups—Betty Grable's iconic World War II–era bathing suit shot—shows her from behind, her back flexed in a slight lordosis as she solicits the viewer over her shoulder. Marilyn Monroe's unforgettable *Seven Year Itch* publicity pose over the subway grate shows her in a similar lordosis, her buttocks swayed backward while her arms hold down her billowing dress. Slightly less demure lordotic poses include Irina Shayk's cover for the 2011 *Sports Illustrated* swimsuit issue, where she kneels in the sand, her lower back arched as her buttocks tip up and back toward her feet. (Granted, her breasts sort of steal the show, but her back is unmistakably lordotic.) Pop star Katy Perry took feline lordosis to a literal extreme by donning a purple catsuit and mask and getting down on all fours in a classic lordosis posture to advertise her perfume brand, Purrs.

The "sexiness" of lordosis is hardly mysterious. For hundreds of millions of years, animals from big cats to mares to rats have become lordotic to signal receptivity. At an early age, males learn that approaching nonreceptive females can mean getting bitten, scratched, wrestled, or boxed. For human males it can be challenging, too. Far better to pass by the nonreceptive females in favor of the ones who are soliciting and signaling, with, among other behaviors, lordosis.

Knowing about lordosis isn't going to make a woman with HSDD suddenly start having orgasms. But understanding cycles of receptivity and nonreceptivity in animals could provide valuable human insights. At least it could reassure some women that it's okay not to want sex all the time and present a simpler reason for why and when a flattened desire might be normal.

The partners of HSDD sufferers might also consider a survey of comparative foreplay. Stroking, neck biting, vulva licking, and ear tongu-

---

*To create instant lordosis (the posture, if not the hormonal reflex), you can go to your closet and put on a pair of high heels. Whether stilettos or wedges, high heels exaggerate the lower back's normal lordosis. If we didn't compensate by tipping out our buttocks and arching that lower spine, we would topple over. Maybe the forced, if artificial, lordosis is what's enduringly attractive about high heels—and why wearing them both looks and feels sexy.

ing are seen across many species of animals. Cornell professor Katherine Houpt notes that for horses, "an adequate period of sexual foreplay is essential." Stallions will nip and nuzzle a mare, starting at her head and ears and moving backward and down to her perineum. Dogs, too, engage in oral grooming precoital activities. Parasitic wasps and fruit flies stroke each other's antennas. Beetle birds engage in cloacal pecking. Of course, what occurs between humans has distinctive appeal to our species, but studying the foreplay of crustaceans, gulls, bats, and geckos could yield the ultimate suite of erotic moves retained by millions of rounds of natural selection for their ability to facilitate copulation and conception.

Perhaps help for HSDD can be found through study of the true nymphomania seen in some cows and mares. Hypersexual behavior occurs as a consequence of disturbances in ovarian function leading to increased testosterone and other male hormones. In horses and cows, ovarian cysts are the cause. Nymphomaniacal cows (most often dairy, as opposed to beef, breeds) paw aggressively at and try to mount other cows. And they bellow "like a bull," with distinctive masculinization of their voices. Similarly affected mares exhibit stallion-like behavior. They flehmen, compulsively urinate, and mount other mares. Experts suggest removal of the affected ovary in this highly disruptive situation.

Until I learned about nymphomania on the farm, I believed the concept to be less a proper medical diagnosis and more a pornography plot driver. But veterinarians not only make this diagnosis—they worry about it, because a nymphomaniac in a barn can wreak havoc and inflict injuries. Learning that the cause in animals is often cyst growth on the ovaries, I wondered whether the millions of women in the United States with polycystic ovary syndrome (PCOS) also experience increases in sex drive and activity. Intriguingly, some women dealing with this virilizing disorder (the medical term for "masculinizing") do describe increased sexual drive. However, the excessive body and facial hair growth that are also features of the disorder may adversely affect a woman's self-image and discourage her from engaging in sex.

The day after Lancelot struck out because of the three-mount rule, I was watching his stallion colleague Biggie go through the same precopulatory paces. Biggie was led into the barn. He received a whiff of frozen mare urine. He was given a look at a receptive mare. He was led to the phantom. Then, with a practiced flair, Biggie straddled the phantom, pumped four or five times, and climaxed. I was looking for behavioral evidence of an orgasm. What I saw was an unmistakable clenching, shuddering, and gripping, followed by a brief motionless moment before Biggie slid off the phantom. Like many stallions who have just ejaculated, Biggie appeared sleepy and "depressed."* The handlers retrieved the giant tube and took it away to be processed. Biggie was led to his stable, and the barn was prepared for Lancelot, who, on this new day, had no trouble getting back in the game.

Obviously we can't know how a horse experiences pleasure with his own ejaculation. But a Japanese research team has reported on behaviors that suggest shared sensations in other animals. In monkeys, they wrote, "copulation culminates at the moment of male ejaculation with body tenseness and rigidity, possibly accompanied by orgasm." Male rats "show jerky stretching at ejaculation following repetitive intromission, firmly holding the female body." Even salmon, the researchers pointed out, "show convulsive stretching with their mouths open wide at sperm emission and egg spawning." And insects show a standardized sequence of movements during sex. Pressed against a female, a male cricket, for example, "assumes a stretching posture," transfers his sperm packet, and suddenly "falls into a complete motionless state." Their conclusion: "There may be a similar mechanism in the final acts of copulation across species."

After examining the similar function and physiology of erections, ejaculations, and orgasms in many species, it's impossible not to postulate that the feelings are also shared. Sensations of orgasm may reward a marine flatworm's multiple penises as profoundly as they pulse through a human male's single member. The "shudder" that a primatologist observed "cours[ing] through" a female siamang's "entire body" after her genitals were licked by a male may have feelings in common with the "violet flannel, then the sharpness" of the poet Molly Peacock's description of an orgasm. The open-mouthed grimace of a lion climaxing could indicate a roar-gasm; the squeals of a mating tortoise, an expression of pleasure.

This could help explain the lengths to which creatures go to have sex.

---

*Katherine Houpt describes a depressed facial expression following ejaculation in breeding stallions.

An animal version of the opioid-oxytocin rush of melting expansiveness that accompanies the muscular fluttering of human orgasm may serve as a crucial incentive that impels animals to try that behavior again and again. Sexual desire in mollusks, fruit flies, trout, worms, gorillas, tigers, and human beings may be driven by a craving for another hit of the chemical cascade that accompanies ejaculation and orgasm.

A *Homo sapiens*–centric view of sexuality can make orgasms seem one of a kind, special, perhaps even uniquely human. But the push of biological reward forms a stronger argument for shared pleasure across the animal kingdom. If this is the case, then orgasm is not the by-product of sex. It is the promise, the erotic ancestry, the bait.

FIVE

Zoophoria

Getting High and Getting Clean

Against the wall of the lab where I perform heart-imaging procedures stands a beige metal box about the size of an office photocopier. It's got a computer screen on the front and, below that, a keyboard. To the right is a little trapdoor that can spit out receipts, like an ATM. Near the keyboard is a dime-sized, glowing red oval—a fingerprint reader. Once you've pressed your thumb and confirmed your identity, you must enter a series of numerical codes before the box will open. And even then only a small sector will be exposed; you can never gain access to all its contents at once.

This mute machine guards the entrance to a kingdom of euphoria. Locked inside are stacked drawers, each containing an array of highly addictive drugs. There are carousels of morphine vials. Pockets of Vicodin pills. Mini-bins stocked with Percoset and Oxycontin. Clear ampoules of fentanyl. All sit waiting but out of reach in the dark cabinet, unlit and unsparkling, like diamonds on black velvet trays deep in a Cartier safe.

The narcotics contained in this Pyxis MedStation 3500 drug-dispensing apparatus are essential for relaxing patients during medical procedures and for relieving their pain afterward. But the box is there to deter a clever group of highly intelligent and crafty dope fiends: drug-seeking doctors and nurses. Hospitals have learned the hard way that easy on-the-job access can lead to addiction in their personnel. Brilliant colleagues, inventors of life-saving medical devices who rarely fail at anything, would find themselves red-faced, empty-handed, and referred into a career-salvaging "diversion program" if they tried to breach the machine to retrieve an unauthorized Vicodin. The lockbox—and the hospital has dozens of them—protects them from themselves.

That's fine for a white-walled clinical suite where Vicodin tablets don't grow on trees and fentanyl vials don't dangle from vines. But the painkillers and sedatives in that machine are derived from natural opiates that do grow wild—in the *Papaver somniferum* poppy. Imagine the security system you'd need to protect several thousand square miles of poppy fields.

For opium-growing regions, this is a real issue. In Tasmania, a leading producer of medical opium, users sometimes sneak into the fields. Ignoring security cameras, they hop fences and gorge on poppy straw and sap. Dosed on the drug, they flail around in circles, damaging crops. Sometimes they pass out in the fields and have to be carried away in the morning. And there's no way to prosecute these trespassing scofflaws, no rehab to send them to. Because these freeloading opium eaters are wallabies.

I have to admit that the thought of stoned wallabies made me smile. Even the mug shot that accompanied the article in which I read about it was so "wrong": a sweet-faced, gray-brown mini-kangaroo squints before an exotic backdrop of emerald-green poppy stalks. The tableau would be as lovable and cheeky as Peter Rabbit sneaking into Mr. McGregor's garden . . . were it not for the animal's zoned-out eyes and the fact that repeat offenders apparently have a serious drug problem.

Often, what's endearing in animals is detestable in humans. So while we may chuckle at the intoxicated Tasmanian wallabies, we'd be justly horrified if they were Tasmanian children with a heroin habit. And if they were human adults, compulsively eating opium day after day, putting not just their own well-being but that of their families at risk, our horror might turn to disgust.

Indeed, this reaction points to one of the most frustrating, painful, and puzzling aspects of drug addiction. Genetics, vulnerable brain chemistry, and environmental triggers play dominant roles in this illness. But ultimately, on the receiving end of the syringe, joint, or martini glass is a person making the choice, at least in the initial stages of substance use, to shoot up, smoke, or swallow.

To nonaddicts, that choice can be utterly perplexing. Users hemorrhage money, destroy careers, lose homes, and demolish relationships—all in pursuit of a high. Confoundingly, addicts who are parents sometimes make decisions that orphan their children. I've even seen patients stricken off the heart-transplant list—a literal death sentence for them—because they continued to use.

Despite advances in imaging and genetics that clearly characterize addiction as a brain illness, it remains uniquely bewildering. Why *is* it so hard for addicts to "just say no"? Is "can't stop" really just an excuse for "won't stop"? Whether we like it or not, confusion about how we should think about and classify addiction pervades our legal systems, schools, governments—and, frankly, even the field of medicine.* Addicts belong to a set of patients that society, even doctors, judges harshly. So well do addicts know this medical prejudice that they may hide their substance-use histories when they go to a doctor's office or the ER, lest the level of care and compassion decline or disappear entirely. As one doctor I interviewed confided to me, "No one likes an addict."

But nearly everyone likes a cute animal. And so it can be surprising to learn that animals, too—even if they must risk losing their children or, sometimes, their lives—plunder nature's pharmacopoeia. With its vicious war between mind and body, addiction can seem distinctly human. But it turns out that our *Homo sapiens* bodies are not unique in the ways they react to intoxicating substances.

Understanding what drives animals to ingest drugs might help us separate what is inevitable from what is optional about this perplexing disease. The brain chemicals and structures that lead many millions of the world's population to snort, shoot up, or chug are pervasive and powerful. As we'll see, the urge to use has stayed in the gene pool for millions

---

*The U.S. medical community's negative attitude toward addiction can be traced back to the 1914 Harrison Narcotics Act, which criminalized opium use and the doctors who prescribed it. This early legislation defined addiction as a crime, as opposed to an illness, and initiated nearly a century of derision and punishment for the addicted.

of years and for a paradoxical reason. Although addiction can destroy, its existence may have promoted *survival.*

No one issued Flying While Intoxicated citations to the eighty cedar waxwing birds in Southern California who crashed into a reflective glass wall one February day. Drunk on fermented Brazilian pepper tree berries, they all died of spinal fractures and internal bleeding, some of them still clutching the mind-altering fruit in their beaks. The Bohemian waxwings in Scandinavia that sometimes gorge on naturally alcoholic rowan berries and then fall into the snow and freeze to death have no irreverent nickname—unlike Russia's *podsnezhniki,* or "snowdrops," the human drunks who are discovered, dead, in thawing snowbanks every spring. When a horse named Fat Boy nearly drowned in a neighbor's swimming pool after getting sauced on ethanolized apples in a small English village, he made the evening news but didn't have to apologize to the local fire brigade that pulled him out.

These animal encounters with intoxicants, however surprising and even amusing, were probably accidental. But others are not. Some animals show what seems to be more deliberate and chronic drug-seeking behaviors. Bighorn sheep in the Canadian Rockies are reported to scale cliffs to get their fix of a psychoactive lichen and grind their teeth down to the gums scraping it off rocks. In opium-producing regions of Asia, water buffalo (like Tasmania's wallabies) are known to sample a daily dose of the bitter poppies and then show signs of withdrawal at the end of the flower-growing season. The pen-tailed tree shrew that lives deep in the Segari Melintang rain forest in West Malaysia prefers the fermented nectar of the Bertram palm to all other food. The yeasty brew has an alcohol concentration comparable to that of beer (3.8 percent).

When cattle and horses that graze in the chaparral of the western United States lose their sense of direction, go weak in the legs, withdraw from other animals, or suddenly become violent, ranchers immediately suspect locoweed. Several varieties of this legume grow freely throughout the West; the numerous types can be identified by their blue, yellow, purple, or white blossoms that resemble small sweet peas. If the intoxicated livestock don't die from walking off a cliff or blundering up to a predator, "locoed" animals can eventually starve or suffer severe, irreversible brain damage. Despite these dire consequences, some animals actually prefer the plant over their regular foraging options—and, tellingly, one taste of it makes them more likely to try it again. In addition to misadventure and death, locoweed produces another nasty problem that annoys ranchers. Like the cool-kid druggie in homeroom, one locoweed-eating animal can influence others to start. Handlers must be assiduous about removing locoed animals from the herd so the weed-seeking behavior doesn't spread. And locoweed affects wild animals, too. Elk, deer, and antelope have been seen staring dully and pacing nervously after a few nibbles.

A friendly cocker spaniel in Texas once sent her owners' lives into a tailspin when she turned her attention to toad licking. Lady had been the perfect pet, until one day she got a taste of the hallucinogenic toxin on the skin of a cane toad. Soon she was obsessed with the back door, always begging to get out. She'd beeline to the pond in the backyard and sniff out the toads. Once she found them, she mouthed them so vigorously she sucked the pigment right out of their skin. According to her owners, after these amphibian benders Lady would be "disoriented and withdrawn, soporific and glassy-eyed." Soon the neighbors' dogs weren't allowed to come over to play, for fear that they'd pick up Lady's bad habit. Lady's family dreaded the raised eyebrows when they hosted parties and PTA meetings and so started withdrawing from their social obligations because of the dog's new inclination. As amusingly recounted in a story on National Public Radio, one night the dog's human mistress found herself in the backyard at four in the morning, desperately searching for a toad to give to Lady—literally enabling the addiction so the dog would finally come inside and the family could get some sleep.*

Giving alcohol to animals—or watching them imbibe on their own—has entertained humans for centuries. In colonial New England, hogs that got tipsy after eating pomace (the pulpy by-product of cider production) may have provided the sounds that gave rise to a term popular in the day: "hog-whimpering drunk."

Aristotle described Greek pigs becoming intoxicated when "they were

---

*In Australia's Northern Territory, vets have also treated dogs who lick cane toads. After getting "a smile on their face and look[ing] like they're going to wander off into the sunset," many dogs go back "to have a second go. . . . They go on to do it again and again," said one vet.

filled with the husks of pressed grapes." According to the author and alcohol historian Iain Gately, Aristotle also recorded a way to trap wild monkeys by enticing them with alcohol. The technique involved strategically laying out jugs of palm wine for the simians to sample and then simply plucking them up after they got drunk and passed out. Apparently the trick worked just as well in the nineteenth century: Darwin described the same procedure in *The Descent of Man.**

You can see modern-day drunken monkeys in a BBC video shot on the Caribbean island of St. Kitts. The Curious George look-alikes, with their bright, rounded faces, dart among bikini-clad hotel guests. Like teens at a wedding, they wait until no one's looking, then run off with half-drunk daiquiris and mai tais. What comes next is enhanced by the video's quick-cut editing but mirrors what happens to other animals, too, from squirrels drunk on fermented pumpkins to goats sauced on spoiled plums. The monkeys weave. They stagger. They list. They tip over. They try to stand up. They pass out.†

Of course, comparing drug use in animals and in humans has limitations. The superpotent, rapidly addicting, Ph.D.-designed forms of opioids, marijuana, and cocaine peddled to and used by today's human addicts differ significantly from naturally found plant sources of these psychoactive agents. The alcohol available to human consumers is much more refined and intense than what Mother Nature can make on her own. Furthermore, for scientists it's frustrating that most examples of wild animal substance use and its effects are based on observation and anecdote. Indeed, the few papers that do examine wild animal models of intoxication bemoan this fact and call for more stringent field studies. But controlled conditions do occur more frequently in the lab, and animal drug use and abuse have been widely studied in that setting.

Rats, the most examined animal in substance abuse research, have revealed many crossover aspects of intoxication. Like us, in order to

---

*Darwin also detailed a simian hangover: "On the following morning they were very cross and dismal; they held their aching heads with both hands and wore a most pitiable expression: when beer or wine was offered them, they turned away with disgust, but relished the juice of lemons."

†You could argue that the St. Kitts monkeys "choose" to steal drinks. But the Internet abounds with examples of animals being given intoxicants on purpose, for human amusement, a practice that is ethically questionable and in some cases frankly abusive.

start using a substance, they first must overcome an initial aversion. They lose neuromuscular control when under the influence of certain drugs. They seek out and self-administer doses—sometimes to the point of death—of various drugs, from nicotine and caffeine to cocaine and heroin. Once addicted (researchers sometimes say "habituated"), they may forgo sex, food, and even water to get their drug of choice. Like us, they also use more when they're stressed by pain, overcrowding, or subordinate social position. Some ignore their offspring. (Conversely, drug seeking can decrease in lactating female rats.) But rats, although they've become the most popular models for addiction in mammals, are not the only lab animals to be tempted by inebriating substances.

Bees "dance" more vigorously when they're dosed with cocaine. Immature zebrafish hang out on the side of the tank where they were once given morphine. Methamphetamine juices snail memory and performance the way Ritalin might boost a sophomore's PSAT scores. Spiders on a range of drugs from marijuana to Benzedrine spin webs that are overly intricate or nonfunctional, depending on the drug.

Alcohol can make male fruit flies hypersexual and pursue more same-sex matings, perhaps because the ethanol interferes with their reproductive signaling mechanisms. Even humble *Caenorhabditis elegans,* a tiny worm, moves more slowly when exposed to levels of alcohol similar to the ones that make mammals intoxicated. And the females lay fewer eggs when drunk.

Drug seeking. Raised tolerance. Attempts to get stronger and more frequent doses. Begging or jonesing for a drug. If human beings were the only creatures who showed these classic addiction behaviors, then we could say the disease is uniquely human. But clearly we aren't alone. Across the animal kingdom—not just in mammals with highly developed brains—animals react to drugs in comparable, although of course not totally identical, ways.

That we can see parallel effects from intoxicants, whether the organism is a rodent, a reptile, a firefly, or a firefighter, strongly suggests two things. First, animal and human bodies and brains have evolved designated doorways for some of nature's most potent drugs. Called receptors, these doorways are specialized channels on the outsides of cells that allow chemical molecules to enter. Receptors for opiates, for example, have been found in some of Earth's oldest types of fish as well as in humans, and even in amphibians and insects. Receptors for cannabinoids (the intoxicant found in marijuana) have been identified in birds, amphibians, fish, and mammals as well as mussels, leeches, and sea urchins. This introduces the biological likelihood that opiates and cannabinoids—plus many more psychoactive substances—play key roles in maintaining the health and safety of animals. Indeed, these drug-response systems may have evolved and endured because they actually *increase* an animal's survival chances, or "fitness." More on that in a moment.

These animal examples also challenge anyone who would stigmatize addicts or moralize about the disease. What you might see as a personal failing in your no-account uncle who ruins every Thanksgiving with his drunken antics is not a uniquely human impulse. Uncle Bill is not alone in the animal kingdom in seeking and responding to chemical rewards. Maybe knowing that won't make the annual get-together any more pleasant—or his life any easier. But the fact remains that driving his addiction is a chemical reward system shared by other animals, from worms to primates, which has been in existence for millions of years. True, Uncle Bill can choose between a trip to the liquor store and a trip to his AA meeting. But if a fruit fly had the same option, it, too, might sometimes take a rain check on sour coffee in a Styrofoam cup in favor of a warm, soothing hit of ethanol.

Jaak Panksepp never expected to make his name by tickling rats. He'd planned to be an architect or an electrical engineer or, at one point, inspired by his University of Pittsburgh freshman classmate John Irving, a writer. But an internship at a mental hospital when he was an undergraduate set him on a different path. Seeing how the patients there required a wide range of treatments, from short-term stays to padded cells, made him want to understand, he says, "how the human mind, especially emotions, could become so imbalanced as to wreak seemingly endless havoc upon one's ability to live a happy life." And so he became a psychologist and, later, a neuroscientist. He now holds a position that gives him a unique vantage point on how the brains of many species work. As the Baily Endowed Chair of Animal Well-Being Science in Washington State University's College of Veterinary Medicine, Panksepp brings his expertise in human emotional systems to a department devoted to the health of nonhuman animals.

Panksepp specializes in what goes on chemically and electrically in the brains of mammals when they play, mate, and fight, or separate and reconnect. And he is convinced that human addictive behaviors stem from ancient parts of our brains that are shared across species.

Rat tickling came in the mid-1990s, after Panksepp had spent several decades studying play urges in rodents. Using an audio device that measured the ultrasonic vocalizations of bats, Panksepp had discovered that rats make two very different sounds when they're playing. Happily engaged rats emit abundant high-pitched chirps at about fifty kilohertz—much higher than we can hear with the naked ear. To Panksepp it sounded happy, a bit like childhood giggling and laughter. He wondered if the animals would make this sound under other circumstances. One morning, he took a rat accustomed to being held by humans, gently rolled it onto its back, and tickled its belly and armpits. Instantly he heard it: fifty kilohertz vocalizing. He tried another rat. Same thing. Rat after rat, eventually over many years and in many different labs, vocalized at fifty kilohertz when they were tickled in this way.

Panksepp and others found that rats make this "happy" sound in several other specific situations. When they're copulating. When they're about to get food. When a lactating mother is reunited with her offspring. But most especially when two friendly rats are playing with each other.

Their other major vocalization registers at a much lower—but still inaudible to human ears—twenty-two kilohertz. Rats make this very different sound when they're alarmed, anticipating a scary situation, when they're fighting, and especially when they've been defeated in a skirmish. Although not a measure of physical pain, it apparently does reflect psychological distress or psychic pain. Baby rats make a version of it when they're abandoned by or isolated from the warmth of their mothers.

Panksepp says that when you run these sounds through a machine that translates them to a frequency we can hear, the high-pitched notes are roughly analogous to human laughter. The low-pitched calls sound like human moaning. He's found rats make the higher, chirping sound when they're anticipating receiving drugs they desire. They utter the lower, moaning sound when deprived of the drugs and experiencing withdrawal.

Panksepp thinks it's no accident that rats emit the same sound when they're in psychic pain and when they're denied a drug they crave. "Pain" is a word that came up again and again in my interviews with human addicts and the doctors who treat them. Overwhelmingly, addicts report that they need their substances to "dull the pain," "make the pain go away," or "make the suffering disappear."

Rarely do they mean literal, physical pain (although many addictions, especially to opioids, begin with a prescription for relief of bodily pain). The pain that addicts describe is more of an ineffable internal ache—an emotional throbbing or social tenderness.

Panksepp is not the first to wonder whether other animals experience life in a way that could be called "emotionally" painful. This fundamental question has puzzled thinkers for generations: Do animals feel things the way we do?

Charles Darwin tackled the issue in his 1872 book *The Expression of the Emotions in Man and Animals*. Trying to extend his principles of evolution beyond anatomy, he argued that natural selection could be applied to emotion and behavior. The idea didn't catch on. Darwin was up against two centuries of René Descartes's insistence on a dichotomy between body and soul. For Cartesians, only humans—specifically, men—possessed a soul, which was also the seat of intelligence. Having neither soul nor emotions, animals existed in a purely physical realm. Instead of "I think, therefore I am," Cartesians believed that for animals it was more like "I can't think, therefore I can't feel."

Without the tools to track—or even define—emotions in nonhuman species, the behaviorists of the early twentieth century, like J. B. Watson and B. F. Skinner, were obliged to infer what an animal might be experiencing solely by observing its behavior. Here the differences between animals and humans really did get in the way. The facial muscles of most animals don't react in ways that clearly communicate pain to a human observer. Most animals don't vocalize when they're hurt (at least not at frequencies we can hear)—possibly as a self-protective strategy against attracting predators. Many withdraw instead of seeking help. So different are these responses that they supported the behaviorists' idea that animals don't, or can't, perceive physical pain.

Because they couldn't see what was going on inside the cranium, the behaviorists concluded that animal conduct occurred without awareness. If a creature didn't "know" it was in pain, then it couldn't possibly feel pain. Only human brains (and perhaps some other highly developed simian brains), they believed, functioned at a level of cognition high enough to process the unpleasant sensations of pain. Although the behaviorists were trying to reconcile body and mind, they succeeded only in further splitting them. Animals went from being soulless physical entities to boring biological machines. Remarkably, the notion that human consciousness was a prerequisite for feeling pain persisted into the last part of the twentieth century.*

And in some cases tragically, this belief was applied to another group of beings who can't use words to describe their experiences: human infants. The conventional medical wisdom *until the mid-1980s* held that newborns' neurological networks were immature and thus subfunctional. The prevailing doctrine was that babies "couldn't feel" pain the way older humans do.†

Although this view persisted for an uncomfortably long time, pain management is now a priority in both veterinary medicine and human medicine—including, thank goodness, pediatrics.

Advanced brain imaging and other technologies are emerging that allow us to directly study the brain's emotional systems. The techniques are providing evidence for Darwin's view that emotions, like physical structures, have evolved. They are subject to natural selection based on their fitness benefit to individuals. And the reason is pretty simple. What

---

*See the work of Marc Bekoff, Jeffrey Masson, Temple Grandin, and others in the field of animal welfare research for the scientific and compassionate arguments that moved this debate into the twenty-first century.

†In the early 1900s, exploration of whether infants could feel pain led to horrifying experiments in some of the most prominent hospitals in the country. Repetitive pricking of needles into newborns' skin or running their limbs under very cold or hot water to record responses are a few examples. So certain were experts that neonates felt no pain that through the mid-1980s major surgeries on newborn babies were sometimes performed *without anesthesia*. These included major cardiovascular procedures requiring prying open rib cages, puncturing lungs, and tying off major arteries. Though provided with no pharmacologic agents to blunt the pain that cracking ribs or cutting through the sternum might have induced, babies were given powerful agents to induce paralysis—ensuring an immobile (and undoubtedly terrified) patient on whom to operate. Jill Lawson's remarkable story of her premature son, Jeffrey, and his unanesthetized heart surgery provides a heartbreaking account of such a procedure. After Jeffrey's death in 1985, Lawson's campaign to educate the medical profession about the need to treat pain in the young literally changed the field. And likely led to improved awareness of pain in animals, too.

we call "feelings" or "emotions" are not airy, intangible thought-vapors that emanate, auralike, from our brains. Emotions have a biological basis. They arise from the interplay of nerves and chemicals in the brain. And like other biological traits, they can be retained or rejected by natural selection.

Of course, how an animal experiences the world cannot be fully known to a human being. Some scientists, including Joseph LeDoux, an author and neuroscientist at New York University, object to using the word "emotion" when describing the interior world of animals. LeDoux coined the term "survival circuits" to describe the hardwired brain systems that drive animals to defend themselves and promote their well-being.

Randolph Nesse, a University of Michigan psychiatrist and a leader of the growing field of evolutionary medicine, put it this way in a paper in *Science*: "Emotions . . . shaped by natural selection . . . adjust physiological and behavioral responses to take advantage of opportunities and to cope with threats that have recurred over the course of evolution. . . . Emotions influence behavior and, ultimately, fitness." Nesse's view echoes that of E. O. Wilson, who wrote, controversially at the time, "Love joins hate; aggression, fear; expansiveness, withdrawal . . . in blends designed not to promote the happiness of the individual but to favor the maximum transmission of the controlling genes."

Whether or not we use the word "emotion" to describe it, animals seem to be rewarded with pleasurable, positive sensations for important life-sustaining undertakings. These are activities such as finding food, approaching mates, escaping to a hideout, outrunning a predator, and interacting with its kin and peers. The joyful pleasure a young human or animal feels upon reuniting with a caretaking parent encourages bonding, for example. Pleasure rewards behaviors that help us survive.

Conversely, depression, fear, grief, and isolation, among other negative sensations, indicate to an animal that it's in a survival-threatening situation. Anxiety makes us careful. Fear keeps us out of harm's way. Imagine the trouble you'd be in if you didn't feel anxious and fearful when encountering a rattlesnake on a hiking trail or a masked gunman at an ATM.

And there is one thing that creates, controls, and shapes these extremely important feelings: tiny hits of addictive chemicals stashed in microscopic pouches (called vesicles) in our brains.

It's as though we're all born with an internal Pyxis 3500 machine that opens specific drawers in response to our unique genetic "thumbprints" and behavioral "codes." Our personal chemical-dispensing apparatus is stocked with tiny capsules of natural narcotics: time-melting opioids, reality-revving dopamine, boundary-softening oxytocin, appetite-enhancing cannabinoids, and many more—some of which haven't even yet been identified.

Gaining access to one's personal, intracranial lockbox may be one of the most potent motivators in animals, including us. But instead of entering a number, an animal must perform a behavior to release the substances. Behaviors are the codes. Do something that evolution has favored, and you get a hit. Don't do it, and you don't get your fix.

Foraging. Stalking prey. Hoarding food. Searching for and finding a desirable mate. Nest building. These are all examples of activities that greatly enhance an animal's chances of survival, or what biologists call fitness. Pleasant sensations of anticipation and excitement—born in the brain's neural circuitry and chemistry—encourage initiative, risk, curiosity, and discovery in animals.

We humans have a similar suite of life-sustaining activities. We just call them by different names: Shopping. Accumulating wealth. Dating. House hunting. Interior decorating. Cooking.

Indeed, when these activities have been studied in humans and other animals, they are associated with rises in the release of certain chemicals, mostly dopamine and other similar stimulating compounds. Nesse notes that "from slugs to primates," dopamine mediates the search for and consumption of food. Ancient dopaminergic systems have been found in fruit flies and honeybees, suggesting that similar reward experiences may be at play in their behavior. Bees have increased levels of octopamine (their version of dopamine) when they are foraging. Tellingly, their drive to find food appears to come not from personal hunger but rather from a desire for the reward.

Finding safety can also trigger these chemical rewards. Imagine the tremendous relief you felt when the biopsy came back benign, or when the creepy person behind you on the sidewalk turned down another street. That flood of relief is actually a chemical dump within your brain.

Opioid receptors and pathways (the same pathways used by heroin, morphine, and other narcotics) have been found in jawed vertebrates that lived 450 million years ago—well before mammals came on the scene. That means that, from barracudas to wallabies, Seeing Eye dogs to homeless heroin addicts, animals have an ancient and intimate response to opiates.

Researchers working with Panksepp have found that opiates regulate separation and distress calls in dogs, guinea pigs, and domestic chicks. His colleagues have also found that when dogs wag their tails and lick each other's or their owners' faces, that behavior, too, is modulated by opioids. Opioids play a role in early suckling behavior in rats. And the proximity of offspring has been shown to trigger a hit of pleasurable chemical reward in the brains of rat mothers.

Besides opiates and dopamine, many other chemicals work constantly in our bodies and brains. Cannabinoids, oxytocin, and glutamate, among others, create a complex system of simultaneous positive and negative sensations. This cacophonous chemical conversation (what Panksepp calls the "neurochemical jungle of the human brain") is the basis of emotion—emotion that creates motivation and drives behavior.

Human feelings powerful enough to launch a thousand ships, build the Taj Mahal, or ignite pleasurable melancholy at Mimì and Rodolfo's parting in act IV of *La Bohème* have emerged from "survival circuits" (to use LeDoux's term) that we share with other animals. In other words, our emotions exist as they do today because their building blocks helped our animal ancestors survive and reproduce.

And this is precisely why drugs can so brutally derail lives. Ingesting, inhaling, or injecting intoxicants—in doses and concentrations far higher than our bodies were designed to reward us with—overwhelms a system carefully calibrated over millions of years. These substances hijack or ignore altogether our internal Pyxis 3500 mechanisms, removing the need for the animal to input a code, in this case, a behavior, before receiving a chemical dose. Nesse writes, "Drugs of abuse create a signal in the brain that indicates, falsely, the arrival of a huge fitness benefit." In other words, pharmaceuticals and street drugs offer a faux fast track to reward—a shortcut to the sensation that we're doing something beneficial.

This is a critical nuance for understanding addiction. With access to external substances, the animal isn't required to "work" first—to forage, flee, socialize, or protect. Instead, he goes straight to reward. The chemicals provide a false signal to the animal's brain that his fitness has improved, although it has not actually changed at all.

Why spend an afternoon in the dangerous and time-consuming task of foraging for a hundred acorns (or bringing in a hundred new clients) when you could achieve a far more intense reward state with one snort of cocaine? Or, to be less extreme, why go through a half hour of awkward small talk at an office party when a martini or two can trick your brain into thinking you've already done some social bonding?

The excessive, seemingly inexplicable behavior of those addicts who forgo the important, life-preserving chores of daily life becomes clearer when viewed this way. Drugs tell users' brains that they've just done an important, fitness-enhancing task—even though they haven't. Their brain receptors don't know whether that opioid molecule came from a hash pipe or from having a conversation with a trusted friend. They don't know whether the dopamine molecule came from a crack spoon or from the rush of getting five phone numbers at a bar or finishing a tough assignment on deadline. The rewarding feelings signal that they *have* gained resources, found mates, and elevated their social status. The awful irony is that these substances so potently imitate these feelings that their users may cease doing the real work of life. Their brains are telling them they already did.

We can condemn addicts and their poor self-control as much as we wish. Ultimately, however, the powerful urge to use and reuse is provided courtesy of honed and inherited brain biology that evolved to maximize an individual's shot at survival. Seen this way, we're all born addicts. That's what "motivates" creatures to do important things.

And that's why Pyxis 3500 machines stand sentry throughout my hospital. They restrict access. As David Sack, the CEO of the drug-rehab program Promises, told me, "You can't become addicted to a drug you don't have access to."

Putting synthetic and plant-derived drugs into our bodies circumvents the personal lockboxes in our brains. But the stashes of natural drugs are still in there. And, as we've seen, the codes to releasing them are elemental behaviors. This reveals an interesting possibility. Even if an animal doesn't have access to external sources of drugs, there may be another way to hack into the internal stores: by punching in code after code . . . of unnecessary but reward-producing behaviors. Maybe addiction can be activated by things we *do,* almost as effectively as the substances we *take.*

.   .   .

As a cardiologist, I mostly encounter substance addictions as they relate to a patient's heart health. But in the late 1980s, I was training to become a psychiatrist and began treating a patient for depression and anxiety. He was handsome and a meticulous dresser. At our weekly sessions he was unfailingly polite and charming, which I interpreted as an openness to the therapeutic process.

At our first meeting I learned the main reason for his anxiety: he was cheating on his wife. Soon I learned that he was cheating on the mistress, too—with her best friend. While maintaining regular relationships with all three women, he was also having frequent one-night stands. As he described the stress and anxiety of juggling each week's sexual appointments, he explained his utter inability to stop doing it. I could sense his excitement about what he was doing—the danger of sleeping around, of hiding it from his family, the thrill of getting away with it. As his psychiatrist, I thought it all just sounded dangerous. He was risking his marriage, his relationship with his child, and his career (the mistress was a subordinate at work). After several months, he quit coming to therapy; he continued his risky behavior and eventually lost his job and his wife.

At that time, psychiatry had a primary approach to treatment: psychodynamic psychotherapy. The basic premise of this method is that our adult selves are formed in large part by our childhood experiences. The entire time I was treating this patient, my professional assumption was that his inability to have a stable sexual relationship with his wife stemmed primarily and perhaps exclusively from attachment issues connected to early childhood traumas. My supervisors confirmed the diagnosis and supported my treatment plan, so I spent many sessions probing his early years, looking for reasons to explain his promiscuous and risky behavior.

Thinking about it twenty-five years later, I realize that my understanding of his reckless sexual behavior was incomplete. The field has now advanced to recognize that early experiences do actively shape genes and the brain, laying the groundwork for susceptibility to addiction later in life. But what I had missed then was the fact that my patient was addicted to the neurochemicals provided to him through his sexual pattern: the spike of thrill-danger-novelty dopamine and perhaps also the feel-good payoff of sex itself. Nowadays he would probably be referred to a sex-addiction program. But that never occurred to me then.

The brain-disease theory of alcoholism was only just emerging at the time. That behaviors like sex or shopping or overeating could be put in the same category as a substance addiction wasn't part of the medical vocabulary. Even today, addictions to the things a person does, instead of the substances he takes, are not completely understood. The debate over whether or not they're "real" addictions divides doctors within and outside the addiction field.

I have to confess that, until recently, I, too, was extremely skeptical. You're "addicted" to buying shoes. Really? Can't stop eating candy corn? Feel physical withdrawal pains when separated from your pornography or video game? Uh-huh. The model of substance addiction as a brain disease made sense to me, but until recently, applying the term "addiction" to behaviors seemed sloppy—a "no-fault," feel-good copout, a lazy, twentieth-century inability to break bad habits. It's not me, Your Honor. It's my *disease.*

However, spending the past several years trying to understand my human patients through a veterinary perspective has led me to a different view and a surprising hypothesis: substance addiction and behavioral addiction *are* linked. And their common language is in the shared neurocircuitry that rewards fitness-promoting behaviors.

When you look at the most-often-treated behavioral addictions from an evolutionary perspective, they are exceedingly fitness enhancing. Sex. Binge eating. Exercise. Working. It's hard to imagine that "in nature" or when tested by natural selection, those behaviors would produce many downsides, even when taken to extremes.

Gambling and compulsive shopping—although they're human variants—work on the same neural pathways as two extremely beneficial animal activities: foraging and hunting. These involve focused and concerted effort and expenditure of energy with a specific goal of gaining resources, typically food but sometimes shelter or nesting materials. Neurochemical rewards reinforce this positive behavior in animals. As Panksepp puts it, "Every mammal has a system in the brain to look for resources."

By following the neurobiology, we can see that gambling is foraging taken to an extreme, where food has been replaced by a financial payoff. Although food and money are certainly rewards in themselves, the true payoff—the addictive part—is the neurochemical mix associated with seeking and risk-taking. Behavior produces a reward that creates dependence, just as external chemicals do.

Connecting brain-rewarding behaviors to increased survival also allowed me to rethink technological "addictions" like video gaming, e-mail, and social networking. The executive who jokes that she's addicted to her BlackBerry probably doesn't think she needs a twelve-step program to quell her itchy thumbs. But many of us find the urge to check that little screen irresistible—even during an important meeting or when we're behind the wheel. Our smartphones, Facebook pages, and Twitter feeds profoundly combine the things that matter most to animals competing to survive: a social network, access to mates, and information about predatory threats. But like drugs, these devices provide the hit without the work. We get a dopamine squirt without seeking a tangible resource. We may get a lovely opiate flood of feeling part of a herd, without the inconvenience of actual herd mates.

Veterinarians I interviewed were reluctant to apply the word "addiction" to animals. As they pointed out, pets generally do not, on their own accord, get hooked on drugs or alcohol.

But there is one thing they seem to crave: reward. It can be as simple as a pat on the head and a murmured "Good boy." A morsel of frozen liver or mouthful of oats. A tummy rub.

Do a behavior, get a reward. Rewards in the form of food or praise have long been used by animal trainers to produce certain predictable behaviors. As Gary Wilson, a professor and trainer at the Exotic Animal Training and Management Program at Moorpark College in California, told me, external treats in the form of food and congratulatory sounds are in effect bridges to the animal's brain. They pair the feel-good neurochemicals produced by anticipation of nutrients with desired behaviors.*

---

*A technique called clicker training pairs a metallic *tick-tock!* with a food treat every time the animal performs a desired behavior. Eventually the animal comes to associate the sound of the clicker with the feel-good neurochemical rewards of the food. When the treat is discontinued, the animal will continue doing the behavior, because its brain has been conditioned to anticipate reward and actually releases dopamine to the sound alone. A human version of clicker training is increasingly being used to train gymnasts and other precision-sport athletes and to reinforce positive behaviors in classrooms and special education groups. Called TAG teaching (teaching with acoustical guidance), to avoid the animal overtones of clicker training, it works on the same principles of associating behavior and reward. "Neurologically, clicker training activates the dopamine centers in the amygdala," Wilson said. The clicker "becomes a marker, the internal reinforcement of the dopamine system."

Seen this way, the unrecognized goal of some animal training may be to create a kind of behavioral addiction, as animals learn to associate the pleasure of reward with new behaviors. David J. Linden, a professor of neuroscience at Johns Hopkins University and author of *The Compass of Pleasure,* connects this pleasure of learning and training in humans to the neurobiology of other addictions.

He notes that learning, along with behaviors such as gambling, shopping, and sex, "evoke neural signals that converge on a small group of interconnected brain areas called the medial forebrain pleasure circuit." Successful dog training creates what we could call a learning addiction, driven by pleasure circuits. Linden notes that these circuits "can also be co-opted by artificial activators like cocaine or nicotine or heroin or alcohol."

Human medicine has only recently started to regard chemical dependency as a physical and chronic illness requiring ongoing (perhaps life-long) care, rather than a condition (like an infection) that we can treat, cure, and quickly put behind us. Understanding the evolutionary origins of addiction can improve how we care for this disease. It may help us be more compassionate to users and addicts and can help us understand that substance use in animals of all kinds is an attempt to get a little more of what we spend our lives seeking.

If you exposed a hundred people to a carcinogen, they wouldn't all get cancer. It's the same thing with drugs. Expose a hundred animals to a chemical molecule, and they're not all going to get addicted to it. Not all cocker spaniels become toad lickers. Not every monkey steals cocktails or wants to drink one every day. Only some wallabies jump the fence to eat opium poppy sap.

The biological term for these differences within populations is "heterogeneity." What heterogeneity means in terms of addiction is that each person, each animal, has a slightly different response to each chemical. An abundance of research backs this up: there's a strong genetic basis for susceptibility to addiction. Recently, families with histories of substance abuse have started educating their children about their particular inborn vulnerabilities. But environmental factors, from the climate in our mother's uterus to the food we eat and the pathogens we encounter, also play key roles in who becomes an addict. It's becoming clearer to scientists that what you eat, where you live, the work you do, and even how you were parented can change how your genes are expressed. The emerging field of epigenetics considers what happens to our personal genetic code when it meets the real world. It explains why nature/nurture is not a divide but, rather, an endless feedback loop.

Genes give an individual high school sophomore a predetermined potential to become addicted to alcohol or drugs. But when and how he encounters those chemical molecules will create the epigenetic effect. For one teen, a Friday night, postgame first exposure to, say, marijuana can activate neural responses that will make cannabis a gateway drug for future use. For that teen's best friend, that first toke might be just another moment in an ordinary get-together at a friend's house, a teenage dalliance that he'll laugh about in a self-deprecating way later in life. Same party; same drug; two different life outcomes. If either teen had encountered that substance as an older adult or as a younger child, again, the outcome might be different.

Like many humans, some nonhuman animals can enjoy the pleasures of substances without apparent adverse effects. The Malaysian pen-tailed tree shrew imbibes copious volumes of fermented palm nectar without observably diminished reflexes or impaired coordination. Zenyatta, a multilaureled and now retired racehorse, traditionally guzzled a Guinness after every race and went on to win the next one.

Heterogeneity stocks every animal's lockbox with different supplies of drugs. Epigenetics calibrates the codes. Those codes are set and changed throughout our lives. But an important period of code setting occurs in childhood—infancy through adolescence. Human and animal data both suggest that the younger the animal is at first exposure to an external drug, the more likely it is to become addicted and responsive to that drug in the future.

This is a very important point. Our behavioral relationship with potentially addictive neurochemicals begins from the minute we enter the world (and quite possibly before). Suckling has been found to produce an internal opioid hit, a chemical reward for this basic, life-sustaining task. Indeed, Panksepp and others believe the suite of "attachment" neurochemicals are many and powerful, and that some of the codes for releasing them are set in earliest infancy. Various elements of a child's young life—including physical health, "wiring," but also, significantly, parenting—influence how their personal lockbox will respond to increasingly challenging environments.

Like younger children, adolescents, too, have highly malleable brains. Pouring external sources of powerful reward chemicals into the brain at precisely the point it's trying to calibrate the system can have lifelong effects. It can influence tolerance levels and response sensitivity. Across species, a zoobiquitous look at addiction suggests that delaying the age of first use could have powerful protective effects. Extensive study of the effect of alcohol exposure on adolescent rodents and nonhuman primates has shown alcohol's long-term effects on the adult brains of these young mammalian imbibers. Along with impaired cognitive function, early use in these animals may increase their risk of alcohol addiction later in life.

In the United States, we've tried prohibition and "just say no" campaigns. We've set the drinking age at twenty-one and the illegal drug use age at never. None of these interventions has completely stopped teenagers from going after what they want.

But the evidence suggests it's wise for parents to try harder to delay their children's first exposures and, perhaps, to teach them natural ways of achieving those chemical rewards: through exercise, physical and mental competition, or "safe" risk-taking, such as performing.

In some individuals, whether cedar waxwings or late-night partiers, intoxication can lead to tragedy. In humans it's linked to higher rates of motor vehicle accidents, suicides, homicides, and accidental injury. In the wild, intoxicated animals, too, are at greater risk. They can more easily be picked off by predators, miss out on opportunities to mate, or fly into walls.

But nature provides its own abstinence program. Access to plants, berries, and other food sources in the wild vary with seasonality, weather, competition, and many other factors, including predation. And these variations automatically reduce access to substances that might otherwise lead to addiction. It's like a wild version of having one's coke dealer leave New York for Miami between November and March. This lack of ongoing access to substances, coupled with the increased risk of death to an intoxicated animal in the jungle, desert, or savannah, makes the possibility of humanlike addiction in the wild unlikely.

Recovery from addiction may involve restoring the integrity of the lockbox we're born with. Substance abusers can learn healthy behaviors that provide the same (albeit less potent) good feelings they used to seek from a bottle, a pill, or a needle. In fact, that may be what makes some rehab programs so effective for certain addicts. If you look at the behaviors these programs encourage—socializing, companionship, anticipation, planning, and purpose—they're all part of an ancient, calibrated system that doles out internal neurochemical rewards.

Ironically, one way to fight addiction may be with addiction, replacing a dependence on heavily refined drugs with the hard work that makes life worth living. The endorphin release of physical work and exercise. The adrenaline rush of healthy competition and risk in games or business. The exquisite anticipation of planning, serving, and at last eating a great meal. The opioid rush of being part of an actual flesh-and-blood social network. Or the warming satisfaction of helping others. The term "natural high" may sound as dated as a John Denver song, but it's not a metaphor. It's the ancient reward that motivates and sustains all animals, including us.

SIX

Scared to Death

Heart Attacks in the Wild

The magnitude 6.7 earthquake struck at 4:31 a.m. on January 17, 1994. I was jolted out of bed, my heart pounding as I waited along with millions of others around Los Angeles for the ground to stop shaking. When it finally did, I drove to the hospital, adrenaline and caffeine blasting away my haze of fatigue. Not knowing whether we'd soon be receiving a few cuts and bruises or facing a large-scale catastrophe, I entered the UCLA emergency room. In that moment, I couldn't have predicted how completely the morning's geological shift would shake my perspective on medicine over a decade later.

At the time I was a "flea," to use the macho, old-school surgeons' derogatory term for overly analytic internal medicine nerds. I embraced the nickname, enthusiastically hopping up and down while spouting medical minutiae and arcane diagnoses whenever my superiors hove into earshot. On cue, I could explicate the cryptic presentation of Behçet's disease. With dorky rectitude, I competed with other fleas to recall the fifth and sixth diagnostic criteria for relapsing polychondritis. Never in the history of medicine, we told ourselves, was anyone as passionate about Churg-Strauss vasculitis or Rasmussen's encephalitis.

As the newly appointed chief resident in internal medicine, I was also treating real patients with more familiar-sounding maladies. But during this year, with the grunt work of internal medicine residency behind me and the rigors of subspecialty still ahead, I threw myself into the exciting and decidedly cerebral pursuit of medical oddities. At a teaching hospital like UCLA, this isn't just tolerated; it's encouraged.

But all that changed when the earthquake shot up from fifteen miles beneath Earth's crust with the fastest ground acceleration ever recorded in urban North America. Apartment buildings slumped. Freeways snapped. The scoreboard at Anaheim stadium toppled onto several hundred (thankfully empty) seats. Thousands of people all over Southern California were injured.

Instantly I was propelled out of the arcane, into the here and now. Throughout the day we treated serious wounds and minor scrapes. Amid the blur and drama of the days right after what became known as the Northridge quake, a curious trend emerged. Although I didn't notice it then, it held a special significance for me as a budding cardiologist.

On the day of the earthquake, and for twenty-four hours right afterward, the heart attack rate around Los Angeles spiked. L.A.'s coroner noted a fourfold increase in cardiac deaths. As later reported in the *New England Journal of Medicine,* nearly five times as many Angelenos suffered what we call a cardiac event that day, compared to the same day in January in the years before and after. The conclusion: at least some Southern Californians had been scared to death when the earthquake struck.

This research, although fascinating, had little day-to-day effect on my practice of medicine. Most of the time I was treating patients who were not extremely fearful. So it sat on my mental shelf of medical curiosities for several years, until the day a wildlife veterinarian showed me a telling video.

The scene opened on a quiet, curved stretch of beach. Waves glinted in the morning sunshine. Suddenly, an explosion cracked the stillness. A flock of shorebirds burst off the water. They flapped madly toward the center of the lake, chased by a giant unfurling rectangle of net detonating from the cannon. Most of the birds escaped and resettled on the mild waves. But two dozen or so didn't make it. Before they could get airborne, the mesh had grabbed them and trapped them in place.

There the video ended, but my vet colleague filled me in on what happened next. A capture team sprinted to the ensnared birds. Working quickly, the biologists plucked the struggling animals one by one from the net, carefully detangling wings, beaks, and claws. Calmly but hurriedly, they placed the animals into plastic crates with perforated lids.

The caught birds would be tagged, recorded, and released, to provide vital information on the species' health and migration routes. But some individuals would never fly again. Startled by the cannon shot, panicked by the confining net, terrified by the grasping human hands, they had died on the spot.

What I realized as I watched the video was that, although separated by place, time, and species, those shore birds connected to the humans who'd died from heart attacks during the Northridge earthquake. And more than that, the bird deaths connected physiologically with a type of cardiac arrest that kills tens of thousands of people every year, called sudden cardiac death. Exploring the overlap between fear-triggered "heart attacks" in animals and humans could expand the scientific understanding of sudden cardiac death. And it could help safeguard patients from an unseen threat within each of their bodies.

Like Angelenos after the Northridge quake, people all around the world take the shock and drama of earthquakes, tornadoes, and tsunamis literally to heart. Admissions to hospitals for chest pain, arrhythmias, and even death are as predictable after natural disasters as power outages, Red Cross tents, and Anderson Cooper's tight T-shirts.

Man-made calamities, too, jerk hearts out of their normal rhythms. In the early days of the Gulf War, in 1991, Iraqi forces began sending Scud missiles into suburban Tel Aviv and other areas of Israel. During that week of bombing, civilians faced the terrifying possibility of being blown up at any moment. Air-raid sirens suddenly and shrilly discharged around the clock. Statisticians combing through the numbers later uncovered a potent piece of data: rates of cardiac events during that frightening week exceeded expected numbers. More Israelis may have died from the physiology of panic and dread than from actual Scud impacts. As a military strategy, the Scud explosions themselves were nearly useless. The much more effective wartime weapon may have been terror.

After the al Qaeda attacks of 9/11, frightened people across the United States hunkered down in their homes, wondering when the next hit might come. According to data collected from heart-disease patients with recording devices implanted in their hearts,* the anxiety of those terrifying days carried grave cardiac risks. The number of life-threatening heart rhythms detected and shocked surged to 200 percent of normal. And this trend was not only seen at the plane crash sites in New York City, Washington, D.C., and Pennsylvania, but in other parts of the country, too. The physical impact of fear touched Americans whose only connection to the disaster came through their eyes and ears—fixed on horrifying TV images and descriptions of planes crashing, buildings collapsing, and human beings leaping from smoke and flames.

As you've probably felt yourself, our hearts react when something startles us, whether it's an unexpected balloon pop or the ground rumbling to life beneath our feet. Our bodies sometimes respond before our brains can sort out the lethal threats from the harmless jolts. And armchair athletes, take note: you don't need to be actually playing a game for your ventricles to act out the agony of defeat.

Take the 1998 soccer World Cup. England and Argentina had clawed their way up the ladder and were facing off for the chance to compete against the Netherlands in the quarterfinals. While international soccer rivalries are always fierce, this pairing had special resonance for the fans. Sixteen years earlier, the two countries had gone to war over the Falkland Islands. Although Britain officially won the skirmish, many Argentines refused to acknowledge defeat. Every time the two teams subsequently met on the soccer pitch, it turned into a grudge match. This game (which featured a young David Beckham, fouling out after kicking another player in full view of the ref) ended in a tie. The winner would be decided by a penalty kick shoot-out.

One by one, the players lined up in front of the goalie to take their shots. The score had reached Argentina 4, England 3 when the English

---

*Implantable cardioverter defibrillators (ICDs) are surgically placed in hearts that are at risk for arrhythmias that can lead to death. These tiny electronic devices read the heart's rhythm 24/7. If it dangerously decelerates or accelerates, the ICD delivers twenty-five to thirty joules of electricity to "jump-start" or "pace" the heart. Patients' descriptions of these jolts range from "heavy hiccups" to "having a donkey kick you on the chest." Previous studies have noted increased ICD activity after emotionally charged events, including arguments.

player David Batty jogged onto the field. He took a few short sharp strides toward the ball . . . made contact . . . and sent it soaring. But between Batty's Puma cleats and the expanse of the goalposts, the ball met the gloved fingers of goalkeeper Carlos Roa—and the winner was Argentina.

The Argentine fans erupted in relieved, joyful mayhem. But English fans watching on TVs in pubs back home gaped in stunned horror. And that day heart attacks across the United Kingdom *increased by more than 25 percent.*

Multiple European studies have corroborated this unusual link between spectator stress and heart health. Interestingly, soccer matches that end in penalty kicks—and, worse, portentously named "sudden-death" shoot-outs—are the deadliest. Richard Williams, a sportswriter for the *Guardian* newspaper in London, has called them sadistic, "the modern equivalent of a flogging in the market square." In fact, penalty kick shoot-outs are so reliably anxiety-producing that soccer groups from the Fédération Internationale de Football Association (FIFA) to the American Youth Soccer Organization (AYSO) have considered banning this form of tie breaking.

Having stood tensely, hands clenched to my chest, blood pressure rising, at my own children's championship bouts (fencing is their game), I appreciate how this proposed change might diminish dangerous heart rhythms in the nervous parents and grandparents squirming in nylon spectator chairs on the sidelines.

Until as recently as the mid-1990s, the relationship between the heart and the mind was only murkily understood. Among many physicians, the idea that emotions could cause actual physical effects within the architecture of the heart was viewed with nearly the same sideways glance as an interest in healing crystals or homeopathy. Real cardiologists concentrated on real problems you could see: arterial plaque, embolizing blood clots, and rupturing aortas. Sensitivity was for psychiatrists.

A shift came in the 1990s. A team of Japanese cardiologists noticed that the hearts of some patients who came to the emergency room with severe, crushing chest pain after moments of extreme emotional stress were not normal. Their EKGs indicated that they were having a heart attack. But when the doctors injected dye into the heart vessels, they found perfectly healthy, "clean" coronary arteries—no signs of blockage. The only unusual finding was a strange lightbulb-shaped bulge at the bottom of the heart. The shape reminded these doctors of the rounded *takotsubo* pots Japanese fishermen use to catch octopuses. So that's what they named it. Takotsubo cardiomyopathy was a new description—direct, physical evidence that severe stress (fear, grief, agony) alone could alter the heart's chemistry, shape, and even the way it pumps blood.*

The condition quickly acquired a nickname, "broken-heart syndrome." When it was newly named and freshly in vogue, takotsubo cases cropped up in ERs all over the country. One young woman had watched her beloved dog run into traffic and she arrived at an ER covered in blood, clutching her limp pet *and* her chest. (Like most takotsubo patients, she was treated and survived, although some do die.) Another patient was thirty minutes into an intense 3-D blockbuster when extreme palpitations, shortness of breath, and repeated vomiting forced her to flee the theater. Her doctors diagnosed takotsubo.

To understand how any intense emotion can physically harm your heart, it helps to know some cardiology basics. Under normal circumstances, your heart is probably the most important thing you never pay attention to. Like the perfect valet, it's toiling away in your chest right now, meticulous yet unseen, as it has since the twenty-third day after your father's sperm met your mother's receptive egg. Every year your heart beats 37 million times, pumping 2.5 million liters of blood.

Like a house, the heart contains both plumbing and electricity. Plumbing moves blood around your body's pipes—its arteries and veins. Like the main lines carrying water into a house and the sewer lines carrying it out, these arteries and veins can get blocked, with devastating results. Sudden myocardial infarction, for example, the classic heart "attack," is caused by a clog in the blood supply to the heart itself. Ruptures and

---

*Before the takotsubo designation came along, we would diagnose this syndrome as "spasm of the coronary arteries." Certain types of people seemed prone: middle-aged women; people with histories of migraine headaches; and patients with Raynaud's syndrome, a circulation anomaly characterized by white, blood-drained fingertips. The ill-defined heart "spasm" was also linked to cocaine use, so any patient who came into the ER with chest pain combined with suspiciously plaque-free arteries was questioned about drug habits.

bursts in the plumbing can be devastating, too. When large arteries tear or rip open, the result is often fatal.

But a whole other category of cardiac woe comes from inborn or acquired damage to the electrical system. Its health can be discerned from an electrocardiogram (EKG)—the jagged, steep slopes that shoot up and down across graph paper or scroll along a computer monitor. You've seen these representations of the heart's steady current countless times on medical dramas and in pharmaceutical ads. And you've heard audio translations of that electricity, too. When wired to an alarm signal, the clockwork electrical current creates that stable *beep . . . beep . . . beep* that indicates all is well. Nothing calms a nervous on-call doctor better than news that her patient's heart is beating in that pattern. We call it normal sinus rhythm.

Tragically, however, this steadfast, pulsing electrical system fatally short-circuits in seven hundred human hearts every day in the United States* and in many thousands more worldwide. The dependable throb suddenly races out of control or becomes floppy and unpredictable. When you listen with a stethoscope, the *lub-dub*s take on an anxious, irregular, and muffled quality. When the beat speeds up—a condition we call ventricular tachycardia, or VT—it's unmistakable on the EKG. The assembly-line predictable peaks and valleys of normal sinus rhythm loosen into rolling, closely spaced "hills." The lopsided, irregular rhythm, on the other hand, is called ventricular fibrillation (VF). It's equally easy to recognize. The jagged visual static lurches across the monitor or graph paper with a sickening randomness.

To those in the know, the sounds and sights of VT and VF instantly convey one thing: the urgent need for someone to place shock paddles on the sufferer's naked chest; call "All clear!"; and send a few hundred joules coursing toward the malfunctioning heart. If this specialized electrical therapy doesn't arrive immediately, the EKG landscape will devolve from the hilly and craggy warning shapes to the infamous horizontal configuration we refer to—with dreaded respect—as a flatline. The shift in rhythm from "life-sustaining" to "malignant" causes the heart's pumping to decrease or stop. With more precision than poetry, doctors call

---

*Imagine one and a half packed 747s crashing every day and you'll appreciate the public health significance of the problem.

this electrical cardiac catastrophe sudden cardiac death (SCD or sudden death for short).*

Cardiac death can come rather predictably after years of plaque buildup in the brittle arteries of an overweight smoker. Or it can strike with a shock as happens when a high school athlete drops dead from a congenital defect he didn't know he had. The "final common pathway" is the same. It's an electrical malfunction, the shift from a life-sustaining, normal rhythm to the death-heralding arrhythmia of VF or VT.

But certain victims of sudden cardiac death have no previously identified heart problems. In these otherwise healthy patients, a massive emotional jolt alone converts the cardiac rhythm from safe and steady to malignant and deadly. Startled, terrified, horrified, or aggrieved, these patients spew stress hormones, including adrenaline, from their highly activated central nervous systems. These catecholamines gush into the bloodstream. Like a chemical cavalry they appear on the scene, ready to boost strength and stamina to aid an escape. But instead of rescuing the patient, this neuroendocrine burst may rupture plaque deposits, lodge a blocking clot in an artery, and cause a fatal heart attack. It might trigger an extra beat at just the wrong moment and send the heart into VT. And in huge amounts and all at once, the chemicals themselves can be enough to poison muscles, including some of the two billion heart muscle cells in a human ventricle. In these patients, the weapon is essentially the reactive nervous system itself, fully loaded with dangerous catecholamines, waiting for terror to pull the trigger.

That's what happens with takotsubo. Whether activated by lost love or war, geological heaving or a ball game, the catecholamine torrent damages heart muscles, creates the octopus pot–shaped bulges, and sometimes causing dangerous arrhythmias.

But takotsubo is only a small part of the story, as I figured out when I started comparing notes with veterinarians.

---

*Many things can cause VF and VT and sudden cardiac death. Some dangerous heart rhythms are inborn, such as long QT syndrome. Others are acquired: electrolyte disturbances, viral infections, antibiotics and other drugs, and aortic ruptures can all result in fatal arrhythmias. Even acts of God, like being struck by lightning, a high-velocity karate chop, or a Little League line drive to the chest that falls at precisely the wrong moment, can cause the heart's valvular structures to vibrate violently and then stop moving (doctors call this *commotio cordis*).

.   .   .

Dan Mulcahy is the kind of guy you'd want along if you ever found yourself stranded in a Category 5 blizzard with the gas gauge on your Arctic Cat F1000 Sno Pro on empty. Part MacGyver, part Davy Crockett, with a full mustache, wire-rimmed glasses, and a basso profundo rumble, Mulcahy occupies that rare and desirable zone on the Venn diagram where superhero and supernerd overlap. At forty-one, after two decades as a microbiologist studying fish diseases, he switched careers and became a wildlife veterinarian. When I met him, he was working in Alaska, tracking and treating walruses, tundra swans, caribou, and other endangered northern creatures. His job requirements ran from the delicate surgery of installing satellite transmitters in spectacled eider ducks to the bravura choreography involved in collaring a half-ton polar bear as part of a worldwide team monitoring and trying to preserve these animals' disappearing hunting grounds.

When we met, we quickly discovered that we shared a professional and personal fascination—with the ways death lurks in the interplay of the heart and the mind. We bonded, as only two doctors can, by trading ghoulish yet exciting tales of fear-based sudden death we'd seen and treated.

Mulcahy's interest had a poignant and frustrating root: every once in a while, an animal—especially, for some reason, certain birds—would silently die in his hands after being chased, captured, and handled. Sometimes they seemed to get through the medical procedure just fine, only to weaken and die the minute they were reintroduced to a new habitat. Mulcahy knew it wasn't anything he was doing wrong. In fact, his supervision made these important surveys carried out by field biologists much safer for the animals.*

Veterinary textbooks describe a heartbreaking yet eerily predictable reality: animals consistently die from the stress of chase and handling. Veterinarians call it capture myopathy. The term describes a syndrome

---

*Like the on-set veterinarians who allow film studios to declare, "No animals were harmed in the making of this movie," Mulcahy is an on-site veterinarian in the field, monitoring animal safety for field biologists observing and tracking wildlife. Working for the United States Geological Survey, Mulcahy has pioneered and enforces protocols to make these studies safer for the animal subjects.

of illness and death seen in animals who are terrified, captured, or running for their lives from predators, hunters, or well-intentioned but underinformed wildlife biologists. Sometimes the affected animals expire on the spot, crumpling to the ground like a maiden in a gothic novel. Sometimes they endure for a few hours after the stressful incident before expiring. Other times they linger for days or weeks, listless and depressed, unable to walk or even stand, refusing food and water, until they die. In any case, postcapture death rates are disturbingly consistent.* It's usually about 1 to 10 percent of a population, sometimes as high as 50 percent, depending on the species.

Capture myopathy was first noticed by human hunters a hundred or so years ago. It was initially thought to be an exclusive syndrome of big prey, like zebra, buffaloes, moose, and deer. These animals often died mysteriously after a hearty chase, even when the hunters' weapons had left no mark on their bodies.

But then ornithologists started noticing capture myopathy's fingerprint on the muscles of birds from tiny parakeets to lanky whooping cranes to brawny, ostrichlike rhea. Marine biologists described cases in dolphins and whales. Fishermen trawling for wild Norway lobsters off the coast of Scotland saw it pinch their bottom line. The meat of chased lobsters was discolored and had an unappealing, watery texture. It looked spoiled—in a way, it was deader than dead. It rotted faster. At the market, it was rejected on sight.†

Soon wildlife veterinarians realized that pursuit—chasing without rest—was killing animals from every corner of the food web. In South Africa, where animals frequently need to be moved around to accommodate national park boundaries and human encroachment, capture myopathy is a serious health threat and a major cause of death. Spe-

---

*Veterinarians divide capture myopathy into four classic presentations: "capture shock syndrome," "ruptured-muscle syndrome," "ataxic myoglobinuric syndrome," and "delayed peracute syndrome." These terms describe various physical manifestations of the syndrome, ranging from weak muscles and unsteady gait to kidney failure and sudden death. A captured wild animal may exhibit one or more of these syndromes during chase and/or capture.

†These are all examples of animals in the wild, but stress before being slaughtered in broiler hens, sows, steers, and lambs damages their muscles, too. The same muscles, when wrapped in plastic on a Styrofoam tray, are sold as meat. Some farmers recognize this and (perhaps for the wrong reasons) strive toward less-stressful slaughter techniques.

cial care is taken when capturing sensitive giraffe, which aren't as a rule accustomed to running long distances and are also known for their anxiety. Deer, elk, and reindeer in North America have capture myopathy mortality rates from relocation and hunting as high as 20 percent. The Bureau of Land Management's yearly helicopter roundups of wild mustangs in Nevada result in the deaths of a certain number of horses every year from capture myopathy.

Fueling an animal's escape from threat is a powerful neurochemical response: a catecholamine dump. Pushed beyond a safe limit, though, catecholamines can overwhelm the skeletal and cardiac muscles and cause them to break down. When significant chunks of skeletal muscle are degraded, massive amounts of their proteins are released into the bloodstream. These proteins can overpower and eventually shut down the kidneys. The medical name for this muscle damage is rhabdomyolysis. "Rhabdo"—as it's widely known in clinical shorthand—can be fatal, but if it is caught early, it can be effectively treated with hydration and supportive care. In people, it's most often seen in extreme cases of trauma and immobility: an earthquake victim pinned under steel beams and rubble, for example, or a motorcyclist thrown from his bike with multiple skeletal fractures and overwhelming soft-tissue injuries. Veterinarians and physicians alike know that a telltale sign of rhabdo is rust-colored urine—tinged that shade by the overflow of toxic muscle enzymes that the kidneys couldn't filter out.

Military physicians from the U.S. Navy and Marine Corps noted as early as the 1960s that during the intense repetitive calisthenics of basic training, recruits sometimes developed the exhaustion, muscle breakdown, and cola-colored urine seen with rhabdomyolysis. Extreme athletes like cyclists, runners, weight lifters, and even high school football players occasionally report similar symptoms after grueling workouts. Animal athletes are also vulnerable to rhabdo, racehorses in particular. Extreme athletes—whether animal or human—push themselves to perform, often through pain, which sometimes results in rhabdo. "Mind over myocardium" can trigger quiet but deadly effects in both humans and animals.

But sometimes wildlife veterinarians finger capture myopathy as the killer even in cases in which there has been no prolonged chase, no skeletal muscle breakdown, no rhabdomyolysis.

A form of capture myopathy can appear when an animal is simply handled, noosed, netted, corralled, crated, penned, or transported. "Running for your life" is terrifying, but at least you have a fighting chance. Being caught, on the other hand, is the step right before the worst-case scenario: "game over."

As Dan Mulcahy put it, for an animal, "the only time they're caught is when something's going to eat them." Restraint usually means one thing: another animal wants you not to move. Capture and restraint, from an evolutionary perspective, spell a single state of affairs: imminent predation or death. Understandably, brains have evolved an all-systems-go response, a massive, last-ditch catecholamine tsunami.

Examples of animals dying when they've been captured or restrained abound. In species as varied as Irish hares, white-tailed deer, cotton-top tamarins, and antelopes, this combination can equal death. Experts on pikas, the rabbitlike "tundra bunnies" of South America, have learned the hard way that holding a captured pika firmly around its middle can scare it to death. A safer approach is to allow it to stand, unrestrained, in your open hands. And the risk exists not just for flighty prey animals. Top carnivores like brown bears, lynxes, wolverines, and gray wolves have died after being trapped.

Loud noises and heat can intensify the hazard of entrapment. Bighorn sheep in California's Mojave Desert that were rounded up for a relocation program fared especially poorly when a thundering helicopter hovered nearby. Pet rabbits have been known to expire in the presence of blaring rock music and even loud arguments between their owners. Fireworks blasts have reportedly startled and killed pets and livestock from parrots to sheep.

In Copenhagen in the mid-1990s, the Royal Danish Orchestra was performing Wagner's *Tannhäuser* in a public park. Abutting the park was the Copenhagen Zoo. As the chorus keened and the soloists belted out their highest pitches, a six-year-old okapi anxiously circled her enclosure and tried to escape. After struggling for several stressful minutes, she keeled over and died. The vets diagnosed capture myopathy.

Loud, frightening noises—not only those passing through a soprano's warbling epiglottis—have recently been shown to be risk factors for adverse cardiac effects in a variety of populations. One study published in the journal *Occupational and Environmental Medicine* found that people laboring in workplaces so persistently noisy they had to shout to have a normal conversation had twice the risk of serious cardiac events as people whose workplaces were quieter. And in some inherited cardiac conditions, startling loud noises can trigger a disturbance in the heart rhythm that leads to death.*

Interestingly, there's a breed of dog that seems to have evolved a defense against noise's jolting effects. Dalmatians born with long QT syndrome, and therefore susceptible to noise-induced sudden death, fortuitously sometimes also carry a genetic mutation that causes deafness. The auditory disability becomes a cardiac blessing in disguise as their muffled experience of sound may protect their vulnerable hearts from fatal cardiac arrhythmias.

Startling sounds and the feeling of confinement signal danger to animals and humans. Like the okapi trapped at the opera, noise and the perception of entrapment can be enough to ignite a fatal brain-heart reaction. Animal and human sensory systems provide information about the outside world that their brains convert to evasive action. But it's not just noise or restraint that can create terror.

In some cases, the mere thought of restraint can induce comparable physiology. Like the anxious and scared humans watching the events of 9/11 on TV, other animals, too, can experience an intense brain-heart reaction simply by seeing a threat.

Four captive zebras at a zoo in Vancouver once died from what was diagnosed as capture myopathy. They hadn't been chased. The stressor was the simple presence of two intimidating Cape buffalo that had been placed in the zebras' enclosure, an enclosure whose fences and moat would prevent their escape.

A sudden and unexpected appearance of predatory threat can also endanger animals. Some bird-watchers described observing a flock of cinnamon-bellied shorebirds called red knots calmly wading along the

---

*In long QT syndrome, abnormalities in ion-channel function cause the period between waves of heart activity to lengthen dangerously. This condition predisposes the patient to a potentially fatal cardiac arrhythmia. Long QT can be inherited (many of the genes have now been identified) or acquired. Many common medications, including some antibiotics, antidepressants, and antihistamines, can bring on long QT episodes. So can some electrolyte disturbances like severe vomiting and diarrhea.

Tellingly, startling patients who have a long QT can actually kill them. The emotional jolt can trigger an extra beat, which leads to a deadly arrhythmia. And that type of jolt can be caused by sudden loud noises, anger, arguing . . . .*or fear.*

water of an Australian beach. Suddenly, a raptor swooped down and grabbed one of the unsuspecting waders with its lethal talons. As the raptor flew off, the observers noticed something interesting. Although untouched by the predator, several nearby birds suddenly became unsteady and weak. Some fell over when they tried to walk.

Ornithologists call this kind of stress-induced muscle disease "cramp." The adrenaline spew would have affected the birds' heart muscles, too. The red knots, like the zebras, succumbed after merely witnessing a terrifying sight.

Circumstances that are not directly life-threatening can induce potent physiologic responses in humans, too. If, while traveling at ten thousand feet, your airplane hits an air pocket and plunges, your adrenal glands and brain release catecholamines. Your heart rate accelerates and your blood pressure rises. You may feel as though you're going to die. And, worse, as with a restrained animal facing predation, your inability to escape heightens your body's physiologic response.

Your brain processes the threat, but your body generates the response. That sickening, activated state you feel is fear. And fear, say veterinarians, is a key factor in capture myopathy. Some say it's the single *most* important factor. This leads us to an internal but dangerous contributing element to capture myopathy: the captured animal's activated emotional state.

We've seen that animal brains, human and nonhuman, respond and in some cases overrespond to entrapment. It's possible that our imaginative human minds take it a step further, triggering heart responses to traps that are not actually physical: an abusive relationship, crushing debt, an impending prison sentence.

Consider disgraced Enron chief Kenneth Lay's cardiac arrest weeks before being sentenced for an embezzlement conviction. Douglas P. Zipes, an expert in sudden cardiac death (SCD) and the editor in chief of the journal *Heart Rhythm,* told a Florida journalist, "We know that stress about which you can do nothing—the death of a spouse, the loss of a job, or facing life imprisonment—can be associated with sudden cardiac death. I can't crawl into [Lay's] head, but there is no question that the head talks to the heart and can have an impact on heart function."

Overwhelming fear responses to entrapment and threat may not be all that different whether you're a zebra facing a glowering Cape buffalo or a white-collar criminal facing life in prison. Indeed, studies have demonstrated that abusive, unfair bosses; negative, conflict-ready spouses; and suffocating debt substantially up the risk of heart-related death.

Given the power of fear and restraint to cause harm in humans and animals, it's surprising there's not a diagnostic term for these types of deaths. Because capture myopathy in animals and fear-induced cardiac effects in humans are related but complex, it might be helpful to find a way to identify cases in which fear and restraint were at fault. More than a decade ago the Harvard neurologist Martin A. Samuels called for "a unifying hypothesis . . . to explain all the forms of sudden death based on the anatomic connection between the nervous system and the heart and lungs."

The takotsubo moment that started my zoobiquitous journey began with placing the features of stress-induced heart failure in humans side by side with capture myopathy in animals and seeing the many similarities. When doctors notice a pattern of symptoms or physical findings, we create syndromes, which we then name. Veterinarians and physicians might consider adopting a new, common term to describe the role of fear in capture myopathy in animals and sudden cardiac death in humans. I propose the acronym FRADE: fear/restraint–associated death events. FRADE is broad enough to describe emotionally triggered fatalities in both animals and humans but narrow enough to filter out nonemotional causes. It would centralize the clinical anecdotes emerging from both human ERs and wild animal field sites. It could link, say, a takotsubo death in a terrified elderly woman to a capture myopathy death in a trapped okapi. In medicine, as in other fields, commonalities go unnoticed until they're named. Eventually, the neuroanatomical and neuroendocrine systems underlying fear- and restraint-related deaths will be fully characterized and better understood. But until then, using a common term to classify this particular kind of death will help veterinarians and physicians compare these sudden fatal events in ways that may lead to prevention strategies.

Many physicians are only now acknowledging the link between fear and cardiovascular events. But across cultures and throughout history, this dangerous connection has been noted. Voodoo curses and overly ominous thoughts, for example, have created deadly outcomes that are hard to explain from a purely physical point of view.

Many surgeons would sooner hang up their scrubs than operate on patients who are certain they will die on the operating table. "Surgeons are wary of people who are convinced that they will die," Herbert Benson, a founder of the Benson-Henry Institute for Mind Body Medicine at Massachusetts General Hospital told the *Washington Post*. Arthur Barsky, a psychiatrist at Brigham and Women's Hospital in Boston, concurs that these patients create a "self-fulfilling prophecy."

It's called the nocebo (Latin for "I will harm") effect, the opposite of the placebo (Latin for "I will please"). Unlike the well-known effects of the placebo, a nocebo is a harmless agent that, when perceived by the patient to have malignant qualities, produces *negative* effects. If you've ever wondered whether voodoo death is real, the nocebo effect offers a reason why the answer may be yes. When the person delivering the hex is convincing enough—and the victim open-minded enough—the heart-mind connection may initiate that deadly cascade seen in stress-induced cardiac death. Some call this "homicide by heart attack." Genetics probably plays a role, since voodoo deaths tend to cluster in specific ethnicities and regions.*

FRADE may connect to these deaths, too. The link between folklore-laced voodoo deaths and straightforward animal capture myopathy lies in the shared biology of animal and human nervous systems.

Over millennia, animals have developed variations—and some improvements—in the ways they can convert sensations of external danger into safety-seeking behavioral responses. Some release toxins or odors or sting with electricity or poison. Sea anemones retract when prodded and release a squirt of seawater. Flies zoom away from the swatter. But the link between threat and catecholamine release is especially pervasive and ancient. Its origins date back two billion years—before the separation of plants and animals. Potato leaves and tubers, for example,

---

*Sudden unexpected noctural death syndrome (SUNDS), for example, largely afflicts young men from the Hmong ethnic group of Laos who die in their sleep. Hmong are wary of a very specific nightmare (*dab tsog*) in which a terrifying evil spirit appears and actually "kills" the dreamer. This effect probably involves an underlying (possibly genetic) electrical problem with the heart. But it needs the cataclysmic (catecholamine-inducing) stress of the nightmare—which takes its imagery from firmly established folklore traditions—to kill. Young Filipino men have also been reported to die from something similar, called *bangungut* (Tagalog for "to rise and moan during sleep"). There is a reason this may sound familiar to current or former horror movie fans. It is the premise of the *Nightmare on Elm Street* movie franchise, in which teenagers who are pursued and killed in their dreams by the movie's villain, Freddy Krueger, die in real life.

respond to stressors like cold, drought, and chemical burns by releasing catecholamines. This seems to increase the plant's resistance to infection and other threats.

For plants, fleeing is not an option. For vertebrates, however, a responsive heart that can accelerate in a "fleeing escape" or slow profoundly for a "hiding escape" has often meant the difference between life and death. But this elegant and effective system has a fatal flaw. Because underestimating danger even just once in a wild animal's life can spell death, the warning system may be calibrated toward overreaction. The evolutionary medicine expert Randolph Nesse explains these overresponses through the analogy of a smoke detector. Although the alarm can sound at the wrong time, many false alarms are better than one missed true one. As the behavioral ecologists Steven Lima and Lawrence Dill wryly note, "Being killed greatly decreases future fitness."*

Overresponse can be found throughout biological systems. Our immune systems may "overreact" in the course of trying to protect us, causing autoimmune disorders such as rheumatoid arthritis and lupus. Eczema and keloid scar tissue are other examples of the body's exuberant response to injury. Fevers that are possibly meant to battle microorganisms can get too high and cause seizures and brain damage. Coughing that is meant to clear vital airways can devolve into bronchospasm or cracked ribs. And in psychiatry, anxiety disorders, panic attacks, and phobias can be conceived of as pathologic overreactions to danger that stem from protective instincts.

FRADE describes another overcalibration. Adaptively, a surge of catecholamines allows a zebra to gallop away full throttle or madly wrestle free from a lion. Maladaptively, that torrent of stress hormones may break down the animal's muscles, destroy its kidneys, or even stop its heart. It's counterintuitive that your brain and heart can sometimes

---

*Sometimes the thing that's supposed to protect you accidentally kills you. An example of this phenomenon is demonstrated in some human-engineered systems. Take air bags. To be effective lifesavers in the split seconds after impact, these scorchproof polyester pillows must deploy at velocities exceeding two hundred miles per hour. Not surprisingly, although tens of thousands of lives have been saved by air bags since their 1997 introduction, thousands of people have died when the force of rapidly opening air bags ruptured hearts, tore pulmonary arteries, and cracked neck vertebrae. This happened mostly to babies and young children riding in the front seat, before legislation made that illegal.

work in concert to kill you. But FRADE is a reminder that safety systems must be powerful and can be calibrated to overrespond—especially in dangerous settings with no "do-overs."

Unless you're a veterinarian, work in a pet store, or just got elected dog catcher, it's a safe bet you don't need to capture an animal very often. And here in the enlightened twenty-first century, we certainly don't capture and pin down human beings all that much. Or do we?

Once when I was on call in the ICU, a young woman was fighting for her life. A staphylococcal infection had attacked many of her organs, including her heart, which was just barely contracting and relaxing. Her kidneys had shut down and her liver was failing. The critical balance of potassium, calcium, magnesium, and sodium was dangerously disturbed. She hadn't slept for days. A month earlier, she had been a vivacious and popular teacher at an area elementary school. But on this night, her life-threatening condition had made her disoriented . . . *and agitated*. This happens in critically ill patients so often that we have a name for it: ICU psychosis.

Thrashing in her bed, she clawed at the nasogastric tube coming out of her left nostril. Her other hand tugged at the arterial line stuck in the soft skin of her thin left wrist. She had a Cordis line in her jugular vein, a Foley catheter in her urethra, and a hemodialysis catheter in her groin. If she pulled any of them out, blood would be everywhere. If she dislodged the intra-aortic balloon pump supporting what little blood pressure she had, she could easily rip the large, high-pressure artery and bleed to death.

To protect her body from her volatile mind, I called for soft physical restraints. The nurses quickly and gently fastened the fleece-lined, nylon-and-cotton, six-inch-wide straps around her wrists.

For a few seconds, all was calm. The heart monitor emitted the regular and comforting *beep beep beep* signifying a safe, normal rhythm.

But then she sensed the straps on her arms. She began pulling against them. I ordered some intravenous sedation as a form of what we call "chemical restraint." But the sick patient kept thrashing, clearly confused and quite likely terrified. And then the beeping coming from the monitor above the bed changed. It sped up and turned slightly less regular. She'd gone into VT. With her already low blood pressure, this rhythm required immediate action.

ICU teams rehearse the lifesaving choreography these moments call for. When they happen, little needs to be said. On the left side of the patient's chest, a nurse placed a large sticky pad the size of a paperback book connected to a wire, rolled the patient to her side, and placed another pad between her shoulder blades. My cardiology fellow dialed the knob of the defibrillator to 150 joules and calmly asked everyone to step away from the bed. The nurses and other staff members backed away, holding their hands up, palms facing out. If they touched any part of the patient or her bed, the electricity that was about to be administered would conduct into their bodies, too. The fellow pushed the raised red button marked SHOCK.

The teacher's body stiffened for a split second as the load of electricity coursed through her 120-pound body, which "jumped" slightly from the bed. All eyes turned to the monitor. Our ears searched for the steady *beep, beep, beep.* After another split second, we got it. Her cardiac rhythm snapped back to a version of normal.

Whether she went into VT at that exact moment because of the addition of wrist restraints is impossible to say. Her acute infection had provided her with several risk factors: myocarditis, electrolyte disturbances, anemia, hypoxia. But recognizing how restraint introduces risk for cardiac arrest in animals, I now have a different perspective on how it affects human patients.

I had always viewed physical restraint as a necessary safety intervention for my patients who required it. It's a fixture of other professions as well—used more often than you might think. It's common in mental health and geriatric care facilities, where modern-day straitjackets and restraints are sometimes used for patients who might be a danger to themselves or others. Law enforcement, military, and prison officers all rely on restraint devices like handcuffs to control unruly behavior.

Admittedly, there are scenarios in which restraint is the best approach for the safety of all involved. I know it can be for the good of the "detainee" as well as for the cops, soldiers, prison guards, orderlies, and nurses involved, not to mention any bystanders.

But before I learned that veterinarians view restraint as a major player in capture myopathy, I'd never even considered whether it might have a physical downside. In human medical circles, the potential risks of restraint are rarely discussed.

Yet FRADE exists all over the animal kingdom. It's only the separation of human and animal medicine that has allowed us to think that it doesn't exist in human beings as well. Physicians ought to be aware of what veterinarians know: that fear, whether produced accidentally by a well-meaning doctor or purposely by a terrorist threat, can be deadly.

As veterinarians have become more enlightened about the dangers of chase, terror, and capture, they've become more adamant about their responsibility to prevent capture myopathy in animals. Whether it's a grizzly bear caught by the foot in a Canadian forest or a pet bunny on a table in a private practice, most vets agree that they can protect animals by following a few simple stress-reducing guidelines: Keep noise and motion to a minimum. Have a small, well-trained crew that can notice early signs of stress-related distress. Develop an approach that emphasizes calmness.

Thinking about the dangers of fear and restraint has changed how I practice medicine. I still sometimes must order patients to be put in restraints, but I am cautious of the dangers and often have the vets' guidelines in mind as I do so.

Unraveling the threads of sudden cardiac death and capture myopathy, noticing how they twine across species, and reknitting them as FRADE led me to consider another potential danger in a quite unexpected setting. It couldn't be farther from the watery home of an Alaskan shorebird, the back of a police car, or the hospital room of an out-of-control ICU patient. It's the snuggly, warm cocoon of a newborn's nursery.

Sudden infant death syndrome (SIDS)—also known as crib death or cot death—is the *leading* cause of infant mortality between one month and one year of age. More than 2,500 babies die of it every year in the United States. International statistics vary, but SIDS is a leading cause of infant death in all countries in which data are available. Strictly defined, it's "the sudden death of an infant under one year of age that remains unexplained after a thorough case investigation, including performance of a complete autopsy, examination of the death scene, and review of the clinical history." The "unexplained" part is what's so frustrating to doctors. How and why these infants silently slide from life into death in many cases goes unanswered.

Theories abound: environmental pollution, secondhand smoke, bottle feeding, prematurity, low serotonin levels. However, so far, one factor has been overwhelmingly correlated with increased risk of SIDS: putting a baby to bed on its stomach. The reason may at first seem obvious. Too small and weak to turn itself over, an infant snuggling facedown into a soft mattress or bedding can suffocate itself. But it isn't that simple. Babies who die of SIDS often show no postmortem evidence of asphyxiation. So examiners have asked, if these deaths weren't respiratory, might they be cardiac?*

When one is lying prone (facedown), the upper chambers of the heart (the atria) become full as blood rushes in from the major veins. But pressure-sensitive nerves (baroreceptors) within the atria sense the increasing volume and activate a suite of autonomic counterresponses. They decrease the urge to breathe. They also decelerate the heart. These reflexes likely share an evolutionary heritage with the ancient diving reflex, an adaptation to underwater oxygen metabolism seen in many species. And this means that putting a baby to bed on its tummy can trigger a reflexive slowing of its heart and breathing.

The heart rates of animals as distantly related as fish and rodents also decrease, sometimes suddenly, when frightened. Loud, startling noises have been demonstrated to induce extremely slow heart rates in fawns and alligators as well as not-yet-born human infants. This heart slowing, called "fear" or "alarm" bradycardia, is a protective reflex that may keep the animal still and silent, making it less detectable to predators. And it can persist for a surprisingly long time—a minute or more. It's especially powerful in juvenile animals and wears off somewhat as the animal matures. (For more on this, see Chapter 2, "The Feint of Heart.")

In the 1980s, a pathbreaking Norwegian physician with a strong knowledge of animal behavior and physiology had an early zoobiquitous moment. Birger Kaada connected the heart-slowing responses in hiding baby animals to the heart-stopping risk in sleeping baby humans. Although there was general recognition that her theory had validity, few in the medical community recognized, as she did, that some cases of SIDS could be explained by a complex overlay of two heart-slowing effects: facedown posture and fear.

---

*Some cases of SIDS may stem from overlapping neurological, respiratory, and cardio-vascular syndromes. One emerging theory connects SIDS to abnormal brain function leading to improper sensing of rising carbon dioxide levels, called hypercapnia.

This is what might be happening in some cases. A baby is placed in the crib on its stomach, which causes a mild slowing of the heart. Then, a sudden, startling noise—a slammed door, car alarm, heated argument, telephone ring—startles and frightens the child. As with juveniles of many species, the human infant's heart rate plummets in response to sudden jolting noises. Researchers have suggested that some infants' immature hearts simply slow to the point of no return. Or, in other cases, the loud noise could trigger a fatal cardiac rhythm in a baby with an already slowed heart rate. In either case, this would mean that some SIDS deaths connect to the physiology of fear.

But SIDS has another important connection to animal capture myopathy, one that suggests it's part of FRADE. Restraint may play a lethal role in SIDS, too. But for human babies, restraint doesn't come as a net, leg trap, or enclosure. And it doesn't involve wrist restraints, as it does for adult psychiatric or ICU patients. Infant restraint takes the form of a centuries-old and newly revived practice: swaddling.

Swaddling human babies is, and has been, a mainstay of parenting practices around the world. It's said to pacify fussy infants, promote sleep, and keep babies from harming themselves—and to make it easier for caregivers to tote them around. Ideally, swaddling mimics the protective shelter of loving arms or even evokes a reassuring sense memory of a snug uterus.

And, interestingly, swaddling offers a slight protective effect against SIDS—but only if the infant is sleeping on its back, according to a study by doctors at the Children's University Hospital in Brussels, Belgium.

These scientists say there's a chilling flip side to swaddling. Put a swaddled baby to sleep on its stomach, play a sudden loud noise, and its risk for SIDS *increases* threefold.

To test this, the Belgian doctors evaluated a group of infants in both swaddled and unswaddled states, their bodies positioned both prone and supine (on their backs). The babies were bound in bedsheets held in place by sandbags. (Be assured: the babies in this 2004 study were monitored constantly; the parents had given their signed consent; and a pediatrician was present the whole time.) The doctors then added a sudden "audio challenge": three seconds of ninety-decibel white noise from a tiny speaker held about an inch from their little ears. (Ninety decibels is about as loud as a blow dryer on "high" or a motorcycle roaring by.)

It turned out that, no matter whether sleeping on their stomachs or their backs, when "restrained" by swaddling, the babies showed earlier and more dramatic heart slowing in response to noise than the unrestrained babies. This suggests that for babies already in the dangerous facedown position, the addition of restraint in the form of swaddling might create a fatal third layer of cardiac slowing, especially with the addition of a loud, unexpected sound.

Swaddling, it must be said, is for the most part safe, playing a role in infant care and physical and emotional security. But if swaddling joins facedown posture and startling sounds, it may be misinterpreted as predatory restraint and further slow the already decelerating heart. Pointing out the ability of noise and restraint to trigger alarm bradycardia in the young of many species adds a zoobiquitous piece to the SIDS puzzle. This calls for direct dialogue among animal physiologists, wildlife biologists, and the primary care pediatricians who can use this information to care for their vulnerable patients.

Like the rhythmic beating of cardiac muscle, the conversation between the heart and the brain starts in the womb and continues until the moment we die. And thank goodness. Because sometimes being surprised, even terrified, can protect us from harm. It fuels a shorebird's escape. It pushes a Californian to seek cover during an earthquake. Powerful yet vulnerable, the heart-brain alliance usually saves lives. But every once in a while, it can also end one.

SEVEN

Fat Planet

Why Animals Get Fat and How They Get Thin

I never, in all my years of counting calories, expected to receive dieting advice from a grizzly bear. But there I was, in a dark conference room with about a hundred zoo veterinarians, raptly listening to a presentation on how Jim and Axhi, two obese Alaskan grizzlies at Chicago's Brookfield Zoo, had dropped hundreds of pounds.

Delivering the secret was Jennifer Watts, Brookfield's easygoing, bespectacled nutritionist, a Ph.D. who oversees the diets of the animals at the zoo. Projected on a screen next to her was a "before" photo. It was just like my favorite moment on every TV makeover show—seconds before the "reveal." The "before" bears' shimmying bellies barely cleared the ground. Rolling billows of flesh rippled along their flanks. Years of overfeeding had ballooned their faces and erased all memory of their necks.

Then Watts flashed the "after" picture. A few chuckles escaped from the animal doctors around me. The difference was huge. Svelte and glossy, the bears simply *looked* healthier. If they'd been my patients, I'd be breathing easier, too, knowing that, along with their weight, their risk of obesity-related health problems had just plummeted.

Although I'm a cardiologist, some days I feel more like a nutritionist. Patients, family, and friends all frequently ask me, "What should I be eating?" We all know by now that choosing the wrong foods and carrying extra weight on our bodies can make us sick. Obesity, weight gain, "eating right": these concerns are at the very core of modern preventative medicine.

And yet, listening to Watts talk about the grizzly bears, I realized something that was at once astounding and obvious: humans aren't the only animals on our planet who get fat. And it turns out that it's not just iconic fatties like hippos and walruses who pack on the pounds. Animals as varied as birds, reptiles, fish, and even insects regularly gain—and then take off—weight. And they do it without ordering dressing on the side or going on slimming crusades involving shots, pills, psychotherapy, and even surgery. Fattening in the animal world has enormous potential lessons for humans—including dieters looking to shed a few pounds and doctors grappling with obesity, one of the most serious and devastating health challenges of our time.

But until that moment it had simply never occurred to me to wonder, *Do animals get fat?*

As you've probably heard many times, we are in the midst of an "obesity epidemic." Millions must cope with this life-threatening disorder. Doctors everywhere are urgently searching for a cure.

What may surprise you about this obesity epidemic, however, is that I'm not talking about overweight humans (not yet, anyway). There's another obesity epidemic going on around us. It afflicts our dogs and cats, horses, birds, and fish. Domestic animals around the world are fatter than ever before, and steadily gaining more weight.

Exact numbers are hard to pin down—in part because pet owners and veterinarians don't always recognize when a beloved Lab or tabby has crossed the line from well fed to positively plump. But studies in both the United States and Australia put the number of overweight and obese dogs and cats somewhere between 25 and 40 percent. (For the time being, animals are still doing better than we are—the proportion of U.S. human adults who are now either overweight or obese is close to a jaw-dropping 70 percent.)

With our pets' excess pounds have come a familiar suite of obesity-related ailments: diabetes, cardiovascular problems, musculoskeletal disorders, glucose intolerance, some cancers, and, possibly, high blood pressure. They're familiar because we see nearly identical problems in obese human patients. And just as in our population, these weight-related diseases among dogs and cats often lead to premature death.

The efforts to combat excess animal girth will also sound familiar. Some dogs are put on diet drugs to curb their appetites. Liposuction has been the treatment of choice for some severely obese canines when the extra flab threatened to snap their spines or splay their hips. Companion felines have been placed on the "Catkins" diet—a veterinary version of the popular high-protein, ultra-low-carb Atkins diet for humans. Veterinarians increasingly treat "portly ponies." They instruct owners not to overfeed chubby fish. They suggest giving husky lizards more exercise to work off their surplus weight. They describe tortoises so fat they can no longer pop in and out of their shells. They've seen so many overweight birds they have a new nickname for them: perch potatoes.

Exotic animals in nonwild settings are getting rounder, too. Concerned about the health effects of extra flab, zoo veterinarians in North America and Europe have placed overweight animals from flamingos to baboons on slimming diets. Many of these regimens borrow strategies from human weight-loss programs. If you've ever tallied daily Weight Watchers points, you understand the routine of the gorillas and cockatoos at Brookfield Zoo, where Jennifer Watts has put the animals on a similar system. In Indianapolis, zookeepers have encouraged rotund polar bears to move around their enclosures by tempting them with calorie-free, artificially sweetened gelatin treats in place of the sugary marshmallows and molasses of yore. In Toledo, plump giraffes have been offered biscuits that are specially formulated with lower salt and higher fiber, in place of the junk crackers they had been getting.

What all these corpulent animals share, what sets them apart from their wild cousins and ancestors, is one thing. We feed them. They are mostly or completely dependent on humans for every meal, and we regulate both the quality and the quantity of everything that passes their lips and beaks. Consequently, we can't really blame them for their weight problems. Of course, a dog will eat pretty much whatever you put in front of her, and still sniff around for more. The very idea of a cat exercising willpower to resist a fattening treat seems ridiculous. And so we're left with one conclusion: we, the species that both manipulates food to make it more unhealthful and has the intelligence to understand that we shouldn't eat so much of it, are to blame. We're responsible not only for our own expanding waistlines but for those of our animal charges as well.

And even just living around us humans can make animals balloon. City rats crawling the alleys in urban Baltimore, for example, grew about 6 percent fatter, per decade, between 1948 and 2006, presumably because their food came almost entirely from human garbage cans and pantries. These rats also showed about a 20 percent increase in the chance of becoming obese. But our fattening leftovers might not have been the sole cause of the rodents' increase in body weight. The researchers who studied these urban rodents found an intriguing parallel weight gain in another group of animals. The city rats' country cousins also became fatter—at nearly the same rates—during some of that same time period. And even though their food supplies were more "natural," rats in the parklands and agricultural areas around Baltimore also showed an increase in the odds they would become obese.

It's pleasing to assume that when animals are in their native environments, eating what they "should" (the unprocessed foods they evolved with), they will stay effortlessly lean and healthy. But that's not necessarily true. I'd long imagined that, in the wild, animals will eat until they are full and then stop. In fact, given the chance, many wild fish, reptiles, birds, and mammals will overindulge. Sometimes spectacularly so—even on wholesome, natural foods. Abundance plus access—the twin downfalls of many a human dieter—can challenge wild animals, too.

Although we may think of food in the wild as hard to come by, at certain times of the year and under certain conditions, the supply may be unlimited. Seeds spill across fields. Larvae cover sand and vegetation. Eggs lie easily available under every leaf. Bushes fill with berries. Flowers ooze with nectar. And when animals' environments are this abundant, they will gorge. Many stop only when their digestive tracts literally cannot take any more. Tamarin monkeys have been seen to eat so many berries in one sitting that their intestines get overwhelmed and they soon excrete the same whole fruits they recently gobbled down. After gorging on abundant prey, carnivorous fish sometimes start excreting undigested flesh. Big felines, like lions, as a matter of course stuff themselves after a hunt until they can barely move. Mark Edwards is an animal nutrition expert at Cal Poly, San Luis Obispo, and was the first nutritionist at the San Diego Zoo and Wild Animal Park. As he told me, "We're all hard-wired to consume resources in excess of daily requirements. I can't think of a species that doesn't." In fact, when presented with unlimited food, domestic species, including dogs, cats, sheep, horses, pigs, and cattle, eat nine to twelve meals per day.

With ready access to superabundance, some wild animals have become impressively fat. A seal with the catchy nickname of C-265 was recently euthanized by the Oregon Department of Fish and Wildlife. His crime: bingeing on more than his share of endangered chinook salmon during the annual run. So enthusiastically did C-265 feast on the lox smorgasbord that he nearly doubled his weight in just two and a half months (going from 556 to 1,043 pounds). His appetite was undeterred by the firecrackers and rubber bullets deployed by rangers to protect the precious salmon stocks. And C-265 was not alone in his gluttony. Dozens of seals have been euthanized since a federal judge's controversial 2008 ruling allowing eighty-five of them to be killed each year to protect salmon reserves.

The weight of blue whales off the California coast fluctuates from year to year, depending on the population of krill, their favorite food. Some years, they're so skinny the individual vertebrae protrude notice-ably from their backs. Other years, as a whale-watching boat captain described it to me, they're "fat, happy, and relaxed." And who can forget the undulating, pendulous, black-and-white bellies in the movie *March of the Penguins*, belonging to birds that could barely waddle after gorging for weeks in the ocean?

In the Colorado Rockies, warmer temperatures since the 1960s have correlated with a body change for yellow-bellied marmots. As Daniel Blumstein, chair of the UCLA Department of Ecology and Evolution-ary Biology, explained to me, "As the snow has melted earlier over the past forty years, marmots have emerged from hibernation sooner, had a longer growing season, and entered hibernation in better condition, which increases both survival and reproductive success." In other words, marmots have gotten fatter. A study Blumstein copublished in *Nature* with biologists from Imperial College London and the University of Kansas showed that across generations of marmots, average weights have swelled by more than 10 percent over the nearly five decades of the research period. If this doesn't seem like a lot, consider that CDC data show that over the same five decades, average adult male weights in the United States have also increased by about 10 percent (from about 166 pounds in 1960 to about 186 pounds in 2002). This trend tracks with the human obesity epidemic, although the implications may be different. Says Blumstein: "The marmot population has tripled in the past decade. Chubby marmots are happy marmots."

Slovakians living at the base of the Carpathian Mountains once believed that their local lake contained a unique species of wild carp—bigger and meatier than the fish found in nearby waterways. But upon closer inspection, it turned out that the impressive specimens were *Esox lucius*—exactly the same species as the smaller fish. A flood had swept nutrients off nearby farms into the lake, providing the piscine gluttons with so much extra food that their bodies had swelled beyond recognition. This ability to get hugely fat when there's extra food around is shared by fish in many other regions.

So wild animals can get fat the same way humans do: in environments with unfettered access to abundant food. Of course, animals also fatten normally—and healthily—in response to seasonal and life cycles (and more on that in a moment). But what's key is that an animal's weight can fluctuate depending on the landscape around it.

A zoobiquitous approach gave me a more nuanced appreciation of why and how animals get fat. It reminded me that weight is not just a static number on a chart. Rather, it's a *dynamic, ever-changing reaction* to a huge variety of external and internal processes ranging from the cosmic to the microscopic.

This echoes something I heard a wise colleague say: "Obesity is a disease of the environment." Richard Jackson is the chair of Environmental Health Sciences at UCLA and a former head of the National Center for Environmental Health at the CDC. In an impassioned Internet video recorded in 2010, he explained what he meant:

One of the problems with the obesity epidemic is we too often blame the victim. And yes, every one of us ought to have more *self-control* and ought to exert more *willpower*. But when everyone begins to develop the same set of symptoms, it's not something in their mind, it's something in our environment that is changing our health. And what's changing in our environment is that we have made dangerous food, sugar-laden food, high-fat food, high-salt food . . . and we've made it absolutely the easiest thing to buy, the cheapest thing to buy, and yes, it tastes good, but it's not what we should be eating.

This point is similar to one made by David Kessler, the former head of the U.S. Food and Drug Administration, who pointed the finger at processed foods in his 2009 book, *The End of Overeating*. Excess sugar, fat, and salt, Kessler argued, "hijack" brains and bodies and drive cycles of appetite and desire that make it nearly impossible to resist certain fattening foods. Essentially, even if we can resist one package of chips or plate of cookies, our "environment" now comprises endless mountains of these foods everywhere we look.

And these fattening landscapes present themselves to animals, who then overconsume. Even some animals you'd think would know better.

Early one morning I walked in on this tableau: French fries wilted in their own grease on paper plates holding burger scraps and ketchup smears. A yellow bag of M&Ms gaped open beside a gutted sack of Doritos. Half-empty soda cans stood near a pizza box glistening with rainbow streaks of congealing oil.

This wasn't a frat house on Sunday morning or a bulimic's bedroom. No, this was the on-call room used by the overnight team of a cardiac care unit (CCU). The young doctors who had created this mess were on their cardiovascular medicine rotation; some were deep into their training to become cardiologists. These physicians, handpicked from the best med schools, had spent the past twenty-four straight hours treating some of the deadliest conditions known to modern humans: heart attacks, artery ruptures, strokes, and aneurysms. Their night had been a whirlwind of chest pain, abnormal EKGs, angiograms, defibrillations. And most of this trauma had been caused by their patients' underlying coronary artery disease, the leading killer in the United States, which is strongly linked to diets high in sugar, refined carbohydrates, salt, and certain fats.

Throughout my training, in teaching hospitals around the country, catering departments would lay out what used to be called "midnight meals"—sumptuous spreads of pasta, sandwiches, thick cookies, granola bars, hamburgers, greasy fries, and candy. These spreads were reward and encouragement for our extreme working hours. They were good opportunities to bond with our colleagues. But for many of us, the unfettered access to all that tasty temptation in the middle of the night, with the overlay of constant stress, was precisely the "obesogenic" environment we now routinely tell our patients to avoid.

You don't have to be a cardiologist to know what you *should* eat, or at least that a diet of candy and pizza is problematic. But this is precisely why that CCU on-call room is so illuminating. Cardiologists see with their own eyes and hold in their very hands the diseased body parts that come from eating poorly. Putting aside the CCU interns' and residents' youthful sense of invulnerability, a junk-food–eating cardiologist seems like a medical oxymoron. Along with other subspecialty death wishers, including chain-smoking oncologists and alcoholic hepatologists, they are the living (for now) embodiment of the cognitive disconnect between intention and consumption. We consume the dietary weapons of mass destruction even when all our training and experience tell us not to. A survey of almost 300,000 U.S. physicians conducted in 2012 revealed that 34 percent of cardiologists report being overweight, with 4 percent actually obese. Forces beyond knowledge and free will are clearly at play when we eat.

The evolutionary biologist Peter Gluckman calls contemporary obesity an example of "mismatch," the widening gulf between our genetic inheritance and our environment. (From animal ancestors we've inherited eating behaviors that evolved to keep us alive through feast and famine. But thanks to human culture, we've created a mismatched, fat-promoting environment of Frosted Mini-Wheats and electric skateboards.)

Mismatch explains why that scene in the CCU on-call room, instead of being the embodiment of what's *worst* about the way we eat, may represent the legacy of millions of years of inherited eating strategies that have *worked*. And the young on-call doctors are not alone in preferring cookies and other treats when given the opportunity.

In the dry western United States, red harvester ants have adapted over millions of years to eat seeds. For them, it's an ideal food source. Seeds store well. They provide nutrients—protein, fat, carbohydrates—in good ratios.

Seed eating essentially makes these animals vegetarians. But put a slice of tuna in front of the ants, or a sugar cookie, and watch what happens. Forget the carefully calibrated generations of evolution. Forget millions of years of natural selection favoring prudent food-storage behaviors. Those ants devour the meat and the cookie.

Something similar happens with marmots. These sandy-blond rodents live in alpine areas around the world, including California's Sierra Nevada and Colorado's Rocky Mountains. They're mostly herbivorous, feeding on grasses, although they will eat the occasional spider or insect. Yet biologists who've spent careers studying them say that these preferential vegetarians will wolf down raw meat given the slightest opportunity. So will chipmunks and squirrels, who are vegetarian except when they are lactating; then they become not only carnivores but cannibals, eagerly scarfing down road-killed kin.

The reason, says the UCLA evolutionary biologist Peter Nonacs, is pretty simple. Ounce for ounce, meat and processed sugar offer the most nutrients for the least amount of effort. They provide more calories, and they're more digestible. As he puts it, "You don't have to eat a lot of meat to survive." Harvesting a pile of seeds requires a lot of work. Munching on bales of grass requires energy. If an ant or a marmot can skip all that and get straight to the nutrients, that's what it will do.

Evolutionary biologists think the desire for protein—which includes the taste for fat and salt—is an ancient, long-preserved mechanism. A drive for sugar is probably slightly younger, most likely arising about a hundred million years ago, when plants began flowering and concentrating sugars in their seeds and fruits. As humans, we share ancestors—and we also may share urges—with protein- and sugar-seeking animals.

This suggests that the scene in the on-call room, with its fatty pizza, sugary candy, and salty fries isn't necessarily an example of depraved human eating. It may be more a demonstration of preserved food-class preferences. If, for hundreds of millions of years, animals have shared the urge to snatch protein, fat, salt, and sugar, it's almost naively optimistic to think that hearty advice to "just resist junk food" and "eat a healthy diet" could compete with it.

Modern-day food manufacturers, perhaps cynically, have hitched a ride on these evolutionary urges by amping up those elements in their products. There's a reason you can't eat "just one." In an analogous situation, a marmot can't either.

And sometimes that's okay. Animal weight goes up and down—in some cases dramatically and several times throughout the year. Throughout the animal kingdom, this is a sign of health. Indeed, zoo nutritionists do not set single weight goals for the animals in their care. They establish weight *ranges,* and they worry if animals from giraffes to snakes don't move from one end of their range to the other, depending on the season and life stage. In the wild, males of many species fatten in the weeks prior to mating season. Female animals store body fat to nourish eggs and support milk production or other food provisioning for their young. Seals, snakes, and other animals whose bodies require a calorie-draining molt are obliged to store energy as fat in the days and weeks leading up to it. Hibernation, iconically, requires a tremendous shift in body mass to support a months-long fast. Migration, too, triggers key fattening and thinning cycles. And among the most metabolically taxing moments in any animal's life are its first few hours or weeks after being born. Infancy is a time of peak fatness for many creatures, from nestling birds to newborn humans.

Even insects' body fat goes up and down during critical phases of their lives. Some fatten before metamorphosis or laying eggs. With adequate nutrition, bees produce fat in bulk: honeycomb wax is a form of apian fat. And fat exists in plants, too—as waxy, waterproof coatings on leaves and fuel packs in seeds.

But nature imposes its own "weight-maintenance plan" on wild animals. Cyclical periods of food scarcity are typical. Threats from predators limit access to food. Weight goes up, but it also comes down. If you want to lose weight the wild animal way, decrease the abundance of food around yourself and interrupt your access to it. And expend lots of energy in the daily hunt for food. In other words: change your environment.

This is something many zoos are already doing.

If you happen to find yourself at the Copenhagen Zoo at just the right time, you'll witness something one won't see at many other zoos around the world.

A dead impala lies in the middle of an enclosure. Crawling over it, like flies on a discarded slice of salami, are a dozen or so lions. The full-grown male with his distinctive mane sits high on the beast, tearing at its throat and face. A couple of favored females crouch near him, methodically munching. Two or three others work on the carcass's abdomen, loosening the entrails inside. Young cubs—as supple-limbed and clumsy as puppies—dart in and out between their elders, snagging jawfuls of flesh, their muzzles dripping with blood. There's an eerie hum of contented growly purrs, punctuated by the unique snap of teeth going through bone. The big cats stuff themselves until they can barely move, their eyelids drooping in a satisfied daze.

This human-staged simulation of a feast on the African veldt is known as carcass feeding. Nutritionists at the Copenhagen Zoo and others who carcass-feed their lions, tigers, cheetahs, wolves, jackals, and hyenas choose the prey carefully. They make sure the carrion is free of disease and that it's appropriately nutritious. Often the animal to be eaten is from another part of the zoo, euthanized and "recycled" as a meal for the carnivores. Proponents say this whole-food approach (hooves, fur, eyeballs, and all) gives the meat eaters a figurative and literal taste of how they would consume meals in the wild, the way nature intended.

However, detractors (mostly in North America and some parts of the United Kingdom) say the practice is cruel, not to mention off-putting for visiting families unaccustomed to such natural carnage. So although many of them are privately in favor of carcass-feeding, British and American zoo nutritionists bow to public opinion. They serve meat that's already dismembered or entirely ground up. On the occasions when they do feed an animal, say, a big bloody beef leg or haunch, they do it behind the scenes ("off-exhibit") or after hours.

When I asked Mads Bertelsen, a veterinarian at the Copenhagen Zoo, about carcass-feeding, he was unapologetic.

"It's what the animal is meant to do," he told me. Zoos that avoid it for fear of a public outcry are, he said, "bending to a minority of loud voices." He pointed out that if you feed a tiger a patty of minced horse meat, it's still eating a horse but receiving none of the nutritional benefits of crunching though bone, gnawing on gristle, and digesting fur and hair. Indeed, zoos that allow their carnivores to feed on the whole prey animals they would naturally hunt (Tasmanian devils on kangaroos, lions on elands, cheetahs on gazelles) notice cleaner, stronger teeth, healthier gums, and even positive behavioral changes, like a more relaxed demeanor. Like most vets, who abhor anthropomorphizing the animals in their care, Bertelsen stopped short of saying the lions in Copenhagen experience pleasure while they're eating in this more natural way. But he did grin and say that the felines "seem to be having a good time."*

Reconciling how an animal eats in captivity with how that same animal might eat in the wild is a challenge for the veterinarians who treat them and the nutritionists who formulate the menus. In the wild, an animal ideally has free access to choose and eat the healthiest and best-balanced meal it can get its fangs and claws on. But more important, its food is intricately connected with the many activities—both physical and cognitive—it must undertake to get it. Stomach and spirit are rarely separated in wild meals, whether in the thrilling adrenaline rush before a chase, the reward of a morsel of clam meat after wrestling the shell open, or the relaxing sensation of a full belly after a period of hunger.

For a zoo animal, however, feeding decisions for the most part are made *for* him. What he eats. When he eats. How much and even where he'll eat. Yet while a zoo environment limits the whole fleet of inherited, wild instincts to hunt, forage, and be alert to danger, it doesn't entirely erase them. Carcass-feeding is one way to put feeding decisions back in the paws and snouts of zoo animals. Creatively spreading forage items like string beans around an enclosure is another. It gives an animal more control and more challenge than does simply slurping chow out of a bowl. Modifying an animal's surroundings in order to improve its health or well-being is called "environmental enrichment."

Environmental enrichment as an animal husbandry standard came into its own in the 1980s, largely as a way for zoos to reduce undesirable behaviors, like pacing, in the animals in their care. Settings that allowed for more "natural" or "wild" expressions of behaviors could in some cases make the animals healthier.

At the Smithsonian National Zoo, in Washington, D.C., for example, environmental enrichment for octopuses includes adding shelves, arch-

---

*At Bertelsen's zoo, as at many other institutions that carcass-feed, the carnivores are usually fasted for a few days after a big meal, mimicking the more naturalistic gorge-and-fast patterns of some wild animals. Together with scientists from the pet food corporation Hill's, Joanne Altman of Washburn University in Topeka, Kansas, studied five captive African lions at the Topeka Zoo. The cats were switched from daily feedings to just three meals per week. The gorge-and-fast regimen, they found, improved the cats' digestion and metabolism and decreased the amount they ate. The animals showed fewer restless pacing behaviors.

ways, tunnels, and doorways to their tanks for them to explore. As they do in the jungle, orangutans can swing hand over hand along the Orangutan Transport System, a 490-foot-long aerial cable network strung along eight fifty-foot-high towers. Naked mole rats sometimes find their tunnels blocked by pieces of beet or carrot, left there by keepers who want to encourage the animals to gnaw or burrow their way around the obstruction, as they would a root in the wild.

Besides the animal's physical environment, feeding is the main area where veterinarians, nutritionists, and keepers concentrate enrichment. Nutritionists provide smaller and more frequent meals. They scatter and hide food. They offer live prey. Changing these aspects of the animals' environment makes eating a *process*.

No animals evolved to have food placed on a plate in front of them. They ran. They dug. They schemed. They starved. Eating was the reward for all that "work." Even when human agriculture began to improve the predictability of food supplies, those humans still had to catch or raise the meat they ate. Farming crops is essentially just organized foraging.

Nowadays, like many pets and zoo animals, most of us no longer worry about where our next meal is coming from (although sadly one in seven still does). Yet as we increasingly outsource where and what we eat to agribusinesses, supermarkets, and restaurant chains, we hand over not just the inconvenience of food gathering and preparation but also the challenge, the puzzle, and even the excitement of eating. Like that of captive animals, modern human eating has become more and more detached from the complex physiological and behavior-based impulses and decisions around food that natural selection forced us to develop.

When Richard Jackson calls obesity a "disease of the environment," the setting he's taking issue with is the one we've built with human ingenuity. The food we've tinkered with. The marketing that encourages us to consume it. The activity-lessening conveniences that have allowed us to become more sedentary than ever before. Living with abundant food and ready access to it will cause obesity no matter what species you belong to.

But a zoobiquitous perspective reveals other environmental factors, ones we can't even see and rarely think about that may be playing an unacknowledged role in obesity. It turns out that there are cosmic and microscopic drivers of appetite and metabolism—forces more complicated and unexpected than portion size, calorie count, and exercise levels. And they make the story of animals' weight gain much, much more interesting.

Every autumn, around the second week of October, the two male alligators at Brookfield Zoo abruptly stop eating. For nearly six months, Gaston and Tiboy refuse all food. Come early April, when they start bellowing and trying to charge their keepers, their nutritionist, Jennifer Watts, knows they're ready to resume their diet of rats and rabbits, until October, when they'll go off their feed again.

There's a reason the alligators' feeding schedule is like clockwork: clockwork.

As we all know, yearly life on our planet goes predictably from season to season. The amount of sunlight every day increases and decreases with perfect regularity, depending on the time of year and latitude.

Daily life, too, follows a regular schedule that is simultaneously grand and utterly familiar. Every day, as it has for billions of days, light follows dark in our planet's steady circadian rhythm. For more than three billion years, Earth's living creatures, starting with earliest single-celled organisms, have evolved in concert with this simple fact. Circadian rhythms, together with the diurnal rhythms of Earth's yearly trek around the sun, influence hunger, appetite, ingestion, and even digestion.

When I started medical school thirty years ago, I'd have been laughed out of seminar for suggesting that diurnal and circadian rhythms had anything to do with food choice and nutrition, much less obesity. These forces were like tidbits in the *Old Farmer's Almanac*—intriguingly consistent and predictable, observed in both plants and animals, but folksy and inscrutable enough to make them uncomfortably hard to use in any standard scientific sense.

During the past decade, that has changed. Molecular biologists have identified the underlying basis of circadian rhythms: the actual "clocks" that track time throughout our bodies. We'd been sensing their inaudible "ticking," but suddenly we could see how many and varied and yet how consistent they are.

The cells of all human beings, from those on our scalps to those deep in our hearts, contain oscillators built by what are called clock genes. Oscillators influence everything from how fast you burn your calories to when you want to eat them. Oscillators aren't just found in animals' cells. Primitive and species-spanning, they vibrate away in the cells of plants, bacteria, fungi, and yeast as well. Even cyanobacteria, some of the oldest single-celled organisms on Earth, display circadian rhythms organized by their oscillators.

So-called higher creatures—those with brains—have evolved a "mission control" apparatus that coordinates the messages from all these countless oscillator "droids" in the far-off cells. It's called the suprachiasmatic nucleus, or SCN. In humans, it's a pinecone-shaped collection of cells about the size of a sesame seed that sits at the point where the optic nerves intersect in the hypothalamus. The external signals the body takes in, called zeitgebers, exert a powerful effect on all of our physical functions. Temperature, eating, sleep, and even socializing influence our bodies' clocks. But by far the most influential of the zeitgebers is light. When light comes through the eyes, it hits the SCN, which then syncs up the external time signals with the internal oscillators throughout the body.

New research suggests that when, and how much, light beams through your eyes and hits your SCN may play a quiet and unrecognized role in determining your dress or pants size. Several studies have linked shift work to obesity in humans. One assumption has been that the weight gain can be attributed to a lack of sleep. But studies coming out of the animal world suggest that the culprit may not be the hours of missed sleep but the breaking up of light-dark cycles. A rodent study published in the *Proceedings of the National Academy of Sciences* showed that mice housed with constant light—whether bright or dim—had higher body mass indexes (BMIs) and blood sugar levels than mice housed with standard cycles of dark and light.

Farmers fattening chickens for meat have experimented with manipulating their weight through light exposure. In a study reported by the *World Poultry* newsletter, broiler hens "subjected to dim lighting were around 70g heavier than those in bright light."

And think about the Brookfield alligators. What changes for them in October and April isn't their job. They aren't suddenly being forced to stay awake or work a double shift. And it's not temperature. The alligators are in a temperature-controlled enclosure. It's light that makes them start and stop eating.

Studies have shown that disrupting circadian rhythms by even one hour during the switch to daylight saving time may increase depression, traffic accidents, and heart attacks. These rhythms affect consumption and metabolism in animals—it is hard to imagine that they aren't also playing a role in human appetites as well. Controlling environmental light with lamps, TVs, and computers gives us incredible flexibility and productivity. But it interrupts daily and yearly cycles that were billions of years in the making and are shared by countless creatures on our planet.*

Global factors like circadian rhythms can influence an animal's internal clock and govern when and how much it eats. But another set of even more intriguing and powerful processes are going on, out of sight, deep inside animals' bodies. While silent and unseen, these internal drivers illuminate a mystery of variable weight gain: why the very same piece of food can be processed differently by two neighbors, two relatives, or even by the same animal at various times of the year.

Some animal intestines perform an amazing trick. They expand and contract like accordions. This may not sound all that impressive, but its effect on weight can be profound. It allows the body to absorb varying quantities of calories from the same food, depending on the task at hand.

The mechanism is simple: a ribbon of muscle running the length of the intestine allows it to contract and expand. When guts are clenched, they're shorter, tighter, and smaller. When relaxed, they're elongated.

When intestines are in the longer, stretched-out mode, they expose more surface area to the food passing over them. This allows the cells to extract more nutrients and, therefore, energy. When the intestines shrink back to their shortened state, some of the food passes by essentially unused.

The guts of some small songbirds increase by 25 percent during the weeks right before they migrate, when fattening quickly is crucial to power their journey. Similarly, the intestinal surface area of certain grebes and waders nearly doubles during premigration feeding. When

---

*Of course, the biggest factor in the amount of sunlight any creature receives is where it finds itself on the globe. Latitude does seem to correlate with metabolism trends in mammals as well as sugar production in plants. (In general, the farther from the equator, the lower the sugar concentration in blood or berries.) Whether the effects are direct (from exposure to sunlight or other physics forces like electromagnetism or gravity) or evolutionary (adaptations over generations to available foods in any given area) remains a question for more research. But geographic effects on human weight have been all but completely ignored.

they've fattened enough to fuel a long flight, the birds' intestines shrink back down again.

The ability to lengthen and shorten intestines has also been observed in fish, frogs, and mammals, including squirrels, voles, and mice. Jared Diamond, a UCLA physiologist and author, has studied python guts for clues to how these snakes can go months between meals. Like those of birds and small mammals, pythons' intestines are dynamic, responsive organs, able to dramatically increase in size depending on what and when food is passing through.

Animals may be doing "naturally" what we spend tens of thousands of dollars to accomplish with bariatric surgeries that cut out or bypass parts of the stomach or small intestine. In us, as in other animals, less "gut" means fewer calories and nutrition absorbed. For animals, it isn't surgery but, rather, muscular action—triggered by certain foods, seasonal cues, and other unknown factors that expand and contract the gastrointestinal region.

Could a similar accordion-like lengthening and shortening in human intestines underlie some unexplained weight gain in our species? Unfortunately, there's little direct research on when and whether our guts pull off this same trick. But there are intriguing clues. Our intestines are also lined with smooth muscle. And we know from autopsies that human intestines are some 50 percent longer after death, when smooth muscle control is no longer exerted. Perhaps, during life, dynamic muscle activity allows the human intestine to vary its calorie-absorbing length in response to medications, hormones, and even stress—factors frequently pointed to when weight inexplicably increases even when a patient isn't eating more. Many common drugs cause undesired weight gain through unclear mechanisms. It's intriguing to consider whether the smooth muscle effects of these drugs contribute to a songbird-like intestinal stretch leading to greater calorie absorption and weight gain.

But besides the astonishing physiology that makes our guts dynamic, animal intestines hold another key to the complex issue of weight. Within them is a universe invisible to the naked eye that scientists are just beginning to explore and understand.

Deep inside every animal colon, ours included, thrives an entire cosmos of creatures more strange and wondrous than any dreamed up in a Hollywood special effects lab. There are whip-tailed bacteria and tripod-legged viruses, frilled fungi and microscopic worms. Trillions of these invisible creatures make our intestines their home—a dark, teeming world scientists call the microbiome. Our skin, mouths, teeth (and even areas once thought to be sterile, like the lungs) so swarm with invisible creatures that as few as one out of every ten cells in our bodies may actually be human. The rest are much smaller microbes. So profound is this colonization that some geneticists call adult humans "superorganisms," meaning our cells plus those of all the creatures living within our bodies. Each of us is like a coral reef, an individual microhabitat harboring unique combinations of unseen wild inhabitants.*

In general, we should be grateful that these trillions of minuscule bugs and plants want to live in our guts. Many of them break down our food and prepare nutrients for our cells to absorb—processes human cells cannot do on their own. Microbiologists are only just starting to explore how human gene sequences interact with those of all our microbial residents. They're finding that these colonies of aliens might not only influence how we digest and metabolize but even drive us to choose or crave certain foods.

It turns out that within our microbiomes there are two dominant groups of bacteria: the Firmicutes and the Bacteroidetes. In the early 2000s, geneticists at Washington University in St. Louis, were looking at how these bacteria break down food we can't digest on our own. And the geneticists made an interesting discovery.

Obese humans had a higher proportion of Firmicutes in their intestines. Lean humans had more Bacteroidetes. As the obese humans lost weight over the course of a year, the microflora in their guts started looking more like those of lean individuals—with Bacteroidetes outnumbering Firmicutes.

When the researchers looked at mice, they found the same thing. Obese mice had more intestinal Firmicutes. Interestingly, these fat mice produced feces that had fewer calories left in them than the feces of lean mice—suggesting that the obese mice were somehow absorbing more energy from the same amount of mouse chow. This led the researchers to suspect that the Firmicutes are superefficient at mining calories from

---

*For entertaining and illuminating reporting on the microbiome in particular and microbiology in general, see the work of Carl Zimmer, a *New York Times* science writer and the author of, among many other books, *Microcosm* and *A Planet of Viruses*.

food passing through the digestive tract. As a December 2006 *Nature* article about the study put it, "The bacteria in obese mice seemed to assist their host in extracting extra calories from ingested food that could then be used as energy."

What this means is that a booming Firmicute colony might help harvest, say, one hundred calories from one person's apple. That person's friend may have a dominant Bacteroidetes population that would extract only seventy calories from the same apple. This could be one factor in why your co-worker can eat twice as much as everyone else but never seems to gain a pound.

If our personal "house blends" of gut bacteria influence the amount of energy we extract from food, then diet and exercise may not be the only factors driving weight gain and loss. The effects of the microbiome challenge the once-unassailable calories-in, calories-out paradigm.*

In fact, veterinarians have long recognized the power of the microbiome over an animal's metabolic function.† In ruminants and other so-called gut fermenters, such as horses, turtles, and even some apes, nutrition and digestion simply cannot function without the proper balance of microorganisms. Although I learned almost nothing about the power of gut flora during medical school, Brookfield Zoo nutritionist

---

*Entrepreneurial-minded thin people take note: the teeming bacterial clusters inches beneath your belly button may be fermenting a billion-dollar opportunity. If the dominant species of bacteria in our guts helps determine our BMIs, perhaps a fecal or oral infusion of Firmicutes or Bacteroidetes in certain proportions would speed us to our body-image goals. There may come a day when, instead of logging calories, we could lose weight by purchasing desirable gut flora from the thin (but not squeamish) bacterially blessed.

†In human medicine, so-called fecal therapy is a breakthrough treatment for stubborn and sometimes life-threatening diarrhea and other gastrointestinal problems arising from infections with organisms such as *C. difficile*. Feces are acquired from an individual (frequently a spouse) with normal gut flora, mixed into a slurry in a kitchen blender, and placed on the tip of a specialized endoscope for insertion into the small intestine of the ailing recipient. You may be wrinkling your nose, but it's an extremely effective and low-cost solution to restore human health. And farm veterinarians have been doing it for decades. Bug-rich biliary gastric juice is extracted through a fistula created in a side surface of a healthy donor cow. This "liquid gold" (not to be confused with the urinary liquid gold used by stallion breeders) is extracted and then transferred into other animals to normalize their gastrointestinal flora. Zoo veterinarians routinely use fecal therapy to normalize their patients' digestive tracts after rounds of antibiotics. It's especially effective for mother-infant pairs.

Jennifer Watts related to me a core principle emphasized to her during her nutrition training: "Feed the gut bugs first, then the animal." She does it by making sure animals are fed a healthy balance of browse (fresh leafy greens) and silage (partly fermented vegetation). Could it be that eating vegetables is good for us not only for the fiber they provide but because they nourish colonies of beneficial microflora in our intestines? Perhaps we're in effect feeding our gut bugs every time we eat a salad.

The power of the microbiome is well known to another group of veterinarians, the ones who oversee the care of animals we make fat on purpose: livestock. Nowadays, it's common for factory farming operations to administer antibiotics to food animals from fifteen-hundred-pound steers to one-ounce baby chicks. The effect of those antibiotics on the living colonies of gut bugs in the animals' intestines may hold a profound clue to the human obesity epidemic.

I'd long known that antibiotics are used in farming to stop the spread of certain diseases, especially under cramped and stressful living conditions. But antibiotics don't kill just the bugs that make animals sick. They also decimate beneficial gut flora. And these drugs are routinely administered even when infection is not a concern. The reason may surprise you. Simply by giving antibiotics, farmers can fatten their animals *using less feed*. The scientific jury is still out on exactly why these antibiotics promote fattening, but a plausible hypothesis is that by changing the animals' gut microflora, antibiotics create an intestine dominated by colonies of microbes that are calorie-extraction experts. This may be why antibiotics act to fatten not just cattle, with their multistomached digestive systems, but also pigs and chickens, whose GI tracts are more similar to ours.

This is a really key point: antibiotic use can change the weight of farm animals. It's possible that something similar occurs in other animals— namely, us. Anything that alters gut flora, including but not limited to antibiotics, has implications not only for body weight but for other elements of our metabolism, such as glucose intolerance, insulin resistance, and abnormal cholesterol. And don't forget the trillions of creatures making up our microbiomes are constantly interacting in complex ways with one another. They have oscillators that respond to circadian rhythms. The dynamic population of that tiny, contained universe exerts more influence over metabolism than physicians have ever suspected.

When the Firmicutes/Bacteroidetes study appeared in *Nature*, it sparked an interest in other obesity risk factors that are less obviously under our control than diet and exercise. Blogs were soon buzzing about a different study showing that having a fat friend increases a person's chance of becoming overweight himself. The Harvard medical sociologist Nicholas Christakis and U.C. San Diego scientist James Fowler were describing a "contagion" of social habits and practices. Your fat friend's bad food choices and exercising habits could influence your own willpower and attitude toward food. Christakis and Fowler were quick to explain that the finding was not literal but symbolic. You couldn't catch the "fat flu" from an ill-aimed sneeze in the waiting room of a lap band clinic. Rather, what was "infectious" was other people's attitudes toward eating.

But when I studied the animal literature, I learned that infectious obesity may not be solely metaphorical. According to some experts, it is altogether literal and real. Nikhil Dhurandhar, a nutrition and food scientist at Wayne State University, in Detroit, explains: "It has been proven that animals became obese when infected with certain viruses." He calls it "infectobesity." Dhurandhar reports that seven viruses and a prion have been linked with obesity in animals as varied as chickens, horses, lions, and rats. That's right: infectious weight gain, spread or facilitated by microscopic pathogens.

On the hottest days between mid-May and late August, alongside one of the many ponds around State College, Pennsylvania, chances are good you'll spy a tall, thin biologist creeping through the cattails in khaki shorts and a battered cap. He'll be crouched, moving in barely perceptible super slo-mo. Suddenly, with an expert forehand swing, he'll swipe a wood-handled net through a stand of reeds or bulrushes. (The move, he explains, is similar to a lacrosse catch or a tennis stroke, which is why he likes to hire grad students who've played these sports before.) Nipping the mesh shut with his free hand, he'll peek inside to see if he captured his quarry: *Libellula pulchella*, the twelve-spotted skimmer dragonfly.

James Marden is an entomologist and professor of biology at Penn State University. For more than two decades at ponds in central Pennsylvania he has studied the flight mechanics of dragonfly wings. He told me that these insects are among the fittest animals on Earth, extraordinarily lean and muscular. Over 300 million years, dragonflies have evolved so perfectly to the acrobatic demands of hovering, bobbing, and looping the loop that Marden calls them "world-class, elite animal athletes."

Usually dragonflies are pugnacious and extremely territorial, always up for a skirmish with another male. When two meet, they zoom at each other in belligerent, balletic aerial combat that ends with the loser being chased off. Some males, however, loiter on the outskirts of the action. Instead of spoiling for fights and flying straight into brawls, they "glide"—easing their way past challengers without coming to blows, as if to say, "I'm just passing through. No problem. Pay no attention to me. I was just leaving."

In the early 2000s, intrigued by this behavior, and whether it might have something to do with muscle performance, Marden collected some of these slower, evasive dragonflies. And when he got them back to his lab, he discovered something shocking. Although on the outside the dragonflies looked perfectly normal—lean and combat ready—Marden's examination showed that they were actually very, very sick. But their disease was peculiar for these "jet fighters of the insect world." They were all medically obese.

Fat was collecting in their body tissues instead of converting into energy to fuel their extraordinary wing muscles. Their blood sugar* concentrations were double that of healthy dragonflies, putting them in an insulin resistant–like state—similar to what's seen in human patients with type 2 diabetes. They were slow, weak, sluggish, and unable to fight for females or defend territory.

That a wild dragonfly could develop a form of metabolic syndrome† has the potential to revise thinking about human weight gain and maybe even the obesity epidemic itself. When Marden looked inside the dragonflies' guts, he found something that surprised him. Freckling their

---

*Dragonfly blood is called hemolymph, and its main carbohydrate is trehalose; Marden calls it blood sugar.

†Metabolic syndrome increases a patient's risk for heart disease and stroke. Also known as insulin-resistance syndrome, it is diagnosed when triglycerides, blood pressure, or glucose are too high or when a patient's "good" cholesterol (HDL) is too low. An apple-shaped body is associated with metabolic syndrome.

intestines were large white parasites. Some of them were so big—up to one-fiftieth of an inch—that Marden could see them without a microscope. Magnified, they looked mild-mannered enough: like plump little grains of rice.

What the parasites caused in the dragonflies, however, was anything but mild. They were gregarines, protozoans from the family that causes malaria and cryptosporidiosis in humans. In the dragonflies they triggered an inflammatory response that interfered with the insects' ability to metabolize fat. That's why it was collecting in their body tissues, particularly around their muscles. Their fat deposits were reducing their muscle performance, causing the dragonflies to relinquish territory and abandon mating opportunities.

By measuring the way the dragonflies' muscles exchanged oxygen and carbon dioxide, Marden and his graduate student Rudolf Schilder could see that the infection was directly causing these changes. He told me it wasn't just that the dragonflies were weakened by the presence of the parasites, making them duller and slower. Rather, "specific components of their metabolism had changed."

The gregarine infections also caused chronic activation of a signaling molecule involved in immune and stress responses, called p38 MAP kinase. In humans, the same molecule is implicated in insulin resistance that can lead to type 2 diabetes.

Intriguingly, the parasites were noninvasive, meaning they didn't chew into or visibly damage the gut walls. Their inflammatory effect seemed to be triggered by substances they secreted and excreted. Eerily, the blood sugars of uninfected dragonflies became abnormal after they simply *drank water* containing trace amounts of the gregarines' excretions or secretions.

At first, the possibility that obesity has an infectious component seemed ludicrous to me. Having been steeped in the simple diet-and-exercise, calories-in, calories-out approach, and knowing that reducing intake and increasing activity does result in at least temporary weight loss, I thought that infectobesity seemed unexpected and, frankly, even unlikely.

But although I had never heard about it, the search for infectious pathogens that promote weight gain has been under way since at least 1965, when a microbiologist at the State University of New York, Syracuse, explored how a certain worm caused mice and hamsters to become obese. He suggested that the worms might be "leaking" a hormone into the bloodstreams of the rodents, causing them to eat more in order to satisfy the chemistry of the parasite.

And, indeed, infections of many kinds influence appetite. Tapeworms make you hungry. Certain viruses put you off your feed. In fact, appetite is one of the first things doctors ask patients about when we're taking a medical history, because it's one of the most sensitive markers of infection. These facts made me consider more seriously the real possibility that microbial invaders might manipulate what, how, and when we eat.

It wasn't that long ago that a serious human intestinal medical condition was unexpectedly found to have an infectious component. For decades, gastric ulcers were believed to occur as a result of our stressed lives and overly reactive psyches. You got them, the medical wisdom went, if you were anxious and couldn't resist greasy, spicy foods. In 2005, the Australians Barry Marshall and J. Robin Warren, a physician and a pathologist, won the Nobel Prize in Medicine for busting that myth. They identified the cause of many ulcers as *Helicobacter pylori,* a contagious bacterium easily treated with a dose of antibiotics. The road to the Nobel was long, however. For years Marshall and Warren had to endure criticism, rejection, and scorn. But now the microbiome is being examined for organisms that may be responsible for irritable bowel syndrome and Crohn's disease. Maybe obesity should be next.

But the research on infectious causes of metabolic syndromes is still at the stage where nutrition scientists and doctors dismiss it—or at least don't seem quite ready to hear it. Marden published his research in a top academic journal, the *Proceedings of the National Academy of Sciences,* and wrote an opinion piece for a human diabetes journal. Yet he told me there "really wasn't much response. I don't think the medical community was swayed by our results or super eager to hear about it. It's been pretty much a 'so what?' from the medical community."

Whether infection will ultimately be proven to play a part in human obesity is still hard to say. However, a cross-disciplinary, zoobiquitous approach—one that connects the knowledge of a dragonfly expert in an agricultural sciences biology department with researchers looking at human obesity—may spark innovative hypotheses and a broadened view of this major health threat. We live in a world teeming with organisms in us, on us, and around us. Our defenses against them drive many of our diseases. It's critically important that researchers charged with understanding and containing obesity's dangerous growth remain open to ecological factors, including lightness, darkness, seasonal shifts, and, yes, even infectious organisms. As Marden put it when his paper came out in 2006, "Metabolic disease isn't some strange thing having just to do with humans. Animals in general suffer from these symptoms . . . [and] it would be irresponsible for us not to point out these possibilities."

To repeat: "Obesity is a disease of the environment." And while Big Gulps and Segways play a primary role, so, too, do these much larger and much smaller forces. An expanded, environmental approach to weight has already cured two obese patients from the Chicago area—those two obese grizzly bears at the Brookfield Zoo.

Whether it was circadian rhythms, imbalanced microbiomes, seasonally confused intestines, an infectious parasite, or just access to too much food that caused Axhi and Jim to pack on the pounds over the years is hard to say. But their pattern of fattening before Watts changed what, when, where, how, and how much they ate resembles our own.

Watts had decided to make a massive change that was both innovative and as old as eating itself. She would approximate the yearly rhythms of a wild diet. In other words, she would let the seasons and the bears' bodies lead the way.

She started with *what* they ate. For years, their food had been abundant, readily available, and largely unchanging throughout the year. It included processed dog food, bread from a local bakery, supermarket apples and oranges, and ground beef. Gradually, Watts challenged the bears' taste buds. She swapped out a serving of lettuce and introduced kale. She traded mango for apple. Then spinach, celery, peppers, and tomatoes subbed in for sweet potatoes and oranges. Although this produce wasn't exactly identical to what they'd find growing on the banks of an Alaskan river, in terms of nutrient range, variety, and seasonality, it was an improvement.

Soon, when the keepers showed up with a meal, the bears were as enthusiastic as human foodies sniffing out the exotic offerings at a new gastropub. Watts also added whole prey like fish, rats, and rabbits, and timed their appearance on the menu with when those foods would be found in the wild. She also ordered boxes of wax worms, which she dumped in the bears' foraging pile—a big, peaty dirt mound—and let them rummage and eat to their hearts' content. With each of these dietary introductions the bears consumed not only new sources of protein and vitamins at seasonally appropriate times of the year, but whatever new and varied microorganisms happened to be making those food items their home. Although she notes this wasn't at first intentional, Watts was following her own motto: "feeding the bugs" in their GI tracts.

Watts also decided to allow the bears to enter a more seasonally appropriate winter torpor. It wasn't a full hibernation (which many bears in the wild don't do anyway). But it was a big change for Axhi and Jim. For the previous decade, the bears had been awakened every day throughout winter for feeding. Sometimes the keepers had to rouse them by shouting or making loud clanging noises. Watts instructed that the bears be allowed to sleep through the winter months. And she ordered that, if they woke, they not be presented with food around the clock but be given "one shot" at a small amount before it was removed. On the surface, this plan would seem to work in favor of weight loss because it reduced the number of overall calories the bears consumed. But its efficacy may have been deeper than that. Sleep and metabolism are interconnected, and the longer periods of fasting may have signaled other physiologic changes to the bears' bodies, such as intestinal lengthening or shortening.

Finally, the bears were moved to a larger home. In this new environment, their food could be presented to them in ways that were "inconvenient," causing them to mimic the foraging and hunting they'd do in the wild and expend more energy getting at a meal.

Yet even with all these changes, Watts was not able to wholly re-create the bears' natural diet. Just as it would be almost impossible for us to eat like our ancestors did even a hundred or a thousand years ago, it's unfeasible for zoos to duplicate a wild diet for each animal. The fruit zookeepers buy from grocers and wholesalers is a far cry from the fruit a wild animal eats.* There are no naturally occurring banana plantations in the Canadian Rockies. No orange groves. No wild watermelon vines

---

*Everything we recognize as wholesome fruit is the product of careful crafting and managed evolution—begun with the earliest human agriculturalists and "improved" over the millennia, intensively over the last few decades. The fruit we find in the supermarket today has been cultivated for human tastes (and transportation convenience).

or mango trees. And even if Watts could have gotten fruit with the same characteristics in the exact proportions to what is found in the wild, the microorganisms on that washed, boxed, refrigerated, and shipped fruit would be completely different from what the animal would be eating in a natural setting.

Fortunately, Watts understood that the fantasy of creating "the perfect wild diet" was just that—a fantasy. She did the best that she could under the circumstances. And it turned out that adjusting the bears' diets in ways that were simply informed by knowledge of their natural ecology was enough. They lost weight. They seemed to feel better and have more energy. In short, they were healthier.

Watts's success carries lessons we can apply to our own lives, whether we want to address the global obesity epidemic or a personal weight-loss effort. Researchers and doctors ought to consider the environment's cyclical periods of abundance and scarcity as well as the seasons' effect on our food-absorbing intestines. We must take seriously the complex universe of the microbiome and the metabolic consequences of infection. We need to think about global forces like day length and light cycles.

Modern, affluent humans have created a continuous eating cycle, a kind of "uniseason." I've started calling this blissful, bountiful, yet static and superfattening environment the "eternal harvest." Sugar is abundant, whether in our processed foods or in beautiful whole fruits that have had their inconvenient seeds bred out of them and that "unzip" from easy-to-peel skins and pop open into ready-to-eat segments. Protein and fat are everywhere available—in eternal harvest the prey never grows up and learns to run away or fight us off. Our food is stripped of microbes, and we remove more while scrubbing off dirt and pesticides. Because we control it, the temperature is always a perfect seventy-four degrees. Because we're in charge, we can safely dine at tables aglow in light long after the sun goes down. All year round, our days are lovely and long; our nights are short.

As animals, we find eternal harvest an extremely comfortable place to be. But unless we want to remain in this state of continual fattening, with its accompanying metabolic diseases, we will have to pry ourselves out of this delicious ease.

Puffed up with extra water and soaring sugar content, commercially grown fruit also has less fiber than fruit found in the wild or in the ancient past.

EIGHT

Grooming Gone Wild

Pain, Pleasure, and the Origins of Self-Injury

Name a medical complaint or affliction, and you'll find an online support group for it. These sites allow people to swap stories, share remedies, and feel less alone. The posts are often heartbreaking. Recently, I was scrolling through some online forums. The threads practically wailed with misery: "I am so worried," "This is tearing me apart," "I am afraid he isn't going to stop," "I am at wits' end," "He's had this problem for years," "Can someone please help me?," "I feel so terrible—like I am a horrible mommy."

The sites were not about human patients. They were about pet birds with the surprisingly common problem known as "feather-picking disorder." While the individual stories varied, the overall theme was the same. Birds with names like Juliet, Zeke, Jubilee, and Ms. Earl were all perfectly healthy, until one day their owners discovered a pile of colorful plumage at the bottom of the cage . . . and a bald patch on their pet's shoulder or chest or tail. The birds were plucking out their feathers, one by one, and sometimes pecking the underlying skin until it bled. Veterinary exams ruled out physical causes of irritation, like mites or infections. The owners installed humidifiers, smoothed aloe vera on the stubbled skin, and invested in higher-quality birdseed. Still the picking continued. One despairing owner of a self-plucking Quaker parrot wrote, "Lately she has been pulling and letting out that little scream like they do when you touch the wrong pinfeather, then she goes right on and does it again, so now she's plucking even if it HURTS . . . and I have seen several small spots of blood on her crop and under her wings."

As a human physician and psychiatrist who's never in my life owned a bird, I nonetheless recognized these symptoms: an inexplicable change in behavior, deliberate actions that caused bodily pain and disfigurement, and confusion and distress in loved ones. They brought to mind a patient I saw a couple of years ago, a twenty-five-year-old woman who presented with heart palpitations. Traversing her inner left forearm was a series of expertly cut incisions, ones that in other circumstances might have been the handiwork of one of my surgical colleagues. Thought had clearly gone into sterilization, cleanliness, and how the cuts would heal. But no doctor had been present when they were made. Instead, my patient had taken a razor in her right hand and sliced into her own skin. She was a "cutter."

Cutting is probably our era's most iconic form of human self-mutilation, seemingly tailor-made for suburban-parent hand-wringing and tabloid ogling. Its name says it all, but in case you don't know: it means taking something sharp—maybe a razor blade, scissors, broken glass, or a safety pin—and purposely slicing it across your skin to draw blood and create wounds. Usually, cutters target parts of their bodies that can be covered up with clothing to hide the evidence—say, their inner arms, thighs, or stomachs. Some do it impulsively, with whatever tool is at hand; others are more ritualistic about it. They might cut at the same time and in the same place every day. Or they create "kits" that hold their favorite cutting implement, along with gauze, Band-Aids, and alcohol wipes for mopping up afterward. As you can imagine, cutters—especially those who do it for many years—develop scars, often parallel lines like crimson ladder rungs up and down their favorite cutting site.

Psychiatrists call cutters "self-injurers" to include the whole range of inventive ways people dream up to hurt themselves. Some burn themselves on purpose with cigarettes, lighters, or tea kettles. Others bruise their skin by banging, punching, or pinching themselves. Those with trichotillomania rub and rip out hair on their heads, faces, limbs, and genitals. Some are swallowers, ingesting objects such as pencils, buttons, shoelaces, or silverware. We see this particular method a lot in prisons.

You may think self-injury occurs only in edgy subcultures or the seriously mentally ill. But my psychiatrist colleagues say it's sweeping through the general population. Therapists and school guidance counselors confirm this.*

Self-injury has gained an unintended endorsement by public figures. I, for one, was shocked by Princess Diana's disclosure to the BBC in 1995 that she cut herself with a lemon peeler and a razor blade. She also engaged in non-blade-related self-injury, including hurling herself into a glass cabinet and throwing herself down a flight of stairs. While Angelina Jolie has restyled herself as a supermom and human rights crusader, she joined such other celebrities as Christina Ricci, Johnny Depp, and Colin Farrell in telling the world about a self-injuring past; their tools included knives, soda can pop-tops, broken glass, cigarettes, lighters, and their own fingers. Cutting's edgy street cred includes feature appearances in angsty teen films like *Thirteen* and *Girl, Interrupted*. And cutting even made a comic turn in the movie *Secretary,* with Maggie Gyllenhaal and James Spader embarking on what is perhaps the most happy-go-lucky sadomasochistic love story ever told.

But confronted by my cutter patient's razor-notched arms, I was still confused. She was a thoughtful, intelligent adult woman with a respectable job, very much like the Maggie Gyllenhaal character in *Secretary.* Why would she cut herself *on purpose*—something a doctor would consider doing only with anesthetic and strict protocols? So although she was in my office for a heart consult, I asked her. She answered matter-of-factly, "My shrink says I'm trying to kill myself. But I'm not. If I wanted to die, I would. Cutting just makes me feel better. It relieves me."

Her answer jibed with what other cutters say. A twenty-two-year-old woman writing on a Cornell University website put it this way: "I began cutting my arms at the age of 12. . . . I think I could best describe the feeling I get as total bliss. It relaxes me."

Bliss? Relaxation? Relief? It "*feels good*"? Even after years of psychiatry training and two decades around a hospital, I still think this sounds

---

*Twenty-five years ago, when I was a medical student on the inpatient psychiatric unit at U.C. San Francisco, self-mutilation was thought to be uncommon in general populations. It was typically diagnosed in conjunction with a developmental or psychotic disability—for example, eye gouging or genital cutting with schizophrenia or head banging with autism. Indeed, self-mutilation occurs in association with certain disorders, including Tourette's syndrome, Lesch-Nyhan syndrome, some forms of developmental delay, and borderline personality pathology.

incredible. But cutters and their therapists say it's true. And they confirm that most self-injurers are not suicidal, though sometimes a cutter will go too deep with the blade and require medical attention.*

But as to *why* they do it, the short answer is that we don't really know. Psychiatrists have linked cutting to adolescence, control issues, emotional imperceptiveness, and an inability to talk about feelings. Self-injury has also been associated with childhood sexual abuse and such other mental conditions as borderline personality disorder, anorexia nervosa, bulimia nervosa, and obsessive-compulsive disorder (OCD). Patients who self-harm report being stressed and anxious, overwhelmed by expectations and choices—or completely isolated and numb.

Childhood traumas and abusive parents have traditionally been blamed for self-mutilating behavior. But this formulation, it turns out, is incomplete. The stereotypical cutter featured in movies and on TV may be a sexually abused, poorly parented borderline girl. But it turns out that rates of self-injury may be roughly similar for men and women. The difference is more in how they do it: men tend to hit or burn themselves, while women usually cut. Some begin the behavior as young adults when they're no longer under the influence of their parents. And many report no history of childhood abuse.†

So the mystery remains. What flicks the switch—changing brooding, hormonal adolescents, and full-grown adults with jobs and responsibilities into self-injurers?

I decided to see what insights a zoobiquitous approach could add. When we find animal behaviors that parallel disturbed human actions, it's an opportunity to look beyond our "hectic modern lives" and "big human brains" for the origins of the symptom. Yet when I first began

---

*Lack of suicidal intent is a relatively new notion when it comes to self-injury—in fact, some of what just twenty years ago we called "hesitation marks" (scars or shallow wounds) on the wrists of a suicide might actually have been evidence of previous cutting.

†In the fourth version of the *Diagnostic and Statistical Manual of Mental Disorders* (DSM-IV), self-harm is listed as a symptom of borderline personality disorder. Other psychiatric texts classify it as an impulse-control disorder, along with exhibitionism, kleptomania, and the compulsive tics and vocalizations of Tourette's syndrome. The much-anticipated *DSM-V* will likely recategorize acts of nonsuicidal self-harm, including cutting, based on our expanded understanding of the neurobiology and genetics that underlie them.

asking whether animals self-injure, the question seemed almost absurd. What would it even mean for an animal to mutilate itself?

The classic image of animal self-harm is a wolf gnawing off a paw to free himself from a hunter's snare. But this purposeful self-injury to escape entrapment (which crops up from time to time in extreme human accounts as well) was not what I was looking for. I sought an animal correlate to the compulsive, trancelike self-injury seen in humans. Needless to say, I could safely assume I would find no evidence of razor blade cuts or cigarette burns among our wild cousins.

And I didn't. But my research quickly turned up an equally fearsome arsenal, usually deployed against enemies. Teeth. Claws. Beaks. Talons. The big question was, do animals ever turn these on themselves? The answer, to my astonishment, was yes, and frequently. Feather-picking disorder in birds was just one of many such examples well known to veterinarians.

A friend of mine once took her cat to the vet assuming it had a skin affliction that was causing all the hair to fall off its legs, revealing red, oozing sores. After some tests to rule out parasites and systemic diseases, her vet said her pet was a "closet licker." It's a common diagnosis for house cats, sometimes called psychogenic alopecia. Out of sight and in secret, the cat was injuring itself with no clear physical trigger, just like a human cutter alone in her room.

Owners of golden retrievers, Labrador retrievers, German shepherds, Great Danes, and Doberman pinschers will probably recognize a condition that often affects those breeds—in which they obsessively lick and gnaw at their own bodies. The open sores they create can cover the entire surface of a limb or the base of the tail. The diagnosis of acral lick dermatitis (sometimes called lick granuloma or canine neurodermatitis) is not linked to an external agent like a fungus, fleas, or infection; the animal is doing it for no apparent physical reason. If you've ever watched a dog chew itself like that, it sometimes seems to be in a kind of trancelike or hypnotic state—eyes glazed, head bobbing, lick . . . lick . . . lick . . . lick . . .

Anyone who's worked in the reptile section of a pet store has seen turtles biting their legs and snakes chewing their tails. And a peek into the stable shows another suffering animal. "Flank biters" are horses that violently nip at their own bodies, drawing blood and reopening wounds.

The owners of these horses, like parents who discover their teenager is cutting, are often confused and heartbroken by the behavior, which can include bursts of violent spinning, kicking, lunging, and bucking.

Behaviors such as flank biting, tail sucking, and feather plucking may be more common than we think, at least in certain breeds. Up to 70 percent of Dobermans, for example, will develop time-consuming and often distressing repetitive actions, including but not limited to self-injury. Nicholas Dodman, a veterinarian at Tufts University, treats and researches compulsive behaviors in horses and dogs. Dodman and his colleagues at the University of Massachusetts and MIT have identified a genetic region on canine chromosome 7 that is associated with an increased risk of a dog's developing what they call canine compulsive disorder (CCD).

Whether OCD in humans and CCD in dogs are the same disorder is hard to say. In human beings we make a diagnosis of OCD when obsessional thoughts drive compulsive behaviors. In contrast, with animals, all veterinarians have in order to make a diagnosis are the behaviors. Without a common language, they have no way of determining whether obsessions underlie the animal's perseverative practices.

When owners bring in pets who circle furniture for hours, do backflips to the point of physical exhaustion, or rub their skin to the point of breakage and bleeding, veterinarians sometimes describe these behaviors as "stereotypies." At the extreme end of the spectrum are head banging, picking, poking, and gouging. In some cases, especially in birds, compulsive vocalization is considered a stereotypy with possible connections to Tourette's syndrome in humans. For veterinarians, any behavior on this spectrum, even the milder end, merits concern and intervention.

Many of the compulsive behaviors seen in horses, reptiles, birds, dogs, and humans share core clinical features, including the potential to cause suffering and profoundly disrupt a patient's life. But many also share an intriguing connection to cleaning activities. You've probably heard about the repetitive hand washing practiced by many OCD sufferers. Similarly, a stressed cat may go overboard with a feline's cleaning tool of choice, its raspy tongue. Veterinarians have come up with a colloquial term that cuts right to the heart of what's going on here. They call it, simply, "overgrooming."

Overgrooming? When I first heard the term, I flashed on countless nature documentaries showing apes combing and nit-picking each other. It surprised me that this benign cleaning and social ritual could escalate into something potentially lethal. I quickly learned that animal grooming covers both a broader range of species and a weirder array of behaviors than I'd ever imagined.

Grooming, to put it plainly, is as basic an activity for many creatures as eating, sleeping, and breathing. Evolution probably favored nature's neat freaks because they were the ones with fewer parasites and infections.

Primates display a wide repertoire of combing and nit-picking techniques. Some chimps pick parasites off each other, place them on a forearm, and kill them with a smack of the hand before eating them. Others use leaves to pinch the bugs out of their partner's fur. Japanese macaques have elaborate finger-and-thumb louse-egg–loosening techniques passed down from one generation to the next (through the maternal line).

But while delousing may be its ultimate purpose, there's a more immediate reason animals groom. Simply stated, grooming feels good, and it plays a vital role in the social structure of many animal groups.

Some groups of chimps scratch each other's backs and clasp hands without the intention of bug removal. Crested black macaques, especially females, embrace each other and rub their sides together. And while most mutual grooming in primates takes place among family members, non-kin also get their fingers in each other's fur—for good reason. When lower-ranking bonnet macaques and capuchins "give" grooming, they gain in return protection, backup in fights, and a chance to hold someone else's baby. Some baboons groom to get close enough to a partner to sniff out whether he or she is aroused and in the mood for mating.

The huge importance of social grooming isn't limited to primates or even land mammals. In the fish world, it sometimes keeps the peace. A tropical reef dweller called the cleaner wrasse operates what are, in effect, underwater spas, where it eats parasites and scar tissue off other fish. These include much larger predators that would normally (and literally) eat the wrasses for breakfast. But in the calming atmosphere of the cleaning station, the wrasses approach the bigger fish without fear, darting around their teeth and even into their gills.

This relationship isn't just a heartwarming example of animal cooperation. Scientists have found that grooming's calming effect is felt not only by the fish receiving the cleaning but also by fish waiting their turn. Both the anticipation and the experience of grooming seem to make the predator fish less likely to chase *any* fish in the area. The scientists who did the study compared this aquatic "safe zone" to a human barbershop in a rough neighborhood where violence is understood to be off-limits.

Grooming's powerful calming effect applies as much to solo as to social cleaning rituals. Cats and rabbits may spend up to a third of their waking hours fastidiously licking themselves. Sea lions and seals spend many hours a day rifling through their own fur. Birds roll in dirt, fluff, preen, and pick with their beaks. Snakes, lacking napkins or hands with which to wield them, often finish a meal by wiping their faces against the ground.

But perhaps no animal has more, or more varied, grooming routines than our own species. Humans primp, wash, and polish . . . alone, in pairs, in groups . . . with and without tools or "product" . . . for free and for price tags verging on the obscene. I'm just one of millions of American women—and, increasingly, men—who find relief at the manicurist or the hair salon when confronted with the stresses of career and family. In fact, I would say that regular and good grooming calms and centers me. The companionship, the care, and, especially, the repetitive tactile stimulation relieve stress and enhance my well-being.

Our species, as a rule, grooms a lot. And as it does for our animal cousins, grooming provides us with physical pleasures: the joy of a warm shower after a week of camping; the satisfying smoothness of a good shave; the indulgence of having attention lavished on us at a salon; and the frisson of looking in the mirror when we're dressed to the nines. (While humans vary in the amount of time and money they choose to spend on grooming, we know that opting out completely carries major social risks.)

It turns out that our sense of well-being is much more than skin deep. Grooming actually alters the neurochemistry of our brains. It releases opiates into our bloodstreams. It decreases our blood pressure. It slows our breathing. Grooming someone else confers some of the same benefits. Even simply petting an animal has been found to relax people.

When I'm sitting in the plush pedicure chair, my feet bathed in warm soapy water, it's hard to believe there is such a thing as *over*grooming. Or that this calming procedure could have anything to do with Princess Diana's slicing lines in her thighs with a razor blade, let alone with a solitary cockatoo in his cage. But grooming encompasses more than just the socially acceptable forms you pay for at the spa.

There's also a more private form of grooming—small behaviors that all but the most virtuous of us engage in all the time and often unconsciously. In general, they're innocent enough, but given the choice, we definitely wouldn't want to show them in public or watch other people do them. Look at your own fingers, the ones holding this book. Are your cuticles smooth or are there some rough edges begging to be picked or nibbled off? Are you twirling a lock of hair around your finger, twisting your eyebrows, stroking your own cheek, massaging your own scalp? Studies looking at hair pulling, scab picking, and nail biting all point to a calm, trancelike state that typically accompanies these small, automatic, self-soothing activities.*

And we unconsciously vary the intensity of these actions. Perhaps the fingers playing with your hair sometimes have the urge to pull a strand out. There's that slight tension as the root clings to the follicle . . . you gently tug harder . . . and a little harder . . . until finally, there's that short, sharp sting and the hair releases.

Or imagine the last time you had a little scab somewhere on your body. Maybe you had the willpower to leave it untouched, but the rest of us probably scraped around its crusty edges with a fingernail, then maybe popped the whole thing off, before it was "ready."

Taking it a bit further, imagine the small satisfaction you get from squeezing a pimple. Those of you who can legitimately claim never to have done this may read the remainder of this paragraph in disgust. But the rest of us know the drill. Feeling along the smooth skin . . . finding a bump, and then—against all advice—squeezing . . . and squeezing . . . feeling the resistance, a prick of pain, until finally it pops and releases. Sometimes there's blood. Once in a while we may even squeeze it again (very much against the dermatologist's orders) and push out still more blood.

Release . . . followed by *relief.* We've all felt it, and even if scabs, pimples, or ingrown hairs aren't your thing, maybe you've bitten your cuticles or scratched your scalp or excavated your nostril a little too hard.

---

*Besides hair twirling and nail biting, many of us chew gum when we're stressed out. Zoobiquitously, some nonhuman primates in the wild pick gum arabic (the springy, saplike base of natural chewing gums) out of trees and chew it. Zoo behaviorists sometimes give their primates this substance as a way of combating stereotypies. Some nonnutritive chewing has, indeed, been shown to have a calming effect, at least depending on which teeth you use (one group of dentists claims that chewing with the rear molars is more relaxing, while using the front teeth or canines perks you up).

In fact, human beings rely on this *release-relief* loop throughout the day. Stroking our hair, picking our toenails, chewing the inside of our cheek—these are powerful self-soothers. We may rub, pull, nibble, or squeeze a little more when we're stressed, but for most of us the behavior never escalates. Folded into our daily lives, these activities help us maintain an activated yet calm state. But for some people the need for that feeling of *release . . . and relief* is so strong that they crave extreme levels of it.

Release . . . and relief—those are the same reasons cutters give for why they cut. The same intensity and promise of sudden relief we might get from pulling a single strand of hair or picking a pimple, dialed *way, way* up, leads cutters to carve lines in their skin with razor blades. If we accept that this behavior is on the same spectrum as less destructive forms of grooming, as my veterinarian colleagues would suggest, then self-mutilation is truly grooming gone wild.

In fact, the addition of genuine pain may even bolster grooming's positive biochemical effects. It turns out that both pain *and* grooming cause the body to release endorphins—those same natural opiates that give marathoners their runner's high. Pain also causes the body to produce catecholamines, which over time damage major organs but in the short term give the body a jolt—spiking the blood sugar, dilating the pupils, and increasing the heart rate. So in a way, self-mutilators are self-medicators, kick-starting their bodies' natural and powerful chemical reactions. Some cutters report a trancelike state combined with an overwhelming need to self-mutilate—not unlike an addict jonesing for a drug, a jogger restlessly anticipating her 5K, or a glassy-eyed German shepherd licking his paw.

As a cardiologist, I was extremely interested to learn that, beyond altering blood chemicals, pain that is self-inflicted can sometimes affect the heart itself. Researchers in Massachusetts outfitted a group of rhesus monkeys known to be self-biters with tiny vests housing heart-rate monitors the scientists could check with a remote control. They found that when the monkeys naturally nibbled at their unfamiliar new ensembles, their hearts showed no significant spike or drop. But when the monkeys bit themselves, their heart rates were markedly elevated for thirty seconds before the behavior, then plunged dramatically the instant their teeth hit fur. A precipitous drop in heart rate—especially one that comes suddenly after it's been elevated by thrill or fear—can create the feeling of calm. Like the self-biting rhesus monkeys, cutters (half-fearfully, half-excitedly) anticipating the moment when the blade hits their skin may be experiencing a mild tachycardia (increased heart rate), followed by a sudden, calming drop once the skin is broken and the blood flows.

So one reason people and animals self-injure may be biochemical: they are caught in a neurotransmitter-based feedback loop in which their bodies reward them with calmness and good feelings after they do something that causes pain. And their hearts may be amplifying the feeling by slowing drastically right after racing with excitement.

What's interesting is how these two opposites—pleasure and pain, grooming and disfiguring—produce similar results in the body. So similar that some people's bodies seem to have mixed them up. Picking, poking, and chewing—which sometimes end up hurting us—stay in the gene pool because they're on the same spectrum as grooming, which calms us down, keeps the peace, maintains our health, and binds our anxiety. But that still leaves us with a problem: whether it's on a normal spectrum or not, self-injury in humans and animals is aberrant, dangerous, and needs to be controlled. Not only is it a sign of psychic distress, but it can cause serious health consequences, starting with nasty infections and ending with death.

It's here that veterinary medicine may offer new insights, or at least new paths, for human doctors to explore. Traditionally, psychiatrists have tried to understand self-harm through a checklist of personality disorders and evidence of past traumas. We might start by looking for a history of sexual abuse or features of borderline personality disorder. But our veterinary colleagues have a more direct approach. Lacking the ability to talk to their patients (and perhaps aided by this as well), they have identified the three most common triggers of self-injury: stress, isolation, and boredom.*

---

*Before addressing these common causes, vets rule out underlying medical conditions. Psychiatrists do this, too, when a new symptom presents. For example, when a patient has a new presentation of depression, the physician may consider hypothyroidism, Cushing's syndrome, or even pancreatic cancer. Similarly, when an animal (human or otherwise) presents with self-injury, organic causes, including physical pain, must be excluded first.

Call a veterinarian to treat a flank biter and she may inquire about the patient's upbringing. (In a canine parallel, a puppyhood spent in a shelter is now recognized as a contributor to disturbed behavior in adult dogs.) Having ruled out a traumatic early "foalhood" and physical causes (say, a twisted bowel or torn ligament), she will then look for acute stress, isolation, and boredom.*

To gauge stress, the veterinarian may investigate the animal's social situation and environment. Is there a bully in the stable? Human or equine? Stress from feeling uncertain or insecure in its environment can lead an animal to self-injure.

Isolation can also provoke it. Simply providing companionship is one fix that veterinarians try. Birds—even ones that seem to want to be alone, that attack and drive out cage mates—have stopped injuring themselves when their enclosures were moved closer to those of other birds. And self-harm plummets in many species of monkeys and apes when they are caged with a companion of the same species. Many stallions stop hurting themselves when they're allowed to graze in the company of mares—their natural grouping—instead of being housed in a stall alone. Animals of many kinds, from cheetahs to racehorses, are sometimes paired with other species, like donkeys, goats, chickens, or rabbits. Part of the reason this works seems to involve the larger animal's fear of stepping on the smaller ones . . . as if a sense of purpose itself diminishes the need to self-harm.

Boredom really rings alarm bells for vets. Free-ranging horses, for example, graze many hours a day. But when a stable hand straps on a feedbag and sates a horse's hunger with an easy hit of tasty, calorie-dense grains, the animal is left with an overfull stomach and time for its idle hooves and teeth to fill.

Boredom is such a risk factor for stereotypies that animal behaviorists in zoos have developed a whole science around it. As mentioned earlier, environmental enrichment creates psychological and physical well-being in animals by fostering behaviors naturally present in the wild. Zoo-

---

*Most reports of animal self-harm come from captive populations, and in some circumstances, captivity itself may exacerbate the triggers. But captive settings aren't the only places animals experience stress, isolation, and boredom. Comparable behaviors may well occur in the wild, too—but given the difficulties and limitations of observing free-ranging animals, wild versions are probably underreported.

keepers excite carnivores with frozen balls of blood and the scent of their favorite prey. Enrichment can be as simple as a new dirt mound to explore; logs, feathers, and pinecones to play with; and different sounds to hear.*

When vets notice animals engaging in stereotypies, they increase or vary the environmental enrichment. When the coyote handler at the Phoenix Zoo watched two coyotes pacing the same pathway in a tight-limbed, ears-back gait, she gave them frozen blood popsicles to play with, hung pigeon wings on branches to encourage them to jump, spread giraffe and zebra urine around bushes to entice them away from their pathway, and filled burlap tubes with peanut butter to get them to work for their treat. After a few weeks, the coyotes were trotting calmly with upright ears.

Trainers give horses a variety of toys to play with, but the most surefire solution for preventing this confirmed herd animal from getting bored and stressed is to . . . give it a herd. After all, horses evolved to live in groups. They generally don't even sleep well unless one of their own stays awake as a sentry. It's no wonder solo living can be stressful for them.

And here's where recognizing our deep connection with other animals may shed some light on the issue of human self-injury, both by putting what we already know into a new context and by suggesting innovative ways to treat the problem. It takes us to a story of a gorilla, some gum, and some nail polish.

Several years ago, a handful of veterinarians wearing face masks and scrubs hunched over a mountainous male gorilla in the gleaming white treatment room of the Birmingham Zoo. Babec was suffering from congestive heart failure, a condition I treat in humans almost every day. It leaves apes of both species weak and lethargic. In the most severe human cases of it, patients feel short of breath and exhausted doing even the simplest activities: walking from the bed to the bathroom, putting on their clothes, sometimes even just talking. The sickest human heart-failure

---

*In 1985, the USDA laid out six elements it deems critical to the psychological well-being of captive animals: social grouping; structure and substrate (meaning the environment of the cage and its flooring, bedding, perches, etc.); foraging opportunities; toys or manipulables; stimulation of all five senses; and training.

patients lose their appetites and drop muscle and weight. Babec, too, had slowed his eating; at 320 pounds, he was a shadow of his former 400-pound self. The ailing gorilla was about to be fitted with a high-tech pacemaker—the same kind that's put into human patients suffering from the most advanced cases of heart failure.

While the technicians anesthetized and intubated Babec, the doctors scrubbed their hands, swabbed disinfectant on his chest, and shaved a large rectangle of silvery-black fur over his heart. Under anesthesia for medical procedures, gorillas can seem uncannily human. Their leathery palms, whorled with familiar-looking fingerprints, relax open at their sides. Their scary bulk and prominent brow ridges, which can look so intimidating when the animal is awake, seem vulnerable, pensive, and even wise when they're under anesthesia.

The doctors made a careful incision with a sterile scalpel and got to work installing the pacemaker. The six-hour operation went well. They closed the wound, bandaged it, and cleared the room so the vet techs could get Babec ready to wake up.

But a couple of things happened during the operation that would send the charge nurse in a human OR into hysterics. In the middle of the procedure, an assistant gave Babec a manicure, painting his normally dark-ish nails Ferrari red. Down by his feet, another zoo staffer shaved little patches of fur from his legs and sewed loose "decoy" stitches into skin the doctors hadn't even approached with the scalpel. Meanwhile, several of the vets did something that is strictly forbidden in a human operating room. Behind their masks, their jaws wrestled large wads of gum. And every so often, they sneaked a marble-sized ball out of their mouths and, inexplicably, worked it into Babec's fur.

The attending veterinarian later explained to me that these human health-code violations were, in fact, clever patient-care strategies. Specifically, they were designed to defend the delicate stitches holding together the true incision in Babec's chest, which, if left unguarded, Babec would pull out in a matter of minutes. But how to protect it? My human patients can generally be cajoled into avoiding the urge to fiddle with their sutures, at least for the thirty-six hours it takes for scar tissue to emerge. But all the lectures in the world won't stop a gorilla from probing at the wound.

So the vets developed an ingenious subterfuge. They would protect the stitches by distracting the patient. And they would do it by harnessing the same instinctive urge that propels the gorilla to pick in the first place: the impulse to groom.

Babec's vets told me he awoke from his surgery the way my human patients often do: groggy, disoriented, and uncomfortable. Peering around the recovery area, he started to move his hand toward his chest, with its new incision, then froze with it in midair. The Ferrari-red fingernails gleamed like hard candies. They held his attention for a good few minutes. When he moved his hand back toward his chest, he didn't get far before his fingers found a wad of gum. He picked and pinched and pulled at the offending material and had only just finished extracting it when his fingers touched another (the veterinarians had heat-treated it after chewing, to kill germs). The fake stitches in his ankles would be next. Every time he finished with one task, another was waiting to grab his attention—distracting him from the most important thing: his chest sutures.

This is a place where human medicine and animal medicine are already converging, although without either side realizing it. Some therapists counsel self-injurers to try a less invasive, distracting "hit" of pain when they get the urge to cut, burn, or bruise. Plunging a finger into a carton of ice cream, squeezing a piece of ice, or snapping a rubber band around the wrist sometimes does the trick. Cutters who crave the bloom of fresh blood can draw a red marking pen instead of a blade across the places they'd like to cut. They can drag ice cubes made with red food coloring over the skin to produce a satisfying crimson trickle. Or they can swipe their fleshy canvas of choice with henna paint (this has the added advantage of drying to a pleasing scablike consistency, which can be picked off the next day). These distractions all deliver the *release . . . relief* response, just in safer ways.

But the vets also point out that animals need both immediate physical distractions and more long-term social changes—in other words, they need solutions to their stress, isolation, and boredom. And when you think about it, that might go for people, too. Young adults in the age of our distant ancestors didn't have anything like the spare time and painless abundance of modern America. The typical middle-class teen is a little like the horse alone in its stall, with most of its needs—especially food but even entertainment and physical activity—provided in easy-to-digest chunks. He's left with lots of extra time and few activities as invigorating as a daily struggle for survival.

The problem may be worsened by technology that isolates even as it entertains and informs. Even those of us who love these activities recognize that watching television, playing video games, and "social" networking alone in a room can leave us feeling disconnected from real people. A survey comparing free-time activities and contentment found that the only pastime that consistently left people of all ages and socioeconomic groups feeling unhappy was watching TV. While bird owners and others with common problems can find solace in online gatherings of people with the same issues, this phenomenon also has a dark side. The Internet provides cutters (and those in other self-injuring subcultures, including anorexics) with the *wrong* kinds of peer groups—ones that enable and support the behavior, offer tips for "improving technique," post poetry praising it, and describe tactics for hiding it.

Zookeepers make animals forage. Should we explore getting teens involved in growing and preparing their own foods, an activity that can produce feelings of profound calmness and satisfaction . . . and purpose? Just as an animal's stereotypies decrease when it has a companion, a pet can provide company, responsibility, exercise, and distraction to a human. Like a lonely horse reintroduced to a herd, isolated cutters could be encouraged to find herds of their own. Whether in more mainstream pursuits (sports, theater, music, volunteering) or more niche passions (medieval reenactment, making YouTube videos, competitive Scrabble), the company of other flesh-and-blood human beings depending on one another can bring a deep sense of belonging.

Psychotherapy, the traditional (and often very effective) treatment for extreme self-harm, may actually combine the two approaches vets use for self-injuring animals. Supportive counseling gives a cutter the beginnings of a herd: a person to talk to, sit near, and be responsible to (by showing up for appointments). Psychotherapy can also be viewed as a form of social grooming, calming and "touching" another person through voice, language, response, and presence. Actual touch and massage therapy with literal physical contact (repetitive tactile stimulation) between healer and patient might also be useful ways to *release . . . and relieve* feelings of isolation and stress.

But zoobiquity also raises a deeper question about self-injuring behavior in humans. If someone is burning himself with cigarettes, we certainly need to find a way to stop him. But can we and should we accept or tolerate less extreme forms? In fact, we already do.

The recent rise in self-injury coincides with the popularization of a form of controlled bodily injury. While examining my patient who showed me her cutting scars, I couldn't help but remark on her nearly head-to-toe collection of tattoos. She told me she'd gotten most of them during a five-year hiatus from cutting. Now she was doing both. "I think the reason I've been doing so much more tattooing is that I really want to cut," she told me. "People say a tattoo doesn't hurt. But it does."

Needless to say, tattooing is not the same as self-injury. It's an ancient and, in many places, sacred cultural art form. But it *is* a kind of grooming that bears many similarities to the practices of our primate cousins. It's an intimate interaction between two individuals. It often confers social status. The pain involved in getting a tattoo releases endorphins.

The zoobiquitous notion that self-injury is a form of grooming gone wild opens up a whole new way of looking at our society's increasingly painful and invasive preening rituals. We subject ourselves to full-body waxing, genital bleaching, acid peels, repeated electrolysis, cuticle shaving, adult orthodontia, ultraviolet tooth whitening, laser microplaning, and Hollywood's injection of choice, Botox.

Whether their tool is a tattooist's ink gun, a plastic surgeon's needle, a cutter's razor blade, or their own talons and beaks, sometimes humans and animals simply cross a line. And when this line is crossed, healthy self-care may shift toward significant self-harm. We may not be able to define exactly where that line begins, but we can easily spot when someone has crossed it.

All of us—from full-blown cutters to secret hair pluckers and nail biters—share our grooming compulsions with animals. Grooming represents a hardwired drive, one that's evolved over millions of years with the positive benefits of keeping us clean and binding us socially.

Parents, peers, physicians, and vets should take notice when stress, isolation, and boredom appear. Combating these triggers with the fellowship of a theater group, the primal satisfaction of backyard gardening, or the grooming challenge of a carefully placed wad of gum does more than create a distraction. It uses an evolutionary tool set to repair an evolutionary short circuit.

NINE

Fear of Feeding

Eating Disorders in the Animal Kingdom

The eating disorders unit of a psychiatric hospital takes on a charged atmosphere around 6 p.m. every day, when leaf-thin patients float anxiously into the dining room. Many are draped in a concealing uniform of baggy sweatpants and oversized shirts with sleeves so long only their fingertips peek out. They glance around warily, eyeing one another and slyly sniffing the air to predict what food they will be challenged to swallow. The meals have been calibrated down to the last calorie and garnished to entice the most reluctant eater. Kind but guarded, the nurses, doctors, and ward assistants (including the janitors) are on high alert for food avoiders, food hiders, and food purgers. Sometimes they lock the bathroom doors before serving time to make sure no one slips away for a mid-meal regurgitation.

As a psychiatry resident in the late 1980s, I spent six months rotating through the eating disorders unit at the UCLA Neuropsychiatric Institute. I remember a meal with one particular patient, a fourteen-year-old I'll call Amber. Pale and gaunt, she sat next to me at the round faux-wood table, fixated on the green plastic plate in front of her. On it sat a simple turkey sandwich and a red apple. She stared at the meal. And she stared.

Finally she looked up at me. I was surprised to see something like terror in her eyes. "I can't do it," she whispered. "I just can't. I am scared to eat this food."

Scared to eat. I remember thinking to myself how disordered that seemed. How unnatural. Even before I had become used to taking a comparative approach to human medical conditions, I thought, here is a mental disorder that is completely antithetical to the principles of evolution. In the wild, animals that starved themselves on purpose would be on a collision course with extinction.

And yet this form of self-starvation, called anorexia nervosa, strikes 1 in 200 American women. It's surprisingly lethal. Killing up to 10 percent of the afflicted, anorexia is considered to be the deadliest psychiatric disorder among young females. Bulimia nervosa, the well known binge-purge disorder, affects some 1 to 1.5 percent of women at some point in their lives and .5 percent of men. And there's also a rapidly expanding category of varied eating disorders that get lumped together in a broad diagnostic category called, simply, "disordered eating." It includes troublesome behaviors such as binge eating, night eating, secret eating, and food hoarding.

Eating disorders are often dismissed as being mild, even trivial, an affliction of the wealthy and privileged. However, because of their worldwide prevalence, the World Health Organization has declared them a priority disorder. And as W. Stewart Agras, a Stanford psychiatrist, points out in *The Oxford Handbook of Eating Disorders,* eating disorders of all kinds are on the rise around the world.

In the two decades since I treated Amber, psychiatrists have learned much about who is at risk for eating disorders and what makes them susceptible. Hormonal states and brain chemistries contribute. Because disordered eating can run in families, genetic factors are believed to play a strong role. Certain personality types are especially vulnerable. Sufferers tend to be fearful and anxious, specifically about gaining weight and being fat. Anxiety disorders are frequently diagnosed along with anorexia nervosa. Some anorexics admit to being perfectionists or wanting to punish themselves. Many say they're addicted—either to food or to the euphoric feeling they derive from starving. They report enjoying exerting control over food and figure, and watching the effect their condition has on people around them. Psychiatric explanations have also pointed to early childhood experiences and family dynamics as the cause or trigger.

Eating disorders are complex and subtle, and seemingly very human. As far as we know, other species don't share our concerns with body image and self-worth—preoccupations that fuel human patients' dangerous disordered eating. And a patient's troubled social relationships and obsessions about getting fat certainly seem to be fixed in a very human matrix of culture and social pressures, media messages and memes.

However, closer inspection of veterinary data yields some surprising, overlapping eating behaviors across species. In animals, binge eating, secret eating, night eating, and food hoarding are common. Anorexia nervosa and bulimia nervosa (or close homologues of those extremes) may indeed be present in certain animals under certain stressful circumstances. And while the animal and human "psychology" of these disorders differ, neurobiologically there may be parallels. A zoobiquitous approach made me see that, like Amber, animals may also sometimes be scared to eat. In fact, for many, both wild and domestic, every meal may feel as risky as Amber's sandwich did to her.

To understand what I mean, we have to put two very different fields of inquiry side by side. First: contemporary psychiatry and the bewildering, ill-defined, yet growing diagnosis of disordered eating. Next to it: wildlife biology and the whimsy and mishap of animals' quest for daily bread.

A typical morning in the wilds of Yellowstone National Park might look something like this: A chipmunk pokes a whiskered nose out of her burrow and scurries over to a scattering of pinecones. With a blur of forepaws, she nibbles at several. Alone, furtively, she stuffs her cheeks, sails back to her den, and stashes the hoard in her secret underground hiding place. Then, although she has just eaten, she heads back out and approaches the food again. Ears pricked, eyes wide, she pauses. Scans her surroundings. Ignores a rustle in the leaves. She eyes the nuts. Suddenly: another crackle. This sound is different—she flees for her den, but too late. *Pounce! Swipe!* A bobcat bites down on the chipmunk's neck and carries her limp body away.

In the tall grass nearby, unnoticed by the bobcat, crouches a silent hare—his heart revved and muscles taut for a now-unnecessary sprint to shelter. He remains frozen for a few more moments until, sensing safety, he can resume feeding. But the patch he had been heading toward when he smelled the cat is now out of the question, too dangerous even though it's only a few hops out of reach. Some tufts of lupine, while less nutritious, will suffice. Jaws grinding, heart racing, the hare stuffs his mouth with the easy and available vegetation.

Deep in the grass patch abandoned by the hare, a grasshopper freezes. It senses, but can't see, a hungry spider nearby. Abruptly, the insect stops munching the protein-rich grass. Moving cautiously, it sidles over to a new plant—sugar-laden goldenrod. Its mandibles start churning rapidly on the sticky yellow flowers.

In an aspen grove on the banks of a river, elk browse in seeming calm, their twitching ears and flaring nostrils the only giveaway that stress hormones are pumping through their veins as they monitor the silent forms of a wolf pack stalking their fawns.

Under the river's turbulent water, a juvenile cutthroat trout hides in the cleft of a rock. In the current around her drift mayfly nymphs, midges, and other nutritious morsels. But, too young and inexperienced to take on a predator in the open water, the cautious young fish stays frozen where she is, camouflaged and protected—sacrificing food for safety.

Hawks and eagles, their stomachs awash in hunger hormones, patrol overhead. Their quarry—the vigilant denizens of the gold-and-green scrub, from sagebrush lizards, partridges, and bull snakes to gophers, deer mice, and skunks—all constantly weigh a risky tradeoff. Nibble food in view of the voracious air force above—or remain hidden and go hungry?

As the sun starts to descend, the animals become even more watchful. Some, driven by hunger, are desperate to kill their calories before dark. Others shovel in food selected from their own larders or pilfered from a neighbor's. Some animals awake with the setting sun and begin the dangerous proposition of finding food by moonlight.

One thing you can say about meals in the wild: they're never boring. Every bite requires a life-or-death focus on two things: getting food and avoiding becoming food. If an animal cannot find and secure consistent meals, he will die of starvation. If he's not vigilant, he will fall to predation. In nature, eating is drenched with danger, risk taking, stress, and fear.

But what if, instead of observing animals in Yellowstone, we were peering into darkened kitchens and dining rooms, past closed office doors and tinted car windows? What if those were human animals who were scurrying, sprinting and storing, hiding and nibbling? People who were spending entire mornings in the pursuit of food, obsessing about it, changing their behavior to get or avoid it? Then the scenario might seem quite different. In fact, if exhibited by a twenty-first-century human, many of those behaviors would trouble my psychiatrist colleagues.

Today we no longer view ourselves as cowering prey. After all, we're the most fearsome predators in the history of our planet. In civilized comfort at the top of the food chain, most of us will go through our whole lives never facing a realistic threat from a nonhuman predator. We can be grateful for that, but it obscures the fact that our DNA has a much longer memory.

In the not-too-distant past, we faced, on a daily basis, the very real threat of becoming someone's lunch. Our genetic legacy of survival has depended on the crucial instincts our forebears developed through millions of years of evolution—the instincts that kept them alive and out of the colons of other creatures. Nowadays Volkswagen-sized eagles aren't poised to drop on us as we exit Starbucks, soy latte in hand. But our back-stabbing office politics, our violence-drenched entertainment, even the very process of growing up can trigger physiological reactions as potent as those handed down from our animal ancestors stalked by ravenous carnivores.

We have one obvious thing in common with other animals and our own animal ancestors: we all have to eat. And echoes of the eating strategies of our animal forebears—*guided by fear, anxiety, and stress*—may remain with us, even today, in ancient, inherited eating neurocircuitry and behaviors. It means there may be a "disordered" animal eater lurking in every one of us.

My daily meetings with Amber took place in various locations inside the treatment unit or somewhere on UCLA's campus: on a bench, under a tree. We delved into her childhood memories (such as they were; she was just fourteen, after all); her thoughts; and her visualizations of the future—all to get to the psychodynamic core of why she was afraid to eat.

Ecologists who study animals are, for obvious reasons, unable to engage in such dialogue. Like psychotherapists, they would never strive to understand a single animal's eating behavior in isolation from the world around it. In fact, scientists who study animal eating know that a lot of an animal's behavior around food depends on factors completely out of its control. Weather, food supply, position in the pecking order, and social hierarchy—all these can mean the difference between a belly that's full and one that's empty. And in the wild, one of the biggest diet determiners is the presence of predators. Biologists call it "the ecology of fear."

To study this idea, scientists at Yale built mesh-and-fiberglass field cages over areas of a meadow containing wild grasshoppers and their primary food source, naturally growing vegetation. In some enclosures, the grasshoppers were left to eat in peace. Those insects mostly munched on protein-rich grasses. But another group of grasshoppers was enclosed with an unpleasant surprise: predatory spiders. To protect the grasshoppers, the spiders had had their mouthparts glued shut.

The presence of the arachnids had a significant and surprising effect. Forced to share space with their mortal enemies, the grasshoppers all but gave up eating grass. But they didn't stop consuming food altogether. They chose instead to eat goldenrod, a sugary, carbohydrate-laden flowering plant, instead. The same preference for sugar over protein was seen when the experiment was repeated and the grasshoppers' choice was between a high-sugar cookie and a protein-rich bar. This means something very interesting, says Dror Hawlena, the ecologist who devised the experiment. When stressed out by the spiders, the grasshoppers binged on sugar and carbs.

The threat of predation speeds up the metabolism of a variety of species, preparing them to react to the danger. Revving the engine burns fuel in the blood and muscles. To keep the engine accelerating, they need quick fuel. Simple sugars and carbohydrates fit the bill. Their chemical bonds come apart more readily than the long-chain fatty acids of leafy greens or the complex molecules of proteins, so they don't require a lot of processing in the gut. The body can utilize their energy quickly.*

---

*As Hawlena explained to me, proteins are also rich in nitrogen, much of which animals must excrete to avoid toxicity. Stressed-out grasshoppers and other animals may avoid protein because the energy required to process nitrogen could be better spent on more urgent activities, such as escape.

Psychiatrists studying eating disorders note that bulimic binge eaters rarely overconsume protein or leafy greens. Like the grasshopper, they focus their eating sprees—sometimes with obsessive intensity—on sugars and simple carbohydrates. (Stress eaters who don't later compensate by vomiting or using laxatives sometimes kick this specific sugar-and-carb focus.)

In the Yale study, the insects' food choices were driven by external factors beyond their control, in other words, the ecology of fear. In the presence of a predatory threat, they chose foods that would accelerate a lifesaving escape. These animals provide an underexplored possible context for a human binge eater's food choices. They suggest an evolutionary origin. A stressed person's decision to forgo the chicken breast and vegetables in his lunch box and consume candy bars instead can seem pointless, weak, and even self-destructive. But knowing that some nonhuman animals prefer high-sugar foods when they're fearful could help a human stress eater better understand his own candy binge. While he knows it's unhealthy for his waistline, blood sugar, and molars, the impossible-to-resist impulse may spring from a hardwired response to threats that for eons has saved animals' lives.

Of course, a college undergraduate shoveling candy into her mouth late at night during finals week or an executive loading up on a sleeve of cookies before a business trip are separated from the grasshopper by genetics, brains, cultures, and, certainly, self-awareness. But as animals, they may share physiologic strategies for coping with stress, one of which could be an attraction to getaway-fueling simple sugars during times of stress.

And the ecology of fear influences not only *what* an animal chooses to eat. It also determines *when*. Light-dark cycles affect animals' sense of safety. In some animals light may inhibit eating; in others it may enhance it. In a study of gerbils, for example, researchers found that when nights were dark, the rodents ate substantially more. On bright nights illuminated by a full moon, making them more visible to predators, they ate less. Another study, on rodents known as Darwin's leaf-eared mice, found that shining a light on their cages was enough to make the mice cut their foraging time in half. They ate nearly 15 percent less than usual and, consequently, lost weight. Scorpions have shown a similar aversion to bright nights. The bigger the moon, the less they foraged. It's known that light therapy can reduce food cravings and overeating in some human bulimics. Animal examples could provide a context for these effects. Perhaps this is the evolutionary basis underlying the folk wisdom that a late-night urge to raid the refrigerator can be squelched by flooding the kitchen with bright light.

The ecology of fear can change an animal's entire approach to feeding—even beyond what and when it eats. In his book about mountain lions, *The Beast in the Garden*, the science journalist David Baron tells an interesting story. Starting around the mid-twentieth century, mule deer around Boulder, Colorado, began behaving oddly. Instead of cautiously venturing out of hiding to feed at dawn and dusk as they used to do, the deer began eating, lounging, and even giving birth in full daylight on the lush, landscaped lawns of Boulder's neighborhoods. This lackadaisical behavior coincided with unusually low numbers of predators in the surrounding region—wolves had been hunted to near extinction in the century before, and the mountain lion populations had been decimated. Baron notes that "with its large carnivores gone, Boulder's herbivores flourished."

Around the same time, something similar was happening in Yellowstone. For fifty years, the landscape had been completely devoid of a fearsome predator: wolves. This had an intriguing and measurable effect on Yellowstone's elk. They relaxed. They began grazing deep down in ravines, near streams and in open meadows, far from tree cover. The elk would never have dared to enter these risky, hard-to-escape-from locations when wolves were near. But, unhindered by the fear of a sudden attack, they were able to browse much longer at a stretch, and they discovered a taste for new menu items. They consumed the leaves of cottonwood and willow in addition to their usual grasses. They ate more than usual. They grew fatter. They had more babies.

However, all that changed in 1995. That's when twenty wolves were released into carefully selected sites in Yellowstone by the National Park Service and the U.S. Fish and Wildlife Service. The wolves' presence affected the elk almost immediately. The animals became more vigilant. Repeatedly lifting their heads to scan their surroundings stole crucial time away from browsing. They changed eating locations, preferring to graze in sheltered forests rather than in open meadows—a pattern elk follow when stalked by human hunters as well.

Nowadays, about one hundred wolves patrol Yellowstone, keeping the elk on edge. The impact of fear has restored their eating to the cautious and restricted pattern so common in the wild. Ecologists have identified other animals around the world who eat less, restrict food choices, and put off eating in response to predator intimidation. Manatee-like dugongs, for example, sacrifice grazing opportunities in the underwater seagrass meadows of Australia's Shark Bay when tiger sharks are prowling nearby. Snails in southern New England tide pools browse less on acorn barnacles and algae when they sense predatory green crabs in the vicinity. Impala and wildebeest increase their vigilance when ravenous lions and cheetahs are lurking in the neighborhood.

What's clear is that when intimidation goes up, animals may restrict where, when, and what they eat. And when threat is reduced, eating behaviors may relax. Fear's ancient connection to feeding could allow physicians to understand eating disorders in a whole new way. What ecologists call "encounter avoidance" and "enhanced vigilance" in wary animals might have psychiatric overlaps with "social phobia" and "perfectionism" in human patients.

Intimidation and fear take many forms in the wild. Usually, they involve claws, fangs, talons, and teeth. But there's one threat that stalks living things without the use of a weapon or even a body. Although no one would say that animals consciously worry about it, starvation is another constant menace in the wild.

A modern supermarket is about as far from Yellowstone's wild eating landscape as you can get. Straight aisles, stocked shelves, temperature-controlled air. I'd never thought much about animal food-storage habits, beyond squirrels pushing nuts into the ground, woodpeckers creating "granaries" in old trees, and bees buzzing around, making communal honey. But the drive behind these behaviors connects to one of the most ominous feeding fears in the wild: starvation.

Animal larders are concealed everywhere we turn, from treetops to roots, branches to grasses, rocks, shrubs, fence posts, and eaves. They're more plentiful and elaborate than I'd imagined, containing not just seeds and nuts but such other delicacies as twigs, lichens, mushrooms, carcasses, nectar, and pollen.

Some moles create worm farms in the walls of their burrows to keep their stash fresh and ready to eat. When they catch a worm, they bite off its head and store the body under the cool soil in special areas of their tunnels. Since moles will continue to injure and stash worms as long as they're available, these so-called fortresses can grow quite large. One I read about weighed more than four pounds, was a yard and a half long, and contained more than a thousand earthworms and grubs. In a lovely bit of recycling, some lucky worms get a reprieve: if they can avoid being eaten until they're regrown their lost heads, they can make an escape, especially in the spring, when the soil warms up.

During nighttime feeding runs, mountain beavers in the Pacific Northwest snip off bits of ferns and other greens and stack them in small bundles under logs and on rocks. They also hang them over the low branches of trees and shrubs. The beavers later move these piles of wilted greens into special chilled storage chambers near their nests and feed out of these "mini-fridges" all year. The moisture-laden vegetation is quick to mold, so every week or so the beavers have to take stock and replace the stores. You might go through your crisper from time to time, too, and throw out the liquefying romaine.

And lest you think food caching is limited to vegetarians or rodents, consider that birds of prey are well known to "overkill" and store food. An American kestrel was once seen slaughtering seven mice and storing their bodies in two adjacent grass clumps. A screech owl once found an empty shelf in a barn and stashed twenty-two dead day-old chicks on it. Bears, foxes, and mountain lions conceal animal carcasses under leaves and dirt, for later meals. Spiders routinely kill more insects than they can eat, package them up in take-out containers of spider silk, and go back later to consume them. Jackals were spotted returning at night to a mud pit to retrieve stored shreds of meat they'd submerged earlier in the day.

Eating alone, in the safe privacy of a personal larder, minimizes the time an animal spends in risky, high-predation situations. And a hoarder has extra energy and time for courtship and mating. But it's hoarding's antistarvation advantages that confer the real payoff.

With provisions to insure them against future famines, animals that hoard have a safety net to protect them from dangerous periods of food shortage. Hoarding behavior literally makes animals safe. Safety and food hoarding connect in humans, too, whether it's a prudent stash of dried beans and powdered milk in an emergency-preparedness kit, a pantry pleasingly stacked with tuna cans, or a freezer stocked with chicken breasts.

But psychiatrists recognize, in some hoarding behaviors, signs of underlying disturbance. For example, food hoarding is often observed in foster children with severe attachment disorders, those whose early sense of safety has been fractured. Even the hoarding of nonfood items connects again to the ecology of fear. Human hoarders' stacks of magazines, plastic bags, and receipts make them feel safe. Parting with these treasured items causes them distress, fear, and anxiety.

Compulsive hoarding—of food, objects, and even pets—is currently believed to be a type of OCD. OCD is linked to several other psychiatric disorders, including anxiety and eating disorders. Clinicians know that the majority of patients with anorexia nervosa suffer from anxiety disorders, including OCD and social phobia. The connection between fear and feeding spans species: from human beings to anxious elk, stressed grasshoppers to cautious gerbils. The ecology of fear also underlies another animal syndrome, one with intriguing parallels to human patients.

It's ironic, but answers for suffering anorexic patients may be hiding in one of the last places they might think or want to look: a pig farm. Under socially stressful conditions, some female pigs may self-restrict their food intake, even as the herd mates around them are eating normally. They continue losing weight until they're emaciated. You can easily spot them by their prominent backbone ridges. Like human anorexics whose hair becomes brittle and patchy, pigs afflicted with a condition called thin sow syndrome grow hair that is abnormally coarse and long. Women who are anorexic often stop having periods (in fact, that's technically part of the definition of anorexia nervosa). Thin sows stop going into heat. Members of both species can go on to starve themselves to death.

The similarities go beyond the physiological. In their article "Intriguing Links Between Animal Behavior and Anorexia Nervosa," psychiatrist Janet Treasure and agriculture professor John Owen explain that while "the affected animals restrict their intake of normal food . . . some consume large amounts of straw." That's similar to an old trick of human anorexics, too: eschewing food that's dense in nutrition (and calories) in favor of stomach-stuffing, "negative"-calorie fillers like lettuce and celery. Even more interesting is something Treasure and Owen learned from observing pigs on farms throughout Europe. Like food-deprived rats running nowhere on their wheels and human anorexics putting in yet another hour on the treadmill, pigs with thin sow syndrome are remarkably restless. As Treasure and Owen note, having observed the animals at Greece's largest pig farm, where a full 30 percent of the females were affected, thin sows "spend more time on nonnutritive hyperactive behavior . . . they move incessantly around their pen."

Trying to figure out why and when certain pigs are at greater risk for developing thin sow syndrome has led to a hunt for the gene sequences that underlie it. And the search has turned up an intriguing culprit. In recent decades, consumer taste has veered away from fatty cuts of meat. Human pork eaters want their chops and loins lean. Even bacon has gotten skinnier. Livestock farmers have responded to the demand by breeding leaner pigs. And that's where a problem seems to crop up. Treasure and Owen report that "pigs, especially those that are bred for extreme leanness, can develop irreversible self-starving and emaciation."

What's happened is that selective breeding for leanness has "led to the uncovering of recessive traits which produces extremes." That these traits came to the fore in pigs in a matter of generations leads Treasure and Owen to suspect that anorexia nervosa has "an analogous genetic basis"—in pigs and humans . . . and other animals, too.* This indicates that gene sequences that code for thinness may exist in many animals, although they remain in the background, essentially unactivated, in wild populations, where breeding is not as controlled.

When we look at humans, we see something similar. Studies of twins and generations of families show that heritability for anorexia nervosa is extremely high. Searching for an "anorexia gene" has led inevitably to questions about why it arises in the first place. Evolutionary psychologists have posited several varied theories to explain why anorexia nervosa may have been selected for in our human ancestors. Their hypotheses have included adaptation to famine, social hierarchy effects, and male preference for certain body types (both heavier and thinner).

---

*The fashion for leaner meat is not limited to pork. Metabolic oddities, like double-muscling in cattle, have cropped up in other farm animals as a result of breeding that selects against fat.

What seems much more likely, says Michael Strober, a UCLA professor of psychiatry and biobehavioral sciences and the editor in chief of the *International Journal of Eating Disorders*, is that the gene sequences that anorexics are passing down the family tree are linked to anxiety. Anxiety, high stress, and fear responses are the main features Strober sees in human anorexics and other people with eating disorders every day in his office at UCLA. "People with anorexia nervosa are nervous when they confront change or any novelty in their environment," he told me.

And change is stressful for thin sows, as well. Even assuming an underlying genetic propensity, what's remarkable is when and why the sows are most vulnerable to the syndrome. It strikes most often during the socially and physically demanding weeks between giving birth and weaning their piglets, a period called farrowing. And it's not just the new porcine moms who get so frightened and stressed that their diet is affected. Weaning is a vulnerable and scary time for the piglets, too. In fact, that's when the less gender-specific variant, wasting pig syndrome, can develop. Like females affected by thin sow syndrome, young pigs with wasting pig syndrome refuse food, become emaciated, and sometimes die. Young male pigs are just as susceptible as female piglets, and it strikes at that fraught yet crucial period when they're moving out from under the protection of their mothers into the competitive world of the herd.

Your typical commercial swine operation is not the bucolic idyll you might imagine from fond rereadings of *Charlotte's Web*. Strict, innate social hierarchies that serve pigs well in the wild lead to dominance displays in crowded conditions, especially around feeding. From their first day on the teat to their later days at the trough, pigs compete for food and may bite each other's tails and ears to be first to chow time. The ones that prevail get fatter and healthier. The more timid ones miss out. In this environment, the pigs with genes that are overly expressed for anxiety, especially social unease, are vulnerable to a phenomenon that every middle school teacher and counselor of young adolescents will recognize: bullying. Farmers keep an eye out for bullying in their herds, knowing it can lead to thin sow syndrome. Psychiatrists, too, increasingly recognize eating disorders' important associations with anxiety and dis-ease in addition to more traditional ways to explain why anorexia nervosa occurs: disordered psychosexual development, intrusive family dynamics, perfectionism, and body-image distortion.

Knowing this, then, could clues for treating human anorexia nervosa be found in a pigsty? Farmers, after all, take a financial hit if they just stand by while their sows and piglets starve themselves. Connecting fearful states to eating behaviors, one study showed that piglets treated with anxiety-relieving drugs were indeed able to overcome their self-starving tendencies and resume eating and gaining normal amounts of weight. But pigs in general afflicted with thin sow and wasting pig syndromes don't fare well. One veterinary website says flatly, "There is no treatment." Psychiatrists might agree: they have yet to find a consistently impressive pharmacological fix for anorexia nervosa once it's fully taken root.

But there may be some preventative measures. Farmers advise making sure the animals are warm. They suggest raising the heat in the animals' pens and giving them more bedding material. Similarly, rodent researchers found that warmer ambient temperatures significantly reduced wheel-running by food-deprived rats. In some cases it actually reversed weight loss. This is likely due to the effects of a tiny brain structure called the hypothalamus, located behind the pituitary gland and above the brainstem. Body temperature, food consumption, and metabolism are regulated by the hypothalamus, which also plays a critical role in stimulating and suppressing appetite. Indeed, early traumatic injury to the hypothalamus (and other brain structures) may lead to anorexia nervosa later in life. Conversely, anorexia nervosa itself can bring on hypothalamic dysfunction.

Pig farmers also recommend immediately increasing feed rations for the whole herd, not just the suffering pigs. Whether this reduces competition for food or catches at-risk sows before they tip into syndrome territory, it seems to improve the health of the entire group.

Could these measures help human anorexics? Although those with full-fledged anorexia nervosa would certainly need more comprehensive treatment,* might people with early signs of this disorder benefit from something as simple as kicking up the thermostat a few degrees during stressful periods? Borrowing further wisdom from veterinarians and farmers, physicians and families might keep an eye out for bullies and

---

*One study of ten anorexics in a clinic showed that having them wear heating vests for three hours a day had no effect on body mass.

social competition during critical life transitions, such as adolescence and new motherhood, to stave off anorexia nervosa in at-risk individuals.

Some eating disorders, say psychiatrists, spread socially among susceptible individuals. It can take just a single "thought leader" to disseminate disordered eating behaviors to many others in a group. Today's aspiring bulimics and anorexics can learn tricks of the trade from anorexia-nervosa–promoting (a.k.a. "pro-ana") websites. Images of skeletal celebrities fill screen after screen, providing visitors with "thinspiration." Comments and blogs give isolated anorexics and bulimics around the world a cyber support group in which to crow about their triumphs: a meal skipped, a parent tricked, chocolate bars and noodles regurgitated, an exercise goal exceeded. Online pals commiserate about laxative intolerance and family meals fake-eaten under the eagle eyes of parents or spouses. Helpful tips include how to mask your breath after a vomiting purge and how to smuggle heavy coins into your pockets to tip the scales at your yearly physical. An added thrill for devotees of these sites may come from a sense of persecution and secrecy. The sites are targeted by web managers and parents' groups and are frequently pulled down, only to sprout up again on a different domain or server.

But what might surprise the aficionado-victims of "the ana lifestyle"—not to mention a bulimic men's cross-country team or varsity cheer squad—is how much they have in common with the gorillas at their local zoo or belugas at their local aquarium. Because some of these animals have a nagging (and mostly secret) habit, too. Zoo vets call it R and R: regurgitation and reingestion.

The technical definition of R and R is "the voluntary, retrograde movement of food or fluid from the esophagus or stomach into the mouth." An affected gorilla causes himself to vomit a bolus of food into his mouth or hand, or sometimes onto the ground. Just before he does it, he goes through some preliminary behavior. Gorillas have been seen poking and massaging their stomachs. Others prepare a special place on the ground. Some hunch over or rock back and forth and shake their heads. After the mouthful comes up—onto the floor or into a hand or mouth—the gorilla reconsumes it. They use their fingers, or lick it up directly, or simply rechew and swallow what's already in their mouths.

Sometimes the process is repeated, with the same material going up and down numerous times.*

Just as with human bulimia nervosa, once R and R behaviors start in one member of a group, they can spread. For example, when the elder gorillas of a troop engage in it, infants and adolescents become fascinated, sneaking up behind silverbacks and females to steal slurps of regurgitant. In one group of gorillas, the young literally aped the adults' stooped body postures after watching them go through their R and R routines. The youngsters spit and reswallowed their own saliva, leading researchers to remark that the behavior "might be socially enhanced, if not learned."

R and R is widely believed not to occur in the wild, or at least not that researchers have reported. But it is common in captive settings in both terrestrial and aquatic mammals. Chimpanzees, dolphins, and beluga whales—all animals said to share our so-called higher cognition—have been seen engaging in R and R outside wild environments. One marine mammal specialist described to me a time she saw a beluga whale gag up a swirling ribbon of white liquid and then balletically, deliberately, reingest it while on display in its underwater tank—in full view of a group of disgusted aquarium visitors.

When veterinarians notice R and R, the first thing they do is take stock of the individual's social environment. Like the pig farmers, they carefully monitor group interactions to see where stressors and fear might be coming from and to minimize opportunities for R and R to be learned by others.†

---

*Animals display a spectrum of regurgitative behaviors, from R and R to cud chewing. For many animals, it's a normal part of their digestive process. One of the reasons R and R makes an intriguing natural animal model for bulimia is its observed association with stress in animals.

†They also might alter the animals' diets. Milk products are associated with both human bulimia and animal R and R. In Georgia, keepers at Zoo Atlanta noticed that R and R peaked every day right after the evening meal. That's when the animals were each given a cup of cow's milk to supplement their nutritional intake. Hoping to reduce R and R, the Zoo Atlanta gorilla team experimentally removed milk from the gorillas' diets. Afterward, the R and R patterns changed significantly. The animals still brought food up, but they reingested it much less frequently. With milk out of the diet, gorillas spent more time eating hay—a more appropriate food item for them. Intriguingly, R and R behaviors in these gorillas also showed a seasonal difference. The behavior was

Veterinarians are careful to point out that R and R differs in some ways from human bulimia. In fact, R and R bears similarities to another human condition called rumination disorder. It's diagnosed in humans who bring up food from their stomachs into their mouths, chew, and then spit out or reswallow repeatedly. One veterinary theory is that R and R behaviors are an animal's way of self-soothing or prolonging feeding pleasure. This may be true, but many human sufferers of rumination disorder also have mental disorders driving the behavior.

Given the relationship between R and R and stress, does this vomiting behavior also connect to the ecology of fear? I believe it does, though it's not predatory fear that underlies R and R but, rather, the dangerous and oppressive anxiety of social stress.

For a scared animal, an emotionally activated digestive tract can be a powerful weapon in its defensive arsenal. The black vultures in McKinney Falls State Park, in central Texas, are notorious regurgitators able to "vomit with a vengeance" when threatened by humans or other animals. Some caterpillars, too, according to lepidopterists, are well-known regurgitators. Some upchuck reflexively at the slightest provocation, while others stoically withstand stressor after stressor until finally they blow. Looking at the opposite end of the digestive tract, some animals defecate as an offensive strategy to drive predators away and to facilitate escape. Others, including many mammals, defecate in response to fear or threat. Perhaps you yourself have felt a stomach-emptying urge— from either end—before a major presentation or during a stressful social encounter.

I found no corresponding term in the human literature to describe it, but wildlife biologists have a great one. They call throwing up when threatened "defensive regurgitation." Although the psychology is vastly different, the effect of stress hormones on the gut may be very similar. Thinking of bulimia nervosa as "defensive regurgitation" may help phy-

---

much more prevalent in winter. In summer, when they were more active, the gorillas were much less likely to vomit on purpose.

As a triggering food for regurgitation, dairy products are preferred by another member of the great ape family. Michael Strober noted, "Yogurt is one of the favorite foods of people with eating disorders. They have this affinity for yogurt. . . . Ask them to list their favorite food; they'd probably say yogurt."

sicians reconsider how they approach and treat this disease. And it may help patients reframe it as well.

I never did get to the bottom of all of Amber's fears. But after a few weeks she left the unit, carrying a few more pounds on her small body and a little less anxiety in her mind. From time to time in the years after, I would see her on campus, home from college. She was recovered and appeared healthy.

But if I could go back to that moment in the dining room when she was scared of a sandwich and change one thing, it would be this. While sorting through her fears—of getting fat, of food, of change—I would help her understand her fear of feeding as a protective physiology gone astray. I would tell her about the ecology of fear and share the story of the Yellowstone elk. How severely they restricted their eating when wolves abounded. And how their eating expanded when the predators went away. We would work together to help her identify the wolves in her life and uncouple her fear from her feeding. Because Amber was a lot like other vulnerable animals venturing out of nests, caves, and burrows. Threat comes not from the food they might eat, but rather from the uncertain and dangerous world in which they must consume it.

TEN

The Koala and the Clap

The Hidden Power of Infection

When monster wildfires scorched southern Australia in 2009, destroying homes and killing nearly two hundred people, one photograph came to symbolize the epic clash of man against nature and the plight of the vulnerable creatures caught in the middle. It showed a firefighter in bright yellow battle fatigues. With smoke rising around him, he crouched on the charred earth, holding a plastic water bottle to the lips of an exhausted koala. As it drank, the animal gripped the firefighter's hand with its small paw. The human—face sooty, hair mussed—gazed intently at the animal, a striking image of compassion and interspecies cooperation.

Around the world, people anxiously followed the story of the koala and the firefighter. At the shelter where the animal's burns were salved and her paws bandaged, she acquired the nickname Sam. This icon of Australia—pulled from the ashes—became the furry face of resilience overcoming adversity, more phoenix than marsupial.

But six months later, Sam hit the blogs again. This time, her story didn't have such a happy ending. In fact, Sam had died. It wasn't the burns that had killed her. The koala was dead of complications from chlamydia.* She had a sexually transmitted disease (STD). Readers learned that chlamydia is an epidemic of such proportions among the thousands of wild koalas in Australia that it threatens extinction of the iconic animals.

Chlamydia and koalas. It's a combination that just seems *wrong*, like a toddler with a heart attack. Small marsupials are innocent and natural, even cute. STDs, let's be honest, are not. Even among physicians accustomed to the sights and smells of the human body, STDs hold little appeal. An international survey of physicians once ranked afflictions by their level of prestige. Brain tumors, heart attacks, and leukemia were the top three. Diseases that strike below the belt were dead last.

And medical advances of the past half century have made it even easier to look away from sexual infections. In developed countries, most of us have the luxury of thinking of STDs as being essentially curable—or, at worst, treatable, chronic diseases requiring daily medication. (Think of antiviral medication for herpes or, in a more extreme case, daily drug cocktails for HIV.) What's more, pervasive and effective "safe sex" education has given the strong message—true in some cases—that barrier methods and abstinence can make you practically impervious to STDs.

But for animals, safe sex isn't a choice. In fact, when you think about it, unprotected sex is the only kind nonhuman animals have. Without access to condoms and abstinence pledges, not to mention antibiotics and vaccines, nonhuman animals have to cope and survive and reproduce somehow, regardless of what infections come their way. When you consider the amount of "unsafe" sex going on, 24/7, in a mere square mile of wilderness, it seems remarkable that animals aren't 100 percent infected at all times with STDs.

Veterinarians, like physicians, often give animal STDs scant attention, compared to other health concerns. Wildlife veterinarians don't regularly count the genital warts on tundra swan penises when they radio-collar them for migration surveys. Nor does yearly population

---

*Technically, the disease affecting the koalas is *Chlamydophila* (usually *C. pneumoniae* or *C. pecorum*). The genome of *Chlamydophila* is slightly larger than that of the closely related genus *Chlamydia*, from which it has been taxonomically split off. While acknowledging this difference, I will use the term "chlamydia" to describe the koalas' infection, as the veterinarians do. Similarly, while "the clap" is a specific and historic reference to *Neisseria gonorrhoeae*, it is used colloquially to refer to STDs in general.

tracking of Yukon caribou end with the females in stirrups and the vet chitchatting while warming up an Arctic-chilled speculum. Even when zoos transport and relocate their animals for breeding, most don't routinely screen for STDs. Among biologists, the handful of professional academic organizations that might discuss animal STDs are loosely organized and spread thinly throughout the world.

Like most patients and physicians, I wasn't exactly clamoring to hear more about STDs. But we all should pay attention, because STDs are remarkably deadly. HIV/AIDS is the world's sixth biggest cause of death. When you combine those numbers with cancer deaths from sexually spread viruses such as human papilloma virus (HPV) and hepatitis B and C, the mortality climbs even higher. STDs are tenacious, ancient, and lethal, and they continue to outfox human attempts to control them. Perhaps physicians can find help for human STD patients in a place they've never thought to look: the genitals of nonhuman animals.

Consider the following: Atlantic bottlenose dolphins sprout cervical and penile warts. Baboons get genital herpes. Copulating whales, donkeys, wildebeests, wild turkeys, and Arctic foxes harbor and transmit warts, herpes, infectious pustular vulvovaginitis, venereal pox, and chlamydia.

Sexually spread brucellosis, leptospirosis, and trichomoniasis cause repeated miscarriages and reduced milk output in cattle. Pig litters can be decimated by bacterial infections acquired during mating. Venereal diseases in farmed geese cause death as well as drops in egg production. Contagious equine metritis so predictably devastates fertility in mares that every stallion of breeding age imported to the United States is required to go into a minimum three-week quarantine to make sure he's not a carrier. Dog STDs can cause abortions and birth failures.

When I first started learning about animal STDs, I was surprised by the range of species they infect. But it wasn't too hard to picture the mechanics of mouse or horse or elephant sex and imagine the genital contact that could lead to the spread of an infection. What was truly eye-opening to me, however, was that the dark, balmy environments favored by sexually transmitted pathogens aren't limited to warm-blooded creatures. Dungeness crabs, for example, are vulnerable to a worm that spreads from males to females when they mate. The worms invade the female looking for her cache of eggs. Once they find it, they start feeding, reducing the number of viable crab offspring.

Even insects carry STDs on their tiny genitals. Two-dot ladybugs, one of the most promiscuous creatures on Earth, can become infected with a sexually spread mite that makes them sterile. A postcoital housefly that lands hungrily on your freshly prepared Dungeness crab chowder may himself harbor a genital fungus, also acquired through copulation. Astonishingly, some of the diseases we humans catch from insects—such as St. Louis encephalitis from mosquitoes and spotted fever from ticks—are actually sexually transmitted among the insects themselves. (If you've never pictured what ladybug or tick or housefly sex looks like, you have an illuminating twenty minutes of Internet image searching ahead of you. Most insects do engage in penetrative sex with genital contact—often in the doggy-style position.)

Indeed, STDs have been found thriving in so many living things, from fish and reptiles to birds and mammals and even *plants,* it's safe to say they're ubiquitous in all sexually active populations. Experts agree that these infections are legion.

And yet you might be whispering to yourself: *So what?* Yes, we want to reduce animal suffering. But in terms of human health, why should we give a moment's thought to diseases of their genitals? To be blunt, since we're not having sex with these animals, why should we care if some koala catches the clap?

The answer is as simple as it is unsettling: because pathogens are always looking for new paths, and they don't differentiate between humans and other animals. For example, rabbit syphilis once spread to trappers in East Yorkshire, who got sores on their hands after handling the animals. There was no sexual contact between the humans and the animals, but the syphilis pathogens didn't care. They were happy to jump the species barrier and curl into warm moist tissue through cuts on the men's hands.

Or think about brucella. In livestock, these nasty bacteria cause spontaneous, late-term miscarriages in females and swollen, bleeding testicles in males. So unforgiving is brucella's attack on the reproductive system that one of its common strains is called *Brucella abortus.* But what's instructive about brucella is how it spreads. Cattle, pigs, and dogs transmit it through sex. So do hares, goats, and sheep. But all these animals can also acquire it nonsexually . . . by eating it. Under the right conditions, brucella organisms can live for several months on many things that might end up in an animal's mouth: feed, water, equipment, and clothing, not to mention manure, hay, blood, urine, and milk.

In a number of animals, the same pathogen has found two different paths of entry into the body—sexual and oral. Through the mouth is the way humans usually acquire brucella infections, too—when they eat contaminated meat, unpasteurized milk, or soft cheeses. Spread this way, from animals to humans, brucellosis is a major public health concern, especially in developing countries, where thousands of cases emerge each year. (In developed countries, it has become mercifully rare, thanks largely to veterinarians who vaccinate animals and monitor the spread of disease.)

Like livestock, humans can become infected with brucella in more than one way. Like the trappers who contracted syphilis infections by touching sick rabbits, zookeepers in Japan got brucellosis when, during the delivery of an infected baby moose, they came in contact with the placenta and mother's vaginal secretions.

And although they're rare, reports do exist of brucella's spreading from human to human—through blood, milk, and bone marrow . . . as well as through sexual intercourse.

Same pathogen. Different paths. Could classifying a condition as "sexually transmitted" be limiting how we consider and understand these infections? After all, bugs are bugs, no matter how they get in. *Streptococcus A,* the common human pathogen that causes strep throat, scarlet fever, and rheumatic heart disease already exploits several routes into the body. Its most common path is respiratory. One person coughs or sneezes droplets containing the bacteria, and another person inhales or picks them up from doorknobs or silverware. But Strep A can, through oral-genital contact, cause penile inflammation and purulent discharge. You can get salmonella from sex with an infected person or by licking raw cookie dough off your finger—either way you'll be down for the count with 104-degree fever, hideous diarrhea, and exhaustion. Hepatitis A, as well, can be picked up during a sexual encounter or by eating at a restaurant where the chef didn't heed the hand-washing instruction sign in the bathroom. No matter which portal it uses to enter your body, the pathogen will give you the same gruesome symptoms: fever, exhaustion, and a complexion the color of Grey Poupon. You might even need a liver transplant.

Studying STDs in animals reminds us that, like all living things, pathogens are constantly evolving. Species suited to one region of the body can change over time, developing new areas in which to live and thrive. Take *Trichomonas vaginalis*. Nowadays, "trich" is one of the least glamorous but most common STDs. In women it causes a fishy-smelling, frothy, yellow-green vaginal discharge. Men infected with trich usually have a slight irritation or burning in the penis, but no other symptoms. But contemporary *T. vag* wasn't always a lowly genital dweller. Ancient, ancestral *T. vaginalis* resided in the digestive tracts of termites. This made it, essentially, a gastrointestinal bug. Changes over trillions of generations (and millions of years), however, allowed it to expand beyond termites and guts into the bodily crannies of many different animals. Eventually a version found its way to human vaginas (and fifteen minutes of microbial celebrity in 2007 when it was featured as *Science* magazine's "cover bug").

Today, cousins of *T. vaginalis* (descendants of that ancient termite-dwelling ancestor) don't limit themselves to human penises and vaginas. Other species of trich have found suitable homes in various parts of human and animal bodies. *T. tenax*, for example, thrives in the dark, moist crevices of rotting teeth. *T. foetus* causes chronic diarrhea in cats and ravages the fertility of cows. *T. gallinae* is practically endemic in the mouths of many birds—voracious raptors and peace-loving doves alike.

T. gallinae (or its close cousin) has, in fact, been colonizing the ancestral birds of Earth for a very long time. Recent research on Sue, the *T. rex* famously on display in Chicago's Field Museum, reveals that she may have died of a raging *Trichomonas* infection that bored holes through her jaw and ultimately left her unable to chew and swallow her food.

Her infection wasn't sexually transmitted, but it shows how, over millions of generations, these microorganisms have deftly adapted to new environments. Like a large family conglomerate where one son controls the real estate holdings, another textiles, and another medical devices, trich has differentiated so that each species specializes and thrives in a specific body region. But regardless of their portal of entry or favorite locale, they are all members of the same genus: *Trichomonas*. So whether it's swabbed from the cervix of a college freshman or collected from the upper esophagus of a carnivorous hawk, under the microscope trich is trich. Again, similar pathogen, different paths.

Gut infection today, genital infection tomorrow. The family albums of ancient pathogens show the evidence of their many migrations around the landscape of our bodies. For example, several hundred years ago, syphilis underwent a major evolution. The pathogen found a new path. Before it discovered its current preference for the human genital tract, the ancestors of the current syphilis microbe caused a horrible skin condition called yaws. It was a disease largely of children, and it spread by skin-to-skin contact. (Yaws still exists, mostly in undeveloped, tropical regions.) But sometime in the last thousand years, yaws somehow found its way into adult genitourinary tracts. Once it discovered the sex superhighway, it morphed into what we now call an STD. But the corkscrew-shaped spirochete that causes it retains the genealogy of its yaws forebear, which was basically a skin disease.

If the same pathogen can be transmitted in any number of ways, and if it can mutate from being a gastrointestinal dweller to a urethral specialist and then change again to become a throat denizen, why do we fixate on sex as the pathway? After all, many organisms can infect us using different routes.

This is a point that physicians—and veterinarians—sometimes overlook. And it's a reason to pay attention to animal STDs. Because pathogens don't discriminate between the warm, moist, nutritious environments they choose to call home, and because they frequently mutate, the animal STDs of today can become the *human* food-borne illnesses of tomorrow. Given chance encounters with human genitals and time to evolve there, those food-borne illnesses can then mutate into the next human STDs.

This is not just an idle theory. It's exactly what happened in the case of the deadliest STD currently stalking our planet. It is now generally believed that HIV evolved from SIV (simian immunodeficiency virus), a pathogen of chimpanzees, gorillas, and other primates. Sex and mother's milk are major transmission routes for SIV within primate populations. Assuming that people were not having sex with chimpanzees or hiring gorillas as their wet nurses, how did SIV jump to humans?

The answer is: the same way brucella infects humans. Through ingestion. The theory is that, by eating the meat of infected monkeys and apes, or getting their blood or other fluids on their hands and faces, hunters in western Africa became unwitting reservoirs of SIV sometime over the last few decades or centuries. Over many years and through many hosts, SIV mutated into HIV and then exploited the same path it had used in the nonhuman primates: sex. What started as an animal disease evolved into a human version we could give to one another. But, of course, sex is not the only way HIV spreads. It can also travel through blood, breast milk, and, on rare occasions, transplantation of infected tissues and organs. Given the way pathogens exploit many routes of entry into a host, it's possible that if another animal were to preferentially feed on humans infected with HIV, the virus could jump into that species and eventually become tailored for sexual spread in that population, too.

But animals—including humans—have not sat idly by as these wily microscopic invaders have launched assaults on our mucous membranes and vulnerable bodily portals. We've evolved fierce infection-fighting arsenals. White blood cells. Antibodies. Fever. Viscous mucous. Thick skin. And, intriguingly, our defenses are not just physical. Animals have also evolved ways of behaving that can reduce the risk of infection. Coughing, sneezing, scratching—even grooming behaviors, like picking, rubbing, and combing—all have an antiparasite benefit at their core. And there are the things we humans do even more deliberately: Washing our hands. Vaccination. Sterilizing dishes. Wearing condoms.

Some behavioral responses protect us once a pathogen has entered our airspace or breached a battlement. But bacteria, viruses, fungi, and worms don't even have to enter our bodies in order to influence our actions. Consider the following automatic behaviors: recoiling from a runny-nosed child in the elevator. Sniffing the opened carton of milk before we pour it onto our cereal. Backing out of a public restroom to avoid grabbing the doorknob. Our behavioral strategies—and immune responses—can be activated by just *thinking* about parasitic infection. (Here, I'll show you: *Bed bugs. Head lice. Pinkeye.* Are you having a reaction?)

Among these reactions are some truly bizarre behaviors that seem to have nothing to do with fighting disease. And, it turns out, they don't. That's because the infections themselves may be steering our actions. Although that may sound like the preposterous premise of a zombie movie, these tiny creatures' ability to influence the behavior of larger animals such as ourselves comes from a billion-year-old game of escalating, coevolutionary cat-and-mouse.

One of the strangest things I've ever seen was a video of a human rabies patient trying to take a drink of water. This patient did not look sick. He was not foaming at the mouth, the way he would have been in a movie. He was not growling like a mad dog or writhing on the gurney with crazy eyes. The man looked perfectly calm and normal. Until a nurse handed him a cup of water. Suddenly, his hands started to tremble. He tried to bring the cup to his lips but couldn't. His head thrashed from side to side as the liquid approached his mouth. It looked as though someone were using a remote control to direct his movements.

Hydrophobia, or fear of water, is a classic symptom of rabies infection. So is aerophobia (fear of moving air) and, as the disease progresses, an uncontrollable urge to bite. These seemingly random behaviors stem from changes the virus causes in the central nervous system of its host. And they may have a fortuitous side effect for the virus itself. The actions may actually help it transmit itself into a new victim. Because the rabies virus is spread through saliva, causing an urge to bite, for example, would be a useful microbial "strategy." So far, however, infectious disease veterinarians haven't found adaptive purposes for causing a fear of water or moving air.

Or consider *Enterobius vermicularis,* a.k.a. pinworms. This common childhood infection alters human behavior by drawing hands away from more productive activities, like homework and setting the table, and redirects them into ferocious anal scratching. This scratching serves two purposes for *E. vermicularis*: It helps burst the gravid females' bodies, releasing the ten thousand eggs they each carry. And it helps those freshly exposed eggs burrow under the child's fingernails, where they wait patiently for the next thumb suck or nail bite to permit them entry into the host's mouth and, from there, his GI tract, where they reproduce.

Or take *Toxoplasma gondii.* Infection with this protozoan has an unusual effect on rodents: it makes them lose their fear of cats. From the rodent's perspective, of course, this is terrible. It makes them easy prey. But from the toxo's point of view, it could not be more clever. That's because the only place on Earth that *Toxoplasma gondii* can reproduce is inside a cat's intestine. By making rodents fearless, the parasite practically gift-wraps and delivers itself to the cats' claws and jaws . . . and from there to guaranteed reproduction.

Humans are "dead-end" hosts for toxo, meaning it can't reproduce in us. But the parasites can still enter our bodies when we eat or touch infected meat, soil, or cat feces. Once inside our brains, the toxo can "encyst," essentially lying dormant and waiting to get back into a cat. The pathogen doesn't know whether it's in a mouse or a mail carrier, a rat or a receptionist. But it continues to produce chemicals and help itself to nutrients in our blood and tissues. In fact, many of us have these encysted toxo infections. And, incredibly, this microorganism may affect our behavior as individuals. Exposure to toxo in the womb may be a contributing factor in developing the often devastating human disease of schizophrenia.

"Brainworms" and other parasites have been shown to spark killing sprees within ant colonies and make crickets and grasshoppers suicidal. One wasp creates a bodyguard for its offspring by infecting a hapless caterpillar which then fights off the wasp's stinkbug predators with powerful swings of its caterpillar head. While toxo, pinworms, and rabies aren't STDs, certain sexually spread ailments work their own microbial puppetry on their hosts. Two STDs—HIV and syphilis—notoriously produce extreme behaviors in people with end-stage infections. HIV dementia compromises judgment and memory. The egomania, impulsivity, and disinhibition that characterize advanced syphilis may have not only propelled the infamous sexual appetites of known syphilitics Al Capone, Napoleon Bonaparte, and Idi Amin but facilitated their power grabs as well. And while patients in the late stages of syphilis are no longer contagious and can't spread the disease, there are diseases where the behavior caused will promote infection.

And this is another way we can learn from animal STDs. Many microbes depend on sex for their transmission. It makes sense that they might induce subtle sex-friendly behaviors, if they could.

But how would a crafty STD microbe get people jumping into the sack with each other? Maybe it would improve the males' pickup lines . . . or confuse normal signaling so that rejections got misinterpreted as come-ons. Maybe it would make the females more alluring. Or increase libido or lower inhibitions to lead to more sex.

This may indeed be exactly what goes on in a range of animals infected by STDs. Male *Gryllodes sigillatus* crickets attract females with intricate symphonies of sound, produced by rubbing their hind legs together. Crickets infected with a certain parasite sing slightly differently than uninfected crickets—but the change seems to increase the males' attractiveness and bring more females their way.

When infected with the sexually transmitted virus Hz-2V, female corn earworm moths start producing excessive amounts of sex pheromones—some two to three times as much as their uninfected sisters and peers. The extra come-hither perfume is believed to attract more male moths—thus aiding the spread of the virus. Intriguingly, these infected females also demonstrate a kind of lepidopteral "no means yes" behavior. Apparently unaware of how politically incorrect it is, they appear to further excite their mates through acts of resistance.

Sexually acquired infection can spur some animals into assertive sex-seeking behaviors. Male swamp milkweed beetles infected with a sexually transmitted mite aggressively move in on nearby mating pairs, busting up the action and pushing aside the other male. When no females are in the vicinity, these infected males approach and attempt to mate with other males.

STDs may even change the "behavior" of plants. Like all living things, plants need to reproduce. For flowering plants, this means getting the sperm-laden pollen from male flowers to the eggs of female flowers. One way floral "sex" is accomplished is by the peripatetic flights and landings of birds, bees, and bats, which carry the pollen from flower to flower when they feed on the blossoms' nectar.

However, the pollen of many flowers teems with microscopic fungi, viruses, and worms . . . all seeking to be transmitted into new hosts. When animal pollinators ascend from a bloom—legs and bellies sticky with what is essentially flower semen—these minute pathogens are often along for the ride. When the bee or hummingbird visits the next flower, it deposits pollen . . . along with a load of these flower STDs.

What's really interesting is that these diseases can make plants, for lack of a better word, promiscuous. The white campion flower, for example, is susceptible to a fungus aptly named "anther smut." A Duke University botanical disease ecologist, Peter Thrall, found that plants infected with anther smut tended to produce larger floral displays. Uninfected plants had punier bunches of flowers. With their big, ostentatious blooms, the flower hussies received (and could accommodate) more visits from more pollinating suitors. By forcing the plant to produce bigger, showier flowers, the fungus was biologically changing its host in a way that made it more attractive to pollinating creatures. This directly benefited the fungus.

A similar "strategy" may be used by the trypanosome that causes an equine disease called dourine. Infected horses, mules, and zebras suffer fever, genital swelling, lack of coordination, paralysis, and even death. Although it's now extremely rare in North America and Europe, dourine once ravaged the cavalries of the Austro-Hungarian Empire and swept across the horse populations of southern Russia and northern Africa. In Canada in the early twentieth century, dourine decimated Indian pony herds.

Dourine spreads when animals mate. Intriguingly, scientists and veterinarians report anecdotally that when dourine is present in a group, the libido of stallions seems to increase.

How this works may be very similar to how anther smut influences the "behavior" of flowers. Full-blown dourine wreaks physical havoc on the animal . . . but the early signs of infection are more subtle. A mare may seem perfectly healthy except for a minor vaginal discharge that reveals itself as a wetness around her tail. Mares infected with dourine often keep their tails slightly raised, presumably to ease discomfort from the increased wetness.

A mare's raised tail is also a signal of sexual receptivity. So is something else, familiar to every horse breeder, that's visible when the tail is up. It's called vulval "winking." Caused by the vulva's contracting and releasing, vulval winking happens when a mare is in heat.

But an under-the-weather, dourine-infected mare, her tail raised and her vulva wet with discharge and perhaps winking with discomfort, may incite randiness in a stallion with her STD-induced false advertising. While the stallion may suffer from the mistake, the pathogen will benefit.

Sometimes the connection between infection and behavior can seem very roundabout. One of the most perplexing endpoints of many STDs is the destruction of their host's fertility. You would think this would be a terrible ploy, for two reasons. If a population can't produce offspring, that usually means the end of the line for the bug. Without a new supply of hosts, where are a bug's descendants to live? And then there's the other problem: If an animal can't have offspring, what would drive it to have sex?

But bugs' success is tied to how much their host *mates*, not to how much their host *reproduces*. (The increasing incidence of STDs in people over fifty illustrates how these infections need sexually active hosts . . . not necessarily fertile ones.) A female animal who is having trouble procreating may in fact try harder—that is, have more sex—than one who is already pregnant. If a pathogen can disrupt the pregnancy cycle by inducing miscarriages or preventing conception, it is likely to enjoy the benefits of increased mating attempts. Is it possible that by hindering reproduction certain STDs are actually driving their hosts to have more sex?

In fact, some veterinary literature supports just that. An STD of deer and other ungulates, for example, puts the females permanently in heat and thus more receptive to sexual overtures. When *Brucella abortus* causes a cow to miscarry her calf, it makes her ready for a new breeding cycle—sooner than if she had carried the calf to term. This revelation suggests that subclinical infections (those percolating below the surface but not actively causing symptoms)—or even as-yet-unidentified pathogens—may play a greater role in unexplained human infertility and repeated miscarriage than we currently suspect.

In other words, even low levels of infection might alter sexual function and behavior. STDs are especially good at going deep undercover once they're inside a body, quietly colonizing it with few overt symptoms. Whether the infections are small and contained or widespread and subclinical, these organisms do affect our bodies and minds in ways that might be unseen to us.

As a medical student at U.C. San Francisco during the height of the AIDS epidemic there, I was instructed to aggressively dispense safe-sex advice. Even if a patient came in with an earache, I brought sex into the conversation. I recommended wearing condoms and avoiding multiple partners. (Remember that quintessential 1984 line "When you sleep with someone, you are sleeping with everyone they've ever slept with"?) I counseled patients to question potential mates ("Do you ever have sex with men?" "Do you use IV drugs?"). A veterinarian can't warn her patient to wear condoms and interview a sex partner before even getting to first base. But one more preventive technique I used to recommend does have an animal correlate. I was trained to advise that potential partners inspect one another's genitals for sores and lesions before engaging in sex.

An animal version of this has been reported in birds. It's called cloacal pecking* and has been described as a male bird pecking inquisitively at the vaginal opening of a female before mounting her. Some researchers speculate that the fluffy white feathers or prominent "lips" around the cloacal openings of many bird species serve as additional aids for assessing health in a potential partner, since ectoparasites and lesions might show up against the pale background. If soiled by diarrhea or other bodily fluids, these structures would also warn a potential suitor of an unhealthy bird.†

Lab studies also show that postsex cleansing might offer a modest level of protection. Rats that are prevented from genital grooming after coitus have higher STD rates than their cleaner counterparts. Many birds preen vigorously after copulation, which some researchers suggest may help do away with bugs trying to hitch a ride on the act. In humans, genital scrubbing does not protect against viral STDs, but it may be slightly effective against bacterial infections. A study of Cape ground squirrels in South Africa showed that the ones having the most sex were also the most frequent masturbators; the researcher speculated that it's a way to flush out the urethra after intercourse in order to protect the animal from sexual infection.

A recent study showed that simply *looking at a photograph* of a sick person caused some people's immune systems to surge. Indeed, animals may have other ways of visually sizing up the health of a mate. For example, in males, red pigmentation—whether in a grouse's comb, a house finch's feathers, or a guppy's skin—may indicate underlying fitness. These animals' bodies cannot create the color red on their own. To impart the bright tint, they have to be healthy enough to source and eat lots of red-pigmented carotenoids, found in fruit or shellfish. Conveniently for any females who might be speed-dating with these males, parasites can interfere with the absorption of these pigments. Animals with paler features are in effect advertising their inferior health status.

But if the thought of colonies of invisible organisms invading your body and controlling your behavior has you reaching for the doxycy-

---

*Birds have a combined reproductive and excretory opening called a cloaca.

†Cloacal pecking may aid sperm competition in birds such as dunnocks, whose precopulatory displays include beak stimulation that induces females to eject sperm from previous partners.

cline, think again. Our best response to the microbial arms race is not necessarily a scorched-earth campaign.

In the 1980s, a British scientist shook up the world of microbiology by asking an outrageous question: *Can we be too clean?* David Strachan was pondering whether hay fever might be related to hygiene and household size. A few years later, a German scientist, Erika von Mutius, was investigating childhood asthma. She was vexed by data that consistently showed that it was most prevalent not in lower-income, more polluted East Germany but, rather, in wealthier, cleaner-living West Germany. The so-called hygiene hypothesis started to circulate, postulating that there are serious consequences to wiping out too many of the microorganisms that have colonized us and our planet for so long. Overusing pesticides, antibacterial agents, and antibiotics, it suggests, kill "good" pathogens along with harmful ones. Plain old better housekeeping and even overly thorough food inspections, the theory goes, create microorganic dead zones. These sterile environments deprive our immune systems, honed over hundreds of millions of years, of necessary daily battles against invaders. And when deprived of external organisms to fight, they sometimes launch an internal attack. An idle immune system will sometimes start attacking itself.

The hygiene hypothesis, while still a matter of debate, is now being used to explain more than just asthma, allergies, and other respiratory diseases. Upsurges in gastrointestinal disorders, cardiovascular disease, autoimmune disorders, and even some cancers are being traced to it, too. However, no one has really looked at the genital environment—and whether it, too, can suffer from being "too clean."

This leads to an intriguing thought. Are some of the pathogens associated with sexual activity beneficial? Most animals have multiple sexual partners—which means that sperm from many different males must duke it out inside vaginas, uteruses, and fallopian tubes to win the conception derby. Conception is not some genteel, quiet pastime; it's a fierce and unforgiving team sport. The swimmers that win sometimes have assists from microscopic wingmen—the *sperm-enhancing microorganisms* that live in semen and may be transferred from penis to vagina to penis to vagina. The sexual act may propel the package of semen, but it is then up to the ejected sperm and its microbiological posse to obstruct and destroy competing sperm. Some of these pathogens aid their sperm's motility, while others act as blockers and killers of competing males' spermatozoa. And if that isn't enough, these teams must also successfully negotiate the vagina's own mix of receptive and defensive microflora.

This means the microorganisms inhabiting an animal's urethra or vagina might make the difference between conceiving and not. Or, when there are multiple male partners, determine which of these males' sperm wins the ultimate prize: fertilization, the chance for his DNA to advance to the next round.*

This made me wonder whether striving for an aseptic genital environment could, in fact, be harmful (beyond the well-known risk of a yeast infection after antibiotic therapy). The human immune system fully matures between the ages of eleven and twenty-five—just when sexual activity is moving into full gear, bringing with it a barrage of new, unfamiliar microorganisms. The hygiene hypothesis demonstrates the risks of underexposure to pathogens of the respiratory and GI systems. Might there be a genital version of the hygiene hypothesis? Could a "just right" mix of microorganisms in your genitals improve your chances for conception or help select the highest-quality sperm for your soon-to-be-conceived child? Might there be a place for a probiotic product to aid conception—similar to the products that improve digestion in the gut microbiome? Or maybe there's an intriguing flip side: Could studying sperm-killing micropathogens in animals lead to new contraceptives?

It's important to stress that, given the threats STDs pose to human health, this is not an argument for unsafe sex. Condoms save lives. Physicians and educators must resoundingly continue to emphasize the absolute necessity of safe-sex practices. But physicians should join veterinarians in considering the long-term ecological perspective of therapies and remain open-minded about unlikely or unexpected consequences of intervention.

As Janis Antonovics, a disease biologist at the University of Virginia, told me, "There is no imperative to cure disease in natural populations. Disease is natural!" Doctors first and foremost have a responsibility to treat individual patients. But ecologists like Antonovics take the

---

*For a lively account of sperm competition tactics across species (the ultimate existential battle), see Matt Ridley's fascinating *The Red Queen*.

pathogen's-eye view of infection. As he explained it to me, every time we perturb a system, by extermination or prevention, there is always a repercussion. An individual may see an immediate benefit from a round of antibiotics, but invariably, necessarily, killing off those organisms causes some unintended side effect, either immediately or down the line. Sometimes it comes back in a more virulent form. Infection (and all the viruses and worms and bacteria and other organisms that create it) is a complex, interconnected, multidimensional web. Tugging out one strand alters the architecture of the entire network.

If Sam the koala had been born a few years later, she might have benefited from the kindness of not just the firefighter but also a biologist named Peter Timms. Timms, along with his colleagues at the Queensland University of Technology, has been developing a vaccine for koala chlamydia. Early trials of the vaccine have cut infection rates slightly and blunted the virulence of the disease. Timms hopes his research will someday not only save koalas but also inform a human chlamydia vaccine.

It's hard to imagine anyone in Australia objecting to vaccinating their national symbol against a disease that causes blindness, infertility, and death. It hardly seems the koalas' fault that the disease that's wiping them out happens to spread via sex. But the development of human vaccines against STDs from chlamydia to HPV to HIV has been hindered by some groups that believe offering protection against these diseases is the same thing as actively encouraging the "immoral behavior" that spreads them.

But here's where a zoobiquitous perspective helps. Looking at these diseases in animals allows us to see infection as infection—independent of the route of introduction. While thinking about a human with chlamydia may make us grimace or blush, koalas with chlamydia likely make us feel sympathetic, or at least impassive. Most of us don't judge the koala for its sexuality. Decreasing the stigma of STDs can improve treatment.

An evolutionary approach could inspire clinical solutions. As we've seen, studying the history of infections may give epidemiologists a head start in identifying those bugs that are getting ready to jump transmission pathways. Maybe there are "good" microorganisms that spread sexually and maintain genital health, the way certain "good" microorganisms maintain intestinal health.

Finally, studying STDs in animals can enlighten us in ways that extend beyond our consideration of the illness, infertility, and death these pathogens can cause. Sexually acquired infections have played an oceanic, though microscopic, role in evolutionary biology. Sam the koala may have succumbed to chlamydia, but not all her sexual partners met with the same fate. In fact, despite their unprotected sexual free-for-all, a small percentage of koalas never became infected. Something enabled them to resist the infection . . . and that something is genetic variation. Every time egg and sperm meet, a new and unique combination of genetic material is created. Every once in a while, the mix is such that the creature that possesses it gains an infection-resistant advantage. This is why, although HIV is complex and deadly to most humans, infectious disease researchers have discovered that about 1 percent of humans (primarily Swedes) seem to be immune to it.*

In populations of clones—those with identical genetics—a single species of virus, bacterium, fungus, or worm can wipe out the whole group. But when individuals within a group each possess a slightly different genetic makeup, chances increase dramatically that some will survive. And nothing provides diversity as predictably and effectively as one particular act: sexual reproduction.

And herein lies a central irony with insights for evolutionary biologists, infectious disease specialists, and sexually active humans. Today we protect ourselves against sex. But over the course of evolution, it has been sex itself that protected us.

---

*A dramatic recent example of genetic resistance to HIV infection occurred when an American man with AIDS developed leukemia while living in Germany. The bone marrow transplant the "Berlin patient" was given for his leukemia came from a donor with a mutation in the genes coding for the CCR5 molecule. The AIDS virus uses CCR5, which usually sits on the surface of the cell, as a "door" to enter and infect the cell. If CCR5 is faulty (there is a mutation), the virus cannot enter. An individual with this mutation is essentially immune to HIV infection. This genetic defect is seen primarily in individuals with European heritage. An estimated 1 percent of people descended from northern Europeans are virtually immune to AIDS infection, with Swedes the most likely to be protected. One theory suggests that the mutation developed in Scandinavia and moved southward with Viking raiders.

ELEVEN

Leaving the Nest

Animal Adolescence and the
Risky Business of Growing Up

A bend in the central California coastline shelters a stretch of soft, white sand ideal for a family day at the beach. The waves shimmer. The sun warms. Kite-perfect, brine-scented breezes gust over sandy dunes, keeping aloft strings of seabirds that glide effortlessly over the gentle breakers.

Go ahead. Slather zinc oxide on the kids. Force them into their swim shirts. Remind them to stay within eyeshot. But before they race, Boogie-boards bouncing behind them, into the water, I should warn you about one thing. Just a few miles out, stretching south of San Francisco toward the Farallon Islands, is a place sea otter researchers call the Triangle of Death.

Great white sharks prowl the chilly waters. Sneaker waves, riptides, and treacherous undertows sweep the shores. The barren seafloor cannot support plant growth, so it's devoid of the sheltering, protective kelp forests found in other coastal regions farther south and north. The depths here teem with higher-than-normal levels of *Toxoplasmosis gondii,* the feared, infection-causing microbe found in some cat feces and uncooked meat.

You will not spot female sea otters in this dangerous location. Otter

pups don't go there, either. Dominant, mature males seem to know better than to venture into these dangerous waters and rarely do. Even scuba
divers, hired by the U.S. Geological Survey to radio-track sea otter movements, refuse to submerge themselves in this perilous location.

But one intrepid type of otter makes frequent forays into the Triangle
of Death—even though shark attacks and unexplained disappearances
occur commonly here. They're adolescent males, the daredevils of the
otter world.

The concept of animal adolescence may surprise you, as it did me.
Of course we've all seen gangly young dogs just out of puppyhood who
haven't quite matched their oversized paws to their less-advanced motor
skills. But the drama, awkwardness, and peril of teen life seems unique
to our species. And, indeed, it probably is one of a kind, if you associate adolescence with teenagers' matchless ability to wound their parents
with a strategic eyeroll or to ruin the family photo with a moody slouch.
But while the details may differ, a larger truth ties human teens to the
vast majority of other species. They all must pass through a fraught transition: the period between leaving the care of adults and becoming adults
themselves.*

We often call adolescence the teenage years, for the obvious reason
that the transition roughly corresponds with that segment of a human
life span. In other animals, the gradual shift from child to adult can last
anywhere from about a week for a housefly to fifteen years for an elephant. For zebra finches, it lasts about two months, starting forty days
after they hatch. In vervet monkeys, the journey from their mother's
side to motherhood (or fatherhood) happens over four years. Even lowly,
single-celled paramecia have an adolescent phase—a don't-blink-or-
you'll-miss-it fifteen to twenty-four hours in which their cell nucleus and
plasm change as well as, believe it or not, their behavior.

We human doctors have dealt with the unique and vexing trials of
this period the same way we have with especially complicated organs or
diseases—by creating a new specialty. "Adolescent medicine" caters to

---

*Parental provisioning can take many forms across species. The kind of parenting we
associate with our own species is seen in many birds, mammals, and other animals. In
fish and other egg-laying animals, parental investment is provided through protective
coatings, shelter, or nutritionally rich eggs that they lay and then abandon. Insects have
a similar strategy.

an in-between population: patients who've outgrown their pediatricians but aren't quite ready for an internist. It addresses the hormonal shifts of puberty and the physical challenges of emerging sexuality. Practitioners in this nascent field work vigilantly to keep at bay a chilling list of threats to young humans: traffic accidents, STDs, alcohol and drug abuse, traumatic injury, teen pregnancy, date rape, depression, and suicide. Much of what we associate with adolescence involves behavioral changes, and lately research has focused on brain changes that help explain those behaviors—like risk taking, sensation seeking, and the somewhat perplexing compulsion to fit in with a group.

Of course, all animals have different things to learn while traversing the arc that takes them from sexually immature, vulnerable child to reproductively capable, developed adult. In our case, those include advanced language skills and critical thinking. But there's one feature that defines adolescence in species from condors to capuchin monkeys to college freshmen. It's a time when they learn by taking risks and sometimes making mistakes.

A surprising and sad fact of life is that just being a human teenager—especially a boy—is very risky, and often deadly. In the United States, once children have survived infancy and early toddlerhood, most will enjoy a brief period of relative safety, until they hit age thirteen.* At that moment, however, the death risk climbs abruptly, mostly because of traumatic injury. The Centers for Disease Control and Prevention (CDC) reports that "among teenagers 12–19, death rates increase with every year of age. This pattern is stronger for males." At about age twenty-five, rates of fatal injuries seen so commonly in adolescents taper off. In adult years, cancer, heart problems, and other long-term diseases emerge as the main health risks.

These stark statistics parallel death trends in the animal world. "Young [animals] suffer higher rates of predator-induced mortality than adults," according to Tim Caro, a UC Davis biologist and author of *Antipredator Defenses in Birds and Mammals*. Risk tapers off as the infant survives

---

*Human infancy is a particularly dangerous phase of life around the world. In a zoobiquitous parallel, the animal newborn is also at increased risk of death, largely from predation, starvation, or accidental injury.

early challenges. But as animals' bodies grow in anticipation of transition, so do the dangers. Consider an adolescent warthog foraging for the first time without the protection of his mother. Because he lacks his full complement of defensive horns and thick hair and doesn't yet have the adult stamina to outrun a predator, his odds of survival would be low if a cheetah came upon him. Since they can't run as fast, fly as high, or otherwise outmaneuver threats as skillfully as adults, young animals fall to predators more often. Less experienced, they misjudge situations and blunder into danger.

Of course, on the whole, modern human teens aren't being picked off by mountain lions or the other hungry predators that threatened our distant ancestors. What kills adolescents disproportionately in many countries around the world is a different lethal presence: motor vehicles. The CDC reports that 35 percent of deaths in the twelve-to-nineteen age group in the United States come from traffic accidents.

Other sudden, violent causes of death threaten teens, too. According to the World Health Organization, interpersonal violence claims the lives of hundreds of ten- to twenty-four-year-olds every day. And gun accidents, suicide, homicide, drowning, burns, falls, and warfare are also leading killers of adolescent humans worldwide.*

Adults recognize this behavior so well that it's enshrined in both law and supposedly forward-thinking parenting strategies. It's why it's harder to rent a car before you're twenty-five and why auto insurance rates are highest for adolescents. Why we set drinking ages and driving ages. Some states and locales dictate how many teens can be in a car at one time. New Jersey forbids all teens, not just the driver, to use electronic devices. And a scarlet rectangle must adorn their license plates, marking them as younger drivers.

Some parents prefer to take safety into their own hands, setting curfews and stocking living rooms with teen bait—gaming consoles, junk food, even alcohol. "If he's going to drink, I'd rather he do it safely in his own home," the thinking goes.

And then there's "choice." A core tenet of teen risk-intervention strategies focuses on teaching teens to make "smart choices." But extensive, new neurological research shows that risk taking at this age isn't really

---

*In some parts of the world, HIV/AIDS is the leading cause of death in all age groups.

a "choice." Profound changes deep in the adolescent brain allow impulsive action to override prudent inhibition. Transitioning teens thrill to novelty. They're drawn to groups of peers. More than adults, they search for ways to stimulate their senses. Their emotional responses are more extreme.

Rest assured that if adolescent rats drove cars, they, too, would have exorbitant insurance rates. Researchers from Rome's Istituto Superiore di Sanità ran a mixed-age group of rats in a maze that ended with a tasty treat. To reach the reward, the rats had to scurry across a narrow plank, suspended high above an open space, with no protective side walls.

Half the rodents flat-out refused to enter that section of the maze. However, of the half that dared, every single one was an adolescent. No babies or elders took the risk.

Adolescent rats display several other common behaviors. When placed in new surroundings, they have lower base levels of anxiety compared to other age groups. They are more impulsive about approaching unfamiliar objects. Novelty doesn't just interest them. It attracts them. They seek it out.

Similarly, when primatologists place unfamiliar items near vervet monkeys, adolescents are the fastest to rush over to investigate. Whether the objects are simply neutral, such as a cardboard box, or unusual but unthreatening, like a tree covered in lights and tinsel, or carry a degree of danger, like a fake tarantula or stuffed snake, adolescents are the ones that eagerly approach, gesture, call out alarms, and try to touch.

Even at great personal risk, adolescent animals seem almost to delight in exploring new things. Preadult zebra finches will approach humans, even sit on an offered finger, when the adults have long since fled. Transitioning sea otters begin venturing into new territories, like the Triangle of Death. Animal behaviorists and human neurologists agree that this sudden lowering of the fear threshold in humans and nonhuman animals stems from specific brain changes.

Said another way: risk taking is normal.

And not only is it normal; it's necessary and serves very specific purposes. For example, to survive on their own, animals need to know how to recognize predators. While the ability to spot threats is to some extent inborn, some of it must be learned during adolescence. For animals, Sun-tzu's classic military advice to "know your enemy" includes studying how he smells, hides, runs, and attacks. And one important way to gain this knowledge is to get close enough to see him in action.

One seemingly suicidal, but actually very effective, way of learning about predators is to gallop, swim, or fly right up to them . . . and live to tell the tale. As Tim Caro, the U.C. Davis biologist, writes in *Antipredator Defenses in Birds and Mammals,* "Young animals may approach and inspect a predator when they see it and perhaps learn about its characteristics, including its motivation and behavior."

For example, instead of hiding from prowling cheetahs and lions, immature Thomson's gazelles often stroll right toward them. Sometimes the young gazelles will trail their hunters for an hour or more, as though the big cats, not the small gazelles, are the quarry. Astonishingly, the discomfited predators frequently slink away, but not before the adolescent gazelles have gotten a good eyeful and whiff of the creature that someday might try to kill them. This practice, however, comes with casualties. According to a Cambridge University study carried out in Tanzania, the curious young gazelles die from feline fangs once in every 417 approaches (compared to one in 5,000 for adult animals). What animal behaviorists call "predator inspection" is seen widely in guppies, gulls, and other fish and birds. While predator inspection often continues into adulthood, the learning begins in animal adolescence—when inexperience also makes it more dangerous.

Luckily for animals, humans aren't the only species whose adults will show the younger generation the ropes. A widespread teaching technique used in species from birds and fish to mammals is something researchers call "mobbing." Moving together in one big group while vocalizing threateningly, a whole group of animals, including both experienced adults and developing adolescents, can intimidate a hunter into seeking its meal elsewhere. Mobbing is an effective antipredation strategy. But, as U.C. Davis animal behavior expert Judy Stamps pointed out to me, it has another crucial, though often overlooked, function.

"Mobbing is a way to impress upon the whole community that something dangerous is near," she told me. "If the whole group is making a huge racket, that helps younger animals learn the community's predators." Mobbing, she continued, is also safer than solo inspection. Young animals, she said, "aren't very good at evading predators." Moving toward danger in the protection of an adult-led mob provides youngsters with a safe and educational close-up view.

When I was in high school, I went through a quintessentially American rite of passage: I took driver's ed. Several decades later, the physical skills of steering, scanning the road, and signaling have become so engrained in my muscle memory, I really don't even remember learning them. But one part of driver's ed remains seared into my mind. Along with generations of neophyte drivers in California, I was made to watch a film called *Red Asphalt.* Produced by the California Highway Patrol, the movie shows scene after gory scene of traffic accidents. Blood rushes down gutters. Bodies lie akimbo under cars. Motorcyclists' limbs are smeared across the pavement. Drivers who spent their teen years outside California might remember being traumatized by other pieces of edu-propaganda, such as *The Last Prom,* which featured a crushed and bloody corsage on the side of the road, or videos with cautionary titles like *Wheels of Tragedy, Mechanized Death,* and *Highways of Agony.*

Movies like these have been terrifying teens for decades. Seen from an animal behaviorist's perspective, though, they may simply be a cinematic tool adults have created to compel human adolescents to inspect their biggest killer: motor vehicles.

Although the threat posed by cars is new, the techniques employed by *Red Asphalt* are age-old. From frightening campfire stories of what lies in the woods to 3-D surround-sound gorefests, human culture routinely uses tales of murder and peril to scare and then instruct. Not only are they age-old, but they're also incredibly popular. And who's drawn to them? Adolescents. Hollywood is populated by wealthy producers who figured out that, like young animals, teenagers will flock to horror movies and gaming worlds their parents have outgrown. A quick glance at the line for a monstrous roller coaster will similarly tell you all you need to know about which age group is drawn to the simulated danger but chemically identical adrenaline rush of a perilous fall. We may not think of these mass entertainments as evolutionarily linked to the antipredation strategies of other animals. But just like mature animals mobbing predators to teach youngsters, human grown-ups write the stories, produce the movies, and build the roller coasters—making money off teens' inherited physiological craving for calculated risk.

Learning to deal with threats doesn't involve only confronting them head-on. It also involves learning when and how to hide from them. For every parent who's been frustrated by a teen who won't meet their gaze, consider what direct eye contact can mean in the wild. It often means you've been targeted. While baby animals often stare at everything around them, adolescents must learn that catching the gaze of the wrong set of eyes can be deadly. Looking-away responses have evolved in many animals, from mouse lemurs to jewel fish. Staring at chickens and lizards causes them to become rigid. House sparrows take flight more readily when eyes are directed at them. Gaze aversion in animals begins during the transition from infancy to adulthood. Studies of humans note a surge of eye gaze aversion in the preteen and teen years.

While young animals are learning how to be vigilant, they can at times be overattentive, identifying threats where none exists. Some overreact to every rustling leaf, looming shadow, or strange smell. I once watched a group of about thirty sea otters startle at a loud noise that turned out to be a false alarm. As the frightened animals raced away to the other side of the lagoon, the adolescents led the way, cutting through the water with full-out swimming strokes. Leisurely pulling up the rear, carefully keeping their heads dry, were the mature otters, who'd had more experience with true danger.

As they test their danger-detection skills, inexperienced yet eager-to-learn vervet monkeys, beavers, and prairie dogs often cry or scream out unnecessary alarm calls. The older members of the group can be surprisingly forgiving toward youths who cry wolf (or jaguar, snake, or owl)—responding with a reassuring return call or simply ignoring the errant signals.

But learning to recognize and avoid predators is really just preparation for a vastly more important and riskier moment in most young animals' lives: leaving the nest.

The young of many species leave their families in adolescence, sometimes for a temporary journey of discovery, sometimes for good. Leaving home, a process behaviorists call "dispersal," varies from animal to animal, by species, and by sex. But whether undertaken by a caterpillar or a zebra, it's an exceedingly dangerous time of life.

Vervet monkeys make an interesting example because their social progression parallels many classic human tales of young men going off to prove themselves. These clever, cat-sized primates, found in sub-Saharan

Africa and on the Caribbean islands of St. Kitts and Barbados, have gray-green fur on their backs, whitish bellies, black faces, and wide, soulful brown eyes. Vervet childhood will sound familiar to many human parents. During an extended infancy that lasts for about a year, a baby vervet sticks close to its mother. At a year, the young monkey's circle widens to include adult members of the group. Yearlings of both sexes play boisterous chase-and-wrestle games.

As they move into their second year—which corresponds roughly to eight to ten years of age in people—the boys' play becomes more frenetic and intense. But the girls drop out of the rough-and-tumble games, their attention suddenly caught by different diversions: playing with infants and figuring out the social hierarchy in which they will spend their lives. Vervet females do not leave their birth group.

Vervet boys follow a different path. Males must strike out into the world on their own, leaving everything behind. Relatives and friends. Familiar foraging territory. Group and adult protection from predators.

But danger comes not just from isolation and the predators they may meet along the way. It's also in the social minefield they're heading into. They must join a new group of monkeys. Approaching and integrating into a vervet group makes our tortured process of applying to college or getting a first job seem almost easy. All alone and newly independent, the adolescent has to first locate a group of strangers. He has to approach them. Then he has to threaten, challenge, try to intimidate, and, finally, fight the dominant, mature males. But diplomacy is critical. If he comes on too strong, he will lose the respect and tolerance of the females in the group—which can be a deal breaker, since vervet groups are matrilineal and females wield the power. Vervet females will not tolerate being threatened. Scaring babies is also strictly taboo. So while he's trying to intimidate the males, the adolescent newcomer has to simultaneously charm the females.

Lynn Fairbanks, a professor of psychiatry and biobehavioral sciences at UCLA, has spent more than thirty years studying wild and captive populations of vervets. She told me that these weeks of male transition are extremely stressful, yet exceedingly critical. How the adolescent performs may affect his status and his access to mates, food, and shelter for the rest of his life. And intriguingly, Fairbanks has discovered that the males who transition most successfully are the ones that show a special willingness to "go for it."

As she told me, in vervets, a degree of impulsivity may be "necessary." It motivates males to leave home and take on the challenges and risks of getting into a new group.

While most vervet immigrants have to settle for second-tier status, the ones who become alphas in their groups share another common trait. Their brashness emerges strongly during adolescence but doesn't stay at peak intensity forever. After they achieve dominance, their impulsivity sinks down to more moderate levels. Fairbanks writes that her findings support "the idea that an age-limited increase in impulsivity in adolescence is not a pathological trait, but instead is related to later social success." In other words, acting a little cocky when you're a teen may not necessarily mean you're going to turn out to be a wild adult. It might even push you up the social ladder.

A similarly lowered risk threshold—indeed, a new *pleasure* in risk taking—likely propels nearly grown birds out of nests, hyenas out of communal dens, dolphins, elephants, horses, and otters into peer groups, and human teens into malls and college dorms. As we've seen, having a brain that makes you feel less afraid enables, perhaps encourages, encounters with threats and competitors that are crucial to your future safety and success. The biology of decreased fear, greater interest in novelty, and impulsivity serve a purpose across species. In fact, it could be that the only thing more dangerous than taking risks in adolescence is *not* taking them.

Linda Spear, a professor of psychology at the State University of New York at Binghamton and the author of *The Behavioral Neuroscience of Adolescence,* agrees. During years of studying the neurology of humans and other species, she has observed "age-specific behavioral characteristics." She explains that, although we notice behaviors in the context of what we call human "culture," adolescent transformations have biological underpinnings and "are deeply embedded in our evolutionary past."

In other words, what we observe as uniquely human adolescent behavior may in fact be shared physiology at work. Admittedly, humans do have a unique ingenuity for amping up the danger. When an adolescent rat or vervet monkey impulsively bursts out to explore something new, he's not also piloting a two-ton SUV full of his friends. A gazelle in thrilling pursuit of a cheetah isn't also tripping on the latest designer drug.

For human parents, knowing that brain and body shifts are causing predictable and universal behaviors is not going to relieve the worry of late nights or the anguish of seeing a tattoo around the ankle of a formerly straight-arrow daughter or son. It certainly won't ease the grief of a parent who's lost a child to what seemed like extreme or unnecessary risk. But putting adolescent impulsivity in a context that sees it as not just normal but physiologically and evolutionarily necessary may make bewildering behavior slightly more bearable.

A number of miles south of the Triangle of Death, near the Moss Landing power plant and slough, there's a sheltered lagoon. Beginning kayakers come here to practice paddling. Ecotourists can board an open-air safari boat to view harbor seals and pelicans. But the sightseers' biggest draw is a group of fifty or so sea otters calmly rafting, grooming, foraging, sleeping, spinning, and occasionally wrestling in the tranquil water.

I spent an overcast August morning observing the Moss Landing otters with Gena Bentall, a research biologist for the Sea Otter Research and Conservation Program at the Monterey Bay Aquarium who has spent thousands of hours documenting the behavior of these marine mammals. As her beagle, Harry, watched from his special bed in the back of her pickup truck, Bentall and I discussed the single distinguishing feature of this otter group: they're all male. Ranging in age from sleek, dark adolescents to grizzled mature adults, these he-otters use the Moss Landing site as a stopover rest area. After swimming long distances along the California coast to breed, explore other territories, or challenge other males, an otter will pull into the Moss Landing lagoon. Some are full-time residents. Some show up only at night. For some, the Moss Landing group is a part-time sanctuary. Food is plentiful; predators are minimal; responsibilities are few. It's a place where territorial males can pass time in a nonterritorial mode and young males can learn the ropes. The relaxed camaraderie of the group reminded me of a men's locker room—a place for growing and grown men to gather, groom, eat, nap, socialize, and play . . . without having to compete for females.

Adolescent male dolphins, elephants, lions, and horses, as well as the teen males of many primates, join so-called bachelor groups like this in the period between leaving their birthplace and starting their own families.

Adolescent African elephants, for example, use them to prepare for the "ritualism of male-male competition," by sparring with other males their own age. According to biologists Kate E. Evans and Stephen Harris, of the University of Bristol, adolescence is an "important learning period" for these young pachyderms, a time of transitioning from "the highly structured breeding herd" to the "much more fluid social system of adult males." Bulls fake-fight with each other as adolescents to determine who is dominant at that moment, and to learn the rules of "bull society."

These groups of young male elephants are especially friendly compared to groups of older animals; they greet one another with gestures like trunk entwining, ear flapping, trumpeting, and joyful defecation. Gena Bentall has cataloged similar familiar greeting behaviors among groups of sea otters, who shove, stroke, nose, and sniff one another. Male wild horses and zebras, which also migrate into all-stallion groups when they leave their birth herd, at about two or three years old, bond through roughhousing and frisky urination.*

Evans and Harris spotted a notable interloper in the adolescent elephant groups: older male bulls. But instead of treating these older males as unwelcome chaperones, the younger elephants seemed to prefer having them around. Evans and Harris write that the elders serve as mentors, socializing the younger elephants and helping them learn to "become dominant males without posing a competitive threat." They also reported that the presence of the mature bulls seemed, in some cases, to suppress testosterone-driven pugnacity in the younger animals.†

Sea otter bachelor groups also include males of mixed ages. Although Bentall did not speculate on whether the presence of the older males affected the younger males hormonally, she said that one of the ways the Moss Landing adolescents find their way to the sheltered lagoon in the first place is by following a mentor male.

---

*Female wild horses disperse from their birth herd, too—either on their own volition or after being chased out by their fathers. Instead of forming all-female groups, however, pre-adult mares integrate themselves into nearby herds, where, as the last to join, they are lowest-ranked.

†Notorious periods of markedly raised testosterone and dangerous behavior in adult male elephants are referred to as "musth." Distinctive physical features of musth include the foul-smelling, tarry sludge that drains from the temporal glands, next to the eyes. Young male elephants may experience "honey musth"—a milder prelude of adult musth, with lighter-colored, sweeter-smelling gland drainage.

For California condors, mentors have played a key role in reeling this endangered species back from the brink of extinction. In 1982, with just twenty-two of these enormous birds left in the world, biologists took emergency action with an accelerated breeding program. By carefully removing eggs from nests as soon as they were laid, the scientists were able to begin building a captive breeding population. By 1992, wildlife conservation teams were ready to reintroduce the condors to their natural habitats in California's redwood and mountain regions.

But they ran into an unexpected problem. They had modeled the release plan on the successful North American reintroduction of the peregrine falcon several years earlier. In that program, biologists had flooded the landscape with fledglings—young birds that were strong enough to fly but still sexually immature and in transition between needing parental care and being able to fend for themselves. The transitioning adolescent falcons had no trouble moving out into the surrounding territory and before long began breeding with one another, reviving the population of the species.

But the condors were different.

As Michael Clark, the head of the Los Angeles Zoo's California Condor Propagation Program, explained to me, unlike peregrine falcons, which are more solitary and don't need mentors, California condors are extremely social. They go through a long preadult phase, during which, by imitation and example, they learn complicated condor conventions for everything from foraging and feeding to resting and nesting. Key to their learning process is living in multiage groups where younger birds can observe older mentors. Hatched in incubators and raised by humans in condor orphanages, the early groups of chicks did not have this experience. Releasing the socially inept preadults created what Clark called a "*Lord of the Flies* situation." The inexperienced birds didn't know what to do when they were out on their own. Some ate garbage and got sick from malnutrition and poisoning. Not knowing better, some landed on telephone poles and electrocuted themselves. Many hung around the release sites before eventually, slowly, moving out into new territory. But perhaps most poignant of all, in the absence of competent adult leaders, some birds followed anything that soared—from eagles to hang gliders. One young bird flew from the Grand Canyon to Wyoming in a single day, dutifully following a false mentor and winding up miles away from home at the end of the day.*

Group life provides animals with many long-term benefits. But sometimes what pulls individual adolescents into groups are short-term brain-based rewards. As Alan Kazdin, a professor of psychology and child psychiatry at Yale and the director of its Parenting Center and Child Conduct Clinic told me, research has shown that simply being physically next to same-aged peers and engaging in activities with them activates pathways for dopamine and other reward neurochemicals.

"Having peers around is a reward and not having them around is felt as the opposite, which begins to explain your 14-year-old's sullen, moody, heedless demeanor around the house," he noted wryly in *Slate* magazine.

Although bachelor groups are seen in many species, adolescent animal groups are not always single sex. Transitioning albatrosses form coed groups called "gams" for several months between fledging and starting their own families. Although they mingle with the opposite sex, they don't mate. Zebra finches, too, congregate in mixed-sex peer groups. Males fine-tune their courtship songs for females and practice outsinging other males. Boy and girl finches preen together; every once in a while the young birds' groups break apart so the kids can fly back to their parents' nests and beg for food—a tendency that may sound familiar to many human moms and dads.

Ancient adolescents also formed groups. One fossil band of dinosaurs was found in a ninety-million-year-old lakebed in Mongolia. They were all between the ages of one and seven—still a few years away from their species' sexual maturity, typically seen at ten years of age. Paleontologists suggested that these two-legged plant eaters may have roamed together in social herds without adult supervision.

Pink salmon, too, grow up entirely without the watchful eyes of parents. A few days after hatching, they emerge from their gravel nests and, under cover of darkness, begin to migrate downstream to the

---

*Thanks to the Los Angeles Zoo, the San Diego Zoo and Wild Animal Park, and Mexico's Chapultepec Zoo, California condor rehabilitation has come a long way since those early days. Hand-reared condor chicks are now exposed to adult mentors in mixed-age groups and are socialized extensively in preparation for release. The wild California condor population now numbers around two hundred and stretches across California, Arizona, and northern Baja California.

open ocean. Before diving into the wide water of the northern Pacific, however, the young fish stop for a week or two in the shallow, calm waters of coastal estuaries. It's here, in this safe environment, that the preadults start figuring out how to swim in a school. First, they group in twos and threes. A few days later, larger groups of five and six eventually combine into a much larger formation. Their daily schedule is much like a human adolescent's. Mornings and afternoons are spent in the school. At nightfall, the groups break up, and the young fish drift individually around the surface of the water before coming together again in the morning. As the salmon learn the choreography and conventions of their fish life, they are also figuring out where they fit into the salmon social hierarchy. With no adult fish to model ideal behavior, the adolescents rely on innate instincts and trial and error to figure out how to get the best feeding spots and secure dominance as adults. It had never occurred to me that fish need to learn the iconic synchronized swim patterns we call schooling. Or that some would be better at it than others.

Schooling, herding, or flocking—moving within and belonging to a group—gives protection to an individual exiting infancy. A group means more lookouts, more eyeballs, more voices to raise danger alarms. But it comes with a price. Individuals coming together to form a group must learn to be inconspicuous. One fin sticking out in an odd direction, a zig when the rest of the group zags, flashing white fur or feathers when all its peers are wearing gray—anything odd or conspicuous makes an animal more obvious to a predator. Adolescent lessons in blending in may serve an animal well for the rest of its life.

We humans don't literally flock or herd or school. But perhaps, if we listen carefully enough to our own adolescents' cries to fit in, we can hear faint echoes of an evolutionary past in which conspicuousness attracted danger. Perhaps this suggests that before parents condemn a child's desperate plea for the "right" Nikes or jeans as materialistic—or dismiss it as overly conforming—they consider a different perspective: the powerful adolescent drive to fit in may represent a precious and ancient protective evolutionary legacy.

Whether it's the Moss Landing otters erupting into a rough-and-tumble wrestling session, Tanzanian gorillas walloping each other in a game of tag, or pink salmon learning to school, peer groups give adolescent animals the chance to practice social behaviors and assess their place in the group. Like high school students figuring out whether they will be jocks and cheerleaders or drama geeks and mathletes, animals go through a similar sorting process. They get a sense of their competition and their community—what it takes to fit in and what it takes to win.

But groups have a troubling flip side. Although they can be safe, pleasant, and necessary, peer groups are not passive sanctuaries that simply shelter young humans and animals until they're ready to burst into the adult world. Groups are elaborate social laboratories, places for young animals to practice adult behaviors. And for social animals, and perhaps especially long-lived social mammals, one of the most important things they're sorting out is social status.

Sometimes the biggest risk for animal adolescents comes not from outside predators but from members of their own species. Susan Perry notes in *Manipulative Monkeys,* her entertaining book about her decades studying capuchin monkeys in the forests of Costa Rica that "the main cause of mortality in capuchins is conflicts with other capuchins." Rival gangs vying for territory, mates, and resources account for much of this violence. But peer groups pose other unique dangers. They tempt, cajole, and shame individuals into doing things that, on their own, they would never do. She told me about monkeys she'd observed that had "extremely high social intelligence" and "great interpersonal skills" whose behavior deteriorated into violent mayhem when they became part of a bachelor group of other adolescents.

She followed one monkey in particular—a youth named Gizmo, who fell in with a gang of seven other males her research team called the Lost Boys. As a young monkey, Gizmo had been appropriately socially deferential and seemed to be headed for a stable, if exactly not illustrious, life in a capuchin troop. But as he emerged from childhood, Gizmo started getting drawn into dangerous situations. Egged on by his socially impulsive brother, Gizmo would end up in brawls with larger, older males. He invariably got soundly thrashed.

As Gizmo accumulated scars and broken bones, his gang also began attracting new members. Soon they totaled eight, each boy more battle scarred and unsuccessful in love than the next. It spiraled out of control. They kept roving and terrorizing the neighborhood, never able to settle down into a coed, mixed-age stable family group. When Perry told me about the Lost Boys, she spoke with the resignation of a high school teacher who can only watch sadly as some of her students inevitably slide into delinquency.

"Their problem," Perry said, "was that their group was just too big. The other troops of capuchins seriously resisted their immigration attempts when they saw eight adolescent male monkeys coming." Perry emphasized that migration by all-male groups is a normal and necessary life stage—there's no safe way to do it, yet they all have to go through it. What was striking about the Lost Boys was that, because their group was so big, they got stuck in the transition stage. Shunned from capuchin society, Gizmo ended up dying a pariah, never having achieved useful social status within the larger group.

And so it goes for human teens, too. "Delinquency and criminal behavior . . . are more likely to occur in groups during adolescence than they are during adulthood," writes Laurence Steinberg, an adolescence expert at Temple University. Drinking, risky driving, sexual risk taking—all are more prevalent, more dangerous, and more likely to occur in groups of adolescents.

For animals and humans alike, falling in with—or afoul of—the wrong crowd can have deadly consequences.

In September 2010, six teens—Raymond Chase, Cody J. Barker, William Lucas, Seth Walsh, Tyler Clementi, and Asher Brown—all died of the same cause. Although they ranged in age from thirteen to nineteen and lived in different states, their deaths were linked by one sad commonality: all six had killed themselves after being bullied.

Their deaths were added to the rolls of the several thousand other teen suicides in the United States in 2010. Suicide is a major adolescent human health threat—among eight- to twenty-four-year-olds nationwide, it's the third most common cause of death.

Like adults who commit suicide, teens who kill themselves usually have an underlying mental illness—in particular, depression or depressed mood. However, one familiar aspect of the adolescent emotional profile may make this age group especially vulnerable to suicide:

their increased impulsivity. With access to physical and pharmacological weapons of self-destruction, an impulsive teen can tip a difficult situation into a deadly one.

Psychological "autopsies," the extensive interviews and investigations conducted by psychiatrists after suicides, have shown that the triggers for teen suicide are remarkably similar across cases. Loss—such as the death of a close friend or family member. Or a best friend's moving out of town, especially for teens who have few friends. Rejection—by a girl- or boyfriend. Deep embarrassment—being kicked off a team, failing an important exam, enduring a humiliating public reprimand by a teacher.

Loss, rejection, embarrassment. The kinds of experiences that are triggers for human suicide also occur within animal groups. But animal behaviorists give them different names: isolation, exclusion, submission, and appeasement. Along with loss, rejection, and embarrassment, these terms describe the complex mixture of reactions and behaviors that contribute to the dynamics of social status within animal groups.

Determining and maintaining status occupies much of the activity within groups of social animals. Aggression by dominants against subordinate members of the group is seen commonly, in animals including sea otters, sea birds, wolves, and chimpanzees. And social hierarchies are in constant flux. A position at the top is never secure. As many animal behaviorists have pointed out, picking on subordinates is a useful, public way for dominant animals to display and preserve their top-dog positions. Although not every animal can be an alpha, top-tier rank carries important benefits, often including exclusive control of mates, food territories, and shelter.

In humans, we see dominants aggressing against subordinates all the time, but we use a more colloquial term: bullying. For years the rap on bullies was that they were insecure, the kids who "feel bad" about themselves. Picking on others, it was believed, momentarily raised their self-esteem. But recent research suggests that bullies feel, on the whole, pretty good about themselves. Their self-esteem is just fine. In fact, bullies often sit comfortably at the top of the social food chain, surrounded by hangers-on, wannabes, and silent bystanders who are more than happy just to be out of the bullies' line of fire.

If animal and human bullying share some common purpose, it may be exactly that: a demonstration of strength and dominance, and a cautionary lesson to anyone who might challenge the status quo. This cross-species perspective on bullying offers insights into why bullies often emerge from the top, not the bottom, of human social hierarchies.

Animals can also help us understand how human bullies choose their victims. In some animal groups, being different can increase an animal's vulnerability.

Not unlike animal predators, human bullies are constantly on the lookout for something that makes their victims stand out a little from the crowd. In North America, a common target of bullies is boys who are—or are perceived to be—gay. In fact, the six September suicides of 2010 had another thing in common besides month and year. All six teens killed themselves after being harassed for appearing to be gay.

How much actual "bullying" occurs among animals is hard to say exactly. If we define bullying as aggression by a dominant animal on a subordinate, then quite a lot. Wildlife biologists and veterinarians frequently characterize male-on-male roughhousing in which no serious injury occurs as "playing." Indeed, when you watch groups of young animals in rough-and-tumble play, whether they're sea otters, dolphins, horses, capuchins, condors, kittens, or puppies, the line between "play fighting" and "bullying" may be unclear. Just as bullying can be invisible to an adult parent, some of the "sparring" or "mock fighting" we see in animal groups may be more intense and purposeful than we have previously thought.

Among animals, peer oppression sometimes comes at the claws or beaks of siblings. The Oxford zoologist T. H. Clutton-Brock described the formative impact of bullying in blue-footed boobies in his *Nature* article "Punishment in Animal Societies." Blue-footed boobies are normally born two to a nest. The first egg to hatch is usually the dominant sibling, a power he or she lords over the second-born by fierce authority displays involving pecking and jostling. Even if the younger chick eventually grows larger than its tyrannical nest mate, their early, in-nest dominance relationship remains for life.

That a propensity to bully may be transmitted across generations was recently explored in another bird, the Nazca booby. When the parents of these Pacific seabirds leave to feed, older, unrelated boobies fly to the unprotected nests and abuse the chicks. Grabbing the youngsters' necks and heads in their orange-and-black beaks, the larger birds squeeze with nutcracker intensity as the small chicks pull away submissively, hiding their bills in their downy chests. Biologists have observed a particularly interesting pattern of abuse: the birds that were attacked the most frequently as chicks were later, as adults, the most likely to attack other youngsters. These Pacific seabirds may be nature's example of "the victimized becoming the victimizer."

The human depression linked to bullying may be uniquely dangerous for impulsive teens. Yet in animals, muted, submissive, perhaps even depressed responses to being picked on may actually make some animals *safer*. Following a violent conflict over hierarchy, the losing animal may be smart to withdraw and not push his luck with repeated challenges. Numerous animal studies have demonstrated that failing to cry uncle results in escalating attacks by the dominant.

While every movie and comic book involving bullies ends with the victim fighting back and often taking the bully down, such revenge fantasies don't often get played out in the animal kingdom. Slinking off to lick your wounds and perhaps find another path often makes more sense than running back and fighting the same bully again and again.

Comparing animal and human behaviors will not bring us to a prescription for "solving" or "curing" complicated human social interactions like bullying. But a species-spanning approach might be able to show us where to start looking.

As far as we know, when an unprotected seabird is victimized, or an unpopular vervet ostracized, or a curious young otter killed during its first solo foraging trip, few adult tears are shed—except, perhaps, by an empathetic field biologist observing through binoculars. But parental care does occur across species. Whether it's a hagfish excreting a protective coating of slime over a clutch of eggs and just swimming away or a Gombe chimp demonstrating a termite-fishing technique to a juvenile, animal parents of all kinds are invested in how their transitioning offspring fare.

Even when they're old enough to live and breed on their own, some animals receive parental care long after they are capable of feeding themselves. The parents of Kloss's gibbons, for example, help defend a child's territory until that offspring can find a mate. Three-toed sloth mothers, like arboreal, insect-eating helicopter parents, vacate a portion of their own territory to assist their offspring in starting its own mature life.

Of course, the parents of an adolescent narwhal, bowerbird, or otter will interact very differently with their transitioning offspring than a human parent will—whether a Japanese *ryosai kenbo* (good wife, wise mom), Russian *mat' geroina* (hero mother), or North American tiger mother. Brains are different. Social structures are different. Development, genes, and environments are all different. Species are different. But a zoobiquitous consideration of parenthood uncovers an embedded reality for mothers and fathers of all species: a parent's genetic legacy depends on its offspring's survival and reproductive success.

For some incredibly unlucky human parents, adolescent risk taking and impulsivity will result in tragedy. Early exposure to alcohol and drugs will put their children in the path of injury, accidental death, and addiction. And the social minefields their children must traverse may exact casualties in the form of severe depression—or even suicide.

If you're a parent, this knowledge won't shrink the lump in your throat while you try to suppress an angry outburst following a missed curfew. It may not stop your fingers from brushing eye-obscuring hanks of hair off your adolescent's face. It's doubtful it could quell the pounding in your chest when you open the e-mail containing your teen's SAT scores. And it's very unlikely to squelch the involuntary screams flying out of your mouth in the final seconds of your child's sports tournament.

But when you find yourself emotionally activated by your teen's behavior, appearance, or prospects, a species-spanning approach might save you a trip to a psychotherapist's office. Instead of blaming "culture" or looking for the early childhood experiences in your own life that contributed to your overreaction, perhaps take a moment to peer much, much further "left" on the evolutionary timeline—and consider the ancient animal roots of your parenting.

And you might take heart from the story of Robert's son, Charley. At sixteen, Charley seemed to be off on a bad course. Bored and unmotivated, he was on the verge of academic collapse. Teachers bemoaned his lack of focus; they said he made no effort unless the topic personally interested him. To make matters worse, Charley preferred riskier pursuits: joyriding and target shooting. When Charley finally enrolled at a university, drinking and smoking became his trademark among his peers.

Robert was in despair. Time after time, he'd tried to get his son to buckle down, to focus on school and on life beyond age twenty. He put together an "emergency plan" to try to salvage his son's future. In one weak but memorable moment, he told Charley, "You will be a disgrace to yourself and all your family."

But don't worry about Charley. He didn't suffer too badly from his risk taking, his rebelliousness, his refusal to accept the world as his elders taught and thought it should be. In fact, he parlayed his iconoclastic nature into one of the most storied careers in the history of science. The mature Charley—Charles Darwin—even later forgave his father's tough parenting, saying, "My father, who was the kindest man I ever knew and whose memory I love with all my heart, must have been angry . . . when he used such words."

Modern parents can take comfort from the fact that most of our teenagers come through adolescence, too—perhaps a little bruised, maybe a little humiliated, but stronger for the journey.

After all, most capuchin monkeys don't go off with a gang and die alone. Most salmon figure out how to school; most vervets get into a new group; most gazelles learn to flee lions and go on to raise young gazelles of their own. And most California sea otters survive, and eventually leave behind, the Triangle of Death.

TWELVE

Zoobiquity

When crows by the hundreds began hobbling around and dropping dead on the sidewalks of Queens, New York, in the summer of 1999, Tracey McNamara felt a stab of dread. Rarely does a single species get sick and suddenly die off without other nearby animals coming down with symptoms, too. A few weeks later, the exotic birds under her care at the Bronx Zoo started falling like flies. McNamara knew an avian killer was on the loose. If she didn't identify it, and fast, it could wipe out her zoo's entire bird population.

McNamara, a veterinarian and the zoo's head pathologist, did two things right away. As a responsible employee, she called New York State wildlife officials to alert them to the alarming appearance of a deadly disease in the Bronx.

But McNamara, a no-nonsense Queens native with a doctorate from Cornell and years of experience analyzing tissues under a microscope, knew a thing or two about bird diseases. With a maverick streak and a love of a good medical mystery, she began her own investigation. Peering at magnified slides late into the night, surrounded by jars of preserved amphibians and exotic reptile fungi, McNamara searched for clues to the mystery of what was killing her birds. One thing was obvious. The killer was swift and ruthless—frying the birds' brains and ravaging other organs. They'd died of massive brain hemorrhages and heart damage. This pointed overwhelmingly to encephalitis—inflammation of the brain—caused by a virus. But which virus?

McNamara knew she had three prime suspects: the viruses that cause Newcastle disease, avian influenza, and eastern equine encephalitis (EEE), all of which notoriously attack birds. With the clock ticking, McNamara began a process of elimination. Newcastle disease and avian flu are highly contagious. Spreading from animal to animal, they can wipe out adjacent flocks in no time. But they couldn't be the culprit: the zoo's exotic flamingos and eagles were dying, yet the chickens and turkeys in the children's petting zoo were fine. McNamara crossed Newcastle and avian flu off the list. That left EEE. But, McNamara realized, the zoo's emus weren't sick. Healthy emus would seem to rule out EEE; the large, ostrichlike birds are particularly vulnerable to this virus and would certainly be showing signs of illness. With three strokes, McNamara had reduced her lineup of suspects to zero.

It had to be a different pathogen, one that didn't spread through bird-to-bird contact. That's when McNamara thought: *mosquitoes.* The petting zoo closed before sundown and opened well after sunrise. The chickens and turkeys were housed safely indoors at dawn and dusk, the prime mosquito feeding times. However, the exotic birds that were dying—the flamingos and cormorants and owls—were housed outside around the clock. This was not a comforting realization. If mosquitoes were indeed spreading this contagion, whatever it was, the birds weren't the only animals at risk. Any warm-blooded creature that provided mosquito meals—like the zoo's rhinos, zebras, and giraffes—was in danger. So, too, McNamara realized grimly, were New York–area human beings.

This was late August. Just a week or so before, emergency room doctors around New York had started tracking a mysterious illness cropping up in elderly people. It appeared to be neurological: patients were presenting with high fevers, weakness, and confusion. Some had signs of swollen brains—encephalitis. When the cluster of sick people reached four, an infectious disease specialist at a Queens hospital raised the alarm, and the Centers for Disease Control and Prevention (CDC) in Atlanta sent a team of epidemiologists to investigate. Because encephalitis was present, the CDC, too, was thinking, "Mosquito vector." As one of the researchers put it, "If you see encephalitis in the later summer, you have to think about viruses spread by mosquitoes." It had been a perfect year for the insect bloodsuckers. A long, dry spring followed by lots of rain and high humidity had created breeding conditions ideal for a population explosion.

After a few days and some tests on the spinal fluid of the sick people, CDC officials triumphantly announced that they had solved the mystery. It was St. Louis encephalitis (SLE). This brain-attacking disease leaves its victims, especially the elderly, with outcomes ranging from bad fever and neck stiffness to death. It has no vaccine, and while reasonably common throughout the South and the Midwest, it hadn't been seen on the East Coast since the 1970s. New York mayor Rudy Giuliani quickly rolled out a $6 million mosquito-abatement plan that included free insect repellant, reams of informational brochures, and a helicopter that sprayed malathion, a potent insecticide, over the city and its alarmed inhabitants.

That might have been the end of that. But there was one big problem with the diagnosis of SLE. As a veterinarian, Tracey McNamara knew it. The virus that causes SLE is passed when a mosquito first bites an infected bird and then, later, a human being. But the birds don't usually get sick from SLE. They don't usually die from it. They're just the carriers, the middlemen. When I spoke with McNamara at Western University of Health Sciences in Pomona, California, where she's now a professor of pathology, she was blunt.

"I had barrels full of dead birds," she told me. "It couldn't be St. Louis encephalitis." She explained that, although the CDC was ready to close the case, she couldn't stop thinking that the dead birds were connected to the sick people. And she knew she was racing the clock. Her avian death toll, especially around the zoo's flamingo pool, was mounting fast. If someone didn't correctly identify the killer, not only would the zoo lose most of its birds, but human beings would be waging a public health battle against the wrong disease. Shortly thereafter came word that two of the human patients had died.

For the rest of the summer McNamara puzzled over the dead street crows, her dead zoo birds, and their possible link to the human deaths from supposed St. Louis encephalitis. Labor Day weekend was the breaking point. Her bird population was ravaged; in rapid succession she'd lost a cormorant, three flamingos, a snowy owl, an Asian pheasant, and a bald eagle. Reports came of a human case of SLE—in Brooklyn. The contagion had spread to a new borough. McNamara stopped following official protocols and called the CDC herself. She offered to share her barrels of downy corpses and the knowledge she had gleaned from all her work in the lab. As she put it, she'd already "ruled out the usual suspects for them"—including St. Louis encephalitis.

Expecting gratitude for the offer of her data set, she was unprepared for what came next. After a terse exchange she calls "condescending," the CDC official she spoke with told her in no uncertain terms that the agency would be sticking with its diagnosis of SLE. She could keep her birds, and her concerns, to herself. The CDC solved human, not animal, outbreaks. McNamara was surprised by the slammed door—she says the official actually hung up on her—and bewildered by the repeated rebuffs when she called back.

McNamara—and, at that moment, the health of animals and humans throughout New York—had fallen victim to that polarizing hypocrisy at the center of medicine and public health. Veterinarians and physicians rarely communicate with each other as equal colleagues.

In her lab in the Bronx, surrounded by dead birds and reports of dying people, and with seemingly no one in the human medical establishment willing to listen, McNamara felt that gulf acutely. Frustrated, yet determined to get to the bottom of the deadly mystery, she began working her other contacts. She sent infected bird tissue samples to a U.S. Department of Agriculture (USDA) lab in Iowa. A different lab in Wisconsin tested bird tissues for SLE and came up negative.

Then the Iowa lab turned up something so decisive and chilling that, McNamara said, it made her "hair stand on end." Whatever this pathogen was, it was only forty nanometers in diameter. And that probably meant it was a flavivirus—related to yellow fever and dengue fever. Working with flaviviruses requires special protective clothing and containment and disposal measures, none of which she had used in her lab while handling the specimens. "That night," she told me, "I went home and wrote my will." The USDA lab contacted the CDC with the latest findings. The CDC remained frustratingly unresponsive.

At two in the morning a few nights later, McNamara sat straight up in bed. She suddenly knew what she had to do. She needed a lab with a higher biohazard safety level. A lab with pathologists who'd seen everything and had a range of experience with infectious agents. "That's when it clicked," McNamara told me. "I had to call the army." The next morning she begged the U.S. Army infectious diseases lab at Fort Detrick, Maryland, to take a look. Within forty-eight hours and in a collaboration McNamara calls the "best of what science can be," the army lab had confirmed McNamara's suspicions. This wasn't St. Louis encephalitis. It *was* a flavivirus.

The virus turned out to be a mosquito-borne pathogen never before detected in the United States—indeed, in the entire Western Hemisphere: West Nile virus. At that point, CDC officials admitted they'd been wrong. They retracted their diagnosis of St. Louis encephalitis and announced the historic and disconcerting arrival of West Nile virus on North American shores. The pathogen quickly migrated across North America, reaching California in 2003. Now it resurfaces each spring and summer throughout the United States, Canada, and Mexico with that year's crop of hungry mosquitoes.

It's hard to say how many lives would've been saved had the human medicine establishment listened to a veterinarian from the beginning. The 1999 West Nile epidemic killed seven people and caused sixty-two known cases of encephalitis. In the years since it first emerged, it's believed to have caused nearly thirty thousand people to get sick and more than a thousand to die. Then there are the animal casualties: thousands of wild and exotic birds—and quite a number of horses—that died from the virus, silently and uncounted.

But the misdiagnosis was a turning point of sorts for public health in the United States. In a report to Congress detailing the outbreak a year later, the U.S. General Accounting Office (now called the Government Accountability Office) admitted that the experience could "serve as a source of lessons" for preparing public health officials to deal with crises of "uncertain causes." (The report, dated exactly one year before the 9/11 terrorist attacks, also suggested that the West Nile event could model how to guard against biological terrorism.)

Tucked beside the usual calls for better communication among government agencies was what at the time was a striking proposal: "The veterinary medicine community should not be overlooked." The CDC heeded the GAO's call and created a new department in 2006: Zoonotic, Vector-Borne, and Enteric Diseases. Charged with monitoring food safety and bioterrorism, it was, significantly and symbolically, headed by a veterinarian, not an M.D. (After only a couple of years, the fledgling department was rolled into a bigger division called the National Center for Emerging and Zoonotic Infectious Diseases.)

Other groups in the United States and around the world have begun to adopt a more species-spanning outlook as well. Bird-watchers, hunters, hikers, and field geologists are invited to upload information about sick or dead animals they encounter onto web-enabled tracking sites that monitor wild-bird and other animal-borne illnesses. The University of Pennsylvania has long had close ties between its veterinary and medical schools, as have Cornell and Tufts. The Canary Database, named for the proverbial sentinels in coal mines and headquartered at the Yale School of Medicine, is a clearinghouse for information about zoonoses (diseases that spread from animals to humans, like West Nile virus and avian influenza), possible bioterrorist attacks, endocrine-disrupting chemicals, and household toxins such as lead and pesticides. The U.S. Agency for International Development (USAID) has funneled hundreds of millions of dollars into the Emerging Pandemic Threats program, which has a mission statement that couldn't be any clearer: "To pre-empt or combat, at their source, newly emerging diseases of animal origin that could threaten human health."*

---

*The program brings together academic institutions, government agencies, and private entities: the University of California, Davis, School of Veterinary Medicine, the Wildlife Conservation Society, the EcoHealth Alliance, the Smithsonian Institution, Global Viral Forecasting, Development Alternatives Inc., the University of Minnesota, Tufts University, the Training and Resources Group, Ecology and Environment Inc., the World Health Organization, the United Nations Food and Agriculture Organization, the World Organization for Animal Health (OIE), FHI-360, the CDC, and the USDA. It's divided into four projects: "PREDICT (to monitor for emergence of infectious agents from high-risk wildlife), IDENTIFY (to develop a robust laboratory network), PREVENT (focusing on behavior change communication to help people avoid high-risk practices that could lead

Jonna Mazet, the U.C. Davis veterinarian who runs a portion of the USAID program called PREDICT, has arguably the most daunting job: like a CIA officer monitoring terrorist activities, she scrutinizes the "viral chatter" coming out of global hot spots in the Amazon, the Congo Basin, the Gangetic Plain, and Southeast Asia. "We don't know what diseases are out there," Mazet says. "We have . . . to try to find the unknown before it spills out and makes the next pandemic. Some people call us virus hunters."

Yet even with government money from many countries around the world, as well as funding from international charities, the discrepancy between what goes into preventing disease outbreaks and what goes into postoutbreak triage is huge. "Over $200 billion of economic losses over the past two decades has gone into *responding* to disease outbreaks," says Marguerite Pappaioanou, an epidemiologist, veterinarian, and former executive director of the American Association of Veterinary Medical Colleges. "The money is clearly there. It's a question of where we decide to spend it." In other words, if an ounce of prevention is truly worth a pound of cure, then strengthening these programs would not only massively decrease suffering and death, but could also save money.

Yet a small but growing number of veterinarians and doctors have realized something in the last few years. The health of all our patients depends on opening a permanent, two-way dialogue. We don't have to leave collaboration to government policy makers and academic institutions—although their work is critical. We can treat the shared diseases of all animals, including humans, by taking a *multispecies*—that is, zoobiquitous—approach in our daily practices.

The effort can be truly low tech and grassroots. Recently a third-year veterinary student on the island of Grenada was holding a free vaccination clinic for neighborhood dogs and cats. She was approached by a local woman, who angrily asked why the animals received free health care while the people were left to fend for themselves. Realizing she had no good answer, the resourceful student, Brittany King, set to work creating a One Health clinic. She recruited students from a nearby medical school and started holding events that offered free

---

to transmission from animals to humans), and RESPOND (to expand family planning services and improve reproductive health in developing countries)."

vision, hearing, blood pressure, and breast exams for people . . . and vaccinations, wound treatment, deworming, and nail trims for animals. The students handed out flyers describing common zoonoses and encouraged people to be on the lookout and to report symptoms they observed in their animals.

A program at Tufts, in Massachusetts, has paired children and dogs with similar heart ailments to help demystify the condition for the children and their concerned parents. Similarly, Winter, the dolphin with the prosthetic tail portrayed in the 2011 movie *Dolphin Tale,* inspires children who have artificial limbs.

My own zoobiquitous journey has utterly changed how I practice and teach medicine. Along with Stephen Ettinger, a pioneer and leader in veterinary medicine, I've started teaching a course on comparative cardiology to UCLA medical students. Recently my cardiology colleagues and I sat, rapt, listening to one of Ettinger's former students present an intriguing case of a life-threatening arrhythmia. It was the sort of medical mystery physicians enjoy as much as reading a chapter in an Atul Gawande book or watching a good episode of *House, M.D.* But in this case, the patient was a rottweiler mix named Shakespeare. The diagnostic strategy we arrived at—from lab tests to medications—for the four-legged patient was all but identical to one we'd recommend to a human patient with a similar disorder.

Along with faculty from the U.C. Davis School of Veterinary Medicine and the veterinarians at the Los Angeles Zoo, I hosted a conference at UCLA School of Medicine in 2011 that brought together physicians and veterinarians who take care of the same diseases in different species. More than two hundred doctors and students from both sides of the species divide spent an academic morning session at UCLA hearing about patients, both animal and human, who were suffering from brain cancer, separation anxiety, Lyme disease, and heart failure. During afternoon "walk-rounds" at the Los Angeles Zoo, veterinarians and physicians compared notes on the animal patients: a rhino recovering from cancer, a lion who'd survived a near-fatal heart condition, condors fighting lead poisoning, and a monkey under treatment for diabetes.

You could say one of the most exciting new ideas in medicine today is something our ancestors took for granted and we somehow forgot—that humans and animals get the same diseases. By working together, physicians and veterinarians may be able to solve, treat, and cure patients of all species.

There is, after all, something truly awe inspiring about seeing the world through genetic and evolutionary connectedness—almost a unified field theory of biology. It reminds us of our shared predicaments; it broadens our empathy and understanding.

And it keeps us safer. Preventive medicine isn't just for people. Keeping animals healthy ultimately helps keep humans healthy. And appreciating these crucial connections readies us to face and fight the next contagion.

Ten years after West Nile hit New York, the world's public health systems mobilized to fight another zoonosis: swine flu, or H1N1.* One of the many headlines from this 2009 pandemic was a disconcerting fact. During its infectious journey around the world, the "human" flu virus had acquired genetic material from pig and bird flu viruses

While the general public may have been surprised by this news, veterinarians and physicians were not. Influenza viruses are notorious shape-shifters. They mutate easily, which is why every year brings a new flu vaccine—each one a variation on preceding themes. But flu viruses can perform another trick. If two different strains, say pig and human, find themselves occupying a single cell in your body at the same time, they can literally trade sections of their genetic code with each other. A new, blended virus can result.

What veterinarians know—and physicians might not—is that flu viruses prowl many animal populations besides pigs and birds. Specific strains of dog, whale, mink, and seal flu have all been identified. Given the opportunity, they could blend with the human strain. Although these volatile viruses haven't, as of this writing, crossed over into human populations, they are being closely tracked by veterinary epidemiologists.

The 2009 swine flu outbreak was but the latest wave in an ocean of diseases emerging from the jungle, the factory farm, the beach, the back-

---

*In fact, swine flu started in human beings; we're the ones who gave it to pigs, so it is technically what is called a reverse zoonosis. But because it was passed back to humans by pigs, after having traveled through the bird population, it's also considered a zoonosis.

yard bird feeder . . . perhaps even the doghouse and the litter box. The avian flu scare of 2005, the severe acute respiratory syndrome (SARS) panic of 2003, the monkeypox eruption the same year, the Ebola worry of 1996, the mad cow terror in Great Britain in the late 1980s—exotic zoonoses are nothing new. Think of a big, infectious killer and it's probably zoonotic, spread or harbored by other animals. Malaria. Yellow fever. HIV. Rabies. Lyme disease. Toxoplasmosis. Salmonella. *E. coli.* These all started in animals and then jumped into our species. Some spread to us via insects like fleas, ticks, and mosquitoes. Others move around in feces and meat. In some cases, the pathogens leave their animal reservoir, mutate, and evolve into bespoke superbugs especially tailored for human-to-human spread.

The *E. coli*–tainted fresh baby spinach that killed three North Americans and sickened more than two hundred in 2006 was traced to the feces of wild pigs in the fields. One of the world's worst outbreaks of the eerily named Q fever struck the Netherlands in the late 2000s.* Thirteen people died and thousands fell ill from the bacterial infection that spread to humans from infected goats on nearby farms.

The threat posed by animal diseases is unnerving enough when they travel among us on their own, without malice or intentional assistance. But, like the loose Soviet nukes we fear may one day end up in the hands of terrorists, zoonoses can also be deliberately wielded against us. Five of the six top organisms that according to the CDC "pose a risk to national security" began as animal diseases: anthrax, botulism, plague, tularemia, and viral hemorrhagic fevers.†

In a world where no creatures are truly isolated and diseases spread around as fast as jets can fly, we are all canaries and the entire planet is our coal mine. Any species can be a sentinel of danger—but only if the widest array of health-care professionals is paying attention.‡

---

*"Q" stands for "query" because when the disease first struck, in the 1930s, its cause was a mystery. Although the *Coxiella burnetii* bacterium was later isolated, the name had already stuck.

†The sixth agent on the list, smallpox, was eradicated by worldwide vaccination programs in part because it's not a zoonosis: it didn't have an animal reservoir.

‡In March 2007, American house pets sounded the alarm. When dogs and cats began getting sick and dying of kidney failure in massive numbers, veterinarians jumped on the case. The problem was traced to tainted pet food, leading to a huge recall across the

Our essential connection with animals is ancient, and it runs deep. It extends from body to behavior, from psychology to society—forming the basis of our daily journey of survival. This calls for physicians and patients to think beyond the human bedside to barnyards, jungles, oceans, and skies. Because the fate of our world's health doesn't depend solely on how we humans fare. Rather, it will be determined by how *all* the patients on the planet live, grow, get sick, and heal.

United States. It turned out that Chinese wheat gluten manufacturers were adding the chemical melamine to their product in order to raise the perceived protein levels and were then selling the gluten to pet food manufacturers. Forewarned by the veterinarians, U.S. food safety and public health officials quickly placed stringent anti-melamine inspections in place for the human food supply. (Unfortunately, Chinese officials didn't put the same measure into effect in time to save hundreds of Chinese babies, who were sickened and in some cases killed by melamine-tainted infant formula.)

Animals can also be sentinels for threats that are not infectious. Animal abuse is very strongly linked to child and domestic violence; British police, for example, have found that when child abuse is suspected in a home, incidents of animal abuse are often reported there first. Mistreatment of animals, particularly cats, strongly presages future antisocial and violent behavior against people. As Melissa Trollinger details in an article about the links between animal cruelty and human abuse, the mass murderers "Jeffrey Dahmer, Albert DeSalvo (the 'Boston Strangler'), Ted Bundy, and David Berkowitz (the 'Son of Sam') all admitted to mutilating, impaling, torturing, and killing animals in their youth."

Acknowledgments

Zoobiquity—the book, the conferences, the research initiative—has been possible only because of the great generosity, support, collegiality, and openness of the hundreds of veterinarians, wildlife biologists and physicians we have met over the course of this project. For sharing their time and tremendous knowledge, and for welcoming us and embracing *Zoobiquity*, we're deeply grateful to each and every one of these doctors. For special support and for their field leadership on the veterinary side, we'd like to thank Stephen Ettinger, Curtis Eng, Patricia Conrad, and Cheryl Scott, as well as the following DVMs: Melissa Bain, Stephen Barthold, Philip Bergman, Robert Clipsham, Vicki Clyde, Lisa Conti, Mike Cranfield, Peter Dickinson, Nicholas Dodman, Kirsten Gilardi, Carol Glazer, Leah Greer, Carl Hill, Malika Kachani, Laura Kahn, Bruce Kaplan, Mark Kittleson, Linda Lowenstine, Roger Mahr, Jonna Mazet, Rita McManamon, Franklin McMillan, Tracey McNamara, Dan Mulcahy, Hayley Murphy, Suzan Murray, Phillip Nelson, Patricia Olson, Bennie Osburn, Marguerite Pappaioanou, Joanne Paul-Murphy, Paul Pion, Edward Powers, E. Marie Rush, Kathryn Sulzner, Jane Sykes, Lisa Tell, Ellen Weidner, Cat Williams, and Janna Wynne.

We are deeply grateful to the many members of the human medical and scientific community who have provided us with support and wise counsel: C. Athena Aktipis, Allan Brandt, John Child, Andrew Drexler, Steven Dubinett, James Economou, Paul Finn, Alan Fogelman, Patricia Ganz, Atul Gawande, Michael Gitlin, Peter Gluckman, David Heber, Steve Hyman, Ilana Kutinsky, Andrew Lai, John Lewis, Melinda Longaker, Michael Longaker, Aman Mahajan, Randolph Nesse, Claire Panosian, Neil Parker, Neil Shubin, Stephen Stearns, Shari Stillman-Corbitt, Jan Tillisch, A. Eugene Washington, James Weiss, and Douglas Zipes. A number of groups also provided access and advice: the Great Apes Health Project, American Association of Zoo Veterinarians, UC Davis

School of Veterinary Medicine, Western University of Health Sciences College of Veterinary Medicine, National Evolutionary Synthesis Center, David Geffen School of Medicine at UCLA, UCLA Division of Cardiology, One Health Initiative, and One Health Commission.

We are indebted to many other friends and colleagues who shared their time and wisdom with us by reading chapters or the whole book: Sonja Bolle—whose keen editorial instincts also brought us together for this project and Daniel Blumstein—whose expertise and kindness provided much needed support and confidence. In addition, our gratitude goes to David Baron, Burkhard Bilger, Emily Beeler, Chris Bonar, and Michael Gisser whose insights and thoughtful suggestions vastly improved the manuscript. Special thanks also to Stephanie Bronson, Susanne Daniels, Beth Friedman, Eric Pinckert, Eric Weiner, Deborah Landau, and Kathleen Hallinan.

The book was greatly enhanced by knowledgeable field experts who carefully read individual chapters for content and accuracy: Kalyanam Shivkumar, Mark Litwin, Tom Klitzner, Deborah Krakow, Greg Fonarow, Laraine Newman, Mark Sklansky, Kevin Shannon, Gary Schiller, Ardis Moe, Daniel Uslan, Mark D'Antonio, Michael Strober, and Robert Glassman.

Our appreciation goes to the team who worked tirelessly to make the first Zoobiquity Conference a tremendous success: Julio Lopez, Cynthia Cheung, Kate Kang, Wesley Friedman, and Meredith Masters. Thanks also to Zachary Rabiroff, Brittany Enzmann, and Jordan Cole for research support.

And with the deepest heartfelt gratitude we thank Jordan Pavlin, our extraordinary editor at Knopf who championed Zoobiquity at every phase, and nurtured it (and us) with her deep experience, sure and steady hand, patience, passion, and vision. Special thanks go also to her assistants, Caroline Bleeke and Leslie Levine for their grace and enthusiasm; to Knopf's Paul Bogaards, Gabrielle Brooks and Lena Khidritskaya for their creativity and energy; to Chip Kidd for designing a beautiful cover; and to the entire Knopf production team for its tenacity and attention to detail.

A stroke of superb good fortune befell us when Tina Bennett became our literary agent. Brilliant and inspiring, incisive, diplomatic and funny, Tina is truly the best in the business, as are the other members of the magnificent team at Janklow and Nesbit, Stephanie Koven and Svetlana Katz.

Singular thanks go to Susan Kwan for taking on the heroic tasks of organizing and formatting the endnotes, bibliographies, and website—and for fact-checking much of the manuscript. Intuitive, inventive and resourceful, Susan has been a privilege to work with. She greatly improved the text and any mistakes that remain are ours.

Finally, this book simply would not have been written without the generosity and forbearance of our families. For their unflagging encouragement, intellectual contributions, and for enduring more than their share of dinner conversations that turned inevitably to the finer points of insect mating or the dread-inducing minutiae of cardiac distress, Kathryn would like to thank Andrew and Emma Bowers, Arthur and Diane Sylvester, Karin McCarty, and Marjorie Bowers. Barbara would like to thank Zachary, Jennifer, and Charlie Horowitz, Idell and Joseph Natterson, Cara and Paul Natterson, and Amy and Steve Kroll.

Notes

Writing *Zoobiquity* has been an exhilarating process of bringing together a tremendous amount of material from many different fields. To make the sourcing as user-friendly as possible, we've divided it into two categories.

For full citations of quotations and references from the text, please see the following endnotes.

For further reading and a complete bibliography of the books, journal articles, popular reporting, and interviews that we cite and which shaped and inspired our thinking, please visit www.zoobiquity.com.

ONE Dr. House, Meet Doctor Dolittle

4 *There was even an article:* A. M. Narthoorn, K. Van Der Walt, and E. Young, "Possible Therapy for Capture Myopathy in Captured Wild Animals," *Nature* 274 (1974): p. 577.

5 *Cardiology in the early 2000s:* K. Tsuchihashi, K. Ueshima, T. Uchida, N. Oh-mura, K. Kimura, M. Owa, M. Yoshiyama, et al., "Transient Left Ventricular Apical Ballooning Without Coronary Artery Stenosis: A Novel Heart Syndrome Mimicking Acute Myocardial Infarction," *Journal of the American College of Cardiology* 38 (2001): pp. 11–18; Yoshiteru Abe, Makoto Kondo, Ryota Matsuoka, Makoto Araki, et al., "Assessment of Clinical Features in Transient Left Ventricular Apical Ballooning," *Journal of the American College of Cardiology* 41 (2003): pp. 737–42.

5 *This distinctive condition presents:* Kevin A. Bybee and Abhiram Prasad, "Stress-Related Cardiomyopathy Syndromes," *Circulation* 118 (2008): pp. 397–409.

5 *But what's remarkable about takotsubo:* Scott W. Sharkey, Denise C. Windenburg, John R. Lesser, Martin S. Maron, Robert G. Hauser, Jennifer N. Lesser, Tammy S. Haas, et al., "Natural History and Expansive Clinical Profile of Stress (Tako-Tsubo) Cardiomyopathy," *Journal of the American College of Cardiology* 55 (2010): p. 338.

6 *Jaguars get breast cancer:* Linda Munson and Anneke Moresco, "Comparative Pathology of Mammary Gland Cancers in Domestic and Wild Animals," *Breast Disease* 28 (2007): pp. 7–21.

6 *Rhinos in zoos:* Robin W. Radcliffe, Donald E. Paglia, and C. Guillermo Couto,

"Acute Lymphoblastic Leukemia in a Juvenile Southern Black Rhinoceros," *Journal of Zoo and Wildlife Medicine* 31 (2000): pp. 71–76.

6 *Melanoma has been diagnosed:* E. Kufuor-Mensah and G. L. Watson, "Malignant Melanomas in a Penguin (*Eudyptes chrysolophus*) and a Red-Tailed Hawk (*Buteo jamaicensis*)," *Veterinary Pathology* 29 (1992): pp. 354–56.

6 *Western lowland gorillas:* David E. Kenny, Richard C. Cambre, Thomas P. Alvarado, Allan W. Prowten, Anthony F. Allchurch, Steven K. Marks, and Jeffery R. Zuba, "Aortic Dissection: An Important Cardiovascular Disease in Captive Gorillas (*Gorilla gorilla gorilla*)," *Journal of Zoo and Wildlife Medicine* 25 (1994): pp. 561–68.

6 *I learned that koalas in Australia:* Roger William Martin and Katherine Ann Handasyde, *The Koala: Natural History, Conservation and Management*, Malabar: Krieger, 1999: p. 91.

7 *In fact, a century or two:* Robert D. Cardiff, Jerrold M. Ward, and Stephen W. Barthold, " 'One Medicine—One Pathology': Are Veterinary and Human Pathology Prepared?" *Laboratory Investigation* 88 (2008): pp. 18–26.

7 *"Between animal and human medicine":* Joseph V. Klauder, "Interrelations of Human and Veterinary Medicine: Discussion of Some Aspects of Comparative Dermatology," *New England Journal of Medicine* 258 (1958): p. 170.

8 *Morrill Land-Grant Acts:* U.S. Code, "Title 7, Agriculture; Chapter 13, Agricultural and Mechanical Colleges; Subchapter I, College-Aid Land Appropriation," last modified January 5, 2009, accessed October 3, 2011. http://www.law.cornell.edu /uscode/pdf/uscode07/lii_usc_TI_07_CH_13_SC_I_SE_301.pdf.

8 *That's when a veterinarian:* Roger Mahr telephone interview, June 23, 2011.

8 *One of the first modern:* UC Davis School of Veterinary Medicine, "Who Is Calvin Schwabe?," accessed October 3, 2011. http://www.vetmed.ucdavis.edu/onehealth /about/schwabe.cfm.

8 *"One Health summit":* One Health Commission, "One Health Summit," November 17, 2009, accessed October 4, 2011. http://www.onehealthcommission.org/summit .html.

10 *"we do not like to consider":* Charles Darwin, *Notebook B: [Transmutation of Species]*: 231, The Complete Work of Charles Darwin Online, accessed October 3, 2011. http://darwin-online.org.uk.

11 *Octopuses and stallions sometimes self-mutilate:* Greg Lewbart, *Invertebrate Medicine,* Hoboken: Wiley-Blackwell, 2006: p. 86.

11 *Chimpanzees in the wild:* Franklin D. McMillan, *Mental Health and Well-Being in Animals,* Hoboken: Blackwell, 2005.

11 *The compulsions psychiatrists treat:* Karen L. Overall, "Natural Animal Models of Human Psychiatric Conditions: Assessment of Mechanism and Validity," *Progress in Neuropsychopharmacology and Biological Psychiatry* 24 (2000): pp. 727–76.

11 *Maybe Lady Diana or Angelina Jolie:* BBC News, "The Panorama Interview," November 2005, accessed October 2, 2011. http://www.bbc.co.uk/news/special /politics97/diana/panorama.html; "Angelina Jolie Talks Self-Harm," video, 2010, retrieved October 2, 2011, from http://www.youtube.com/watch?v=IW1Ay4u5JDE; Angelina Jolie, *20/20* interview, video, 2010, retrieved October 3, 2011, from http:// www.youtube.com/watch?v=rfzPhag_09E&feature=related.

11 *Significantly for addicts and their therapists:* Ronald K. Siegel, *Intoxication: Life in Pursuit of Artificial Paradise,* New York: Pocket Books, 1989.

11 *But examples of what appears:* McMillan, *Mental Health.*

12 *Not so long ago, paleontologists uncovered:* Houston Museum of Natural Science, "Mighty Gorgosaurus, Felled By . . . Brain Cancer? [Pete Larson]," last updated August 13, 2009, accessed March 3, 2012. http://blog.hmns.org/?p=4927.

15 *In 2005,* Nature *published:* Chimpanzee Sequencing and Analysis Consortium, "Initial Sequence of the Chimpanzee Genome and Comparison with the Human Genome," *Nature* 437 (2005): pp. 69–87.

15 Deep homology *is the term:* Neil Shubin, Cliff Tabin, and Sean Carroll, "Fossils, Genes and the Evolution of Animal Limbs," *Nature* 388 (1997): pp. 639–48.

17 *The longest tunnels:* Burkhard Bilger, "The Long Dig," *The New Yorker,* September 15, 2008.

17 *Cockroaches helped solve:* TED, "Robert Full on Engineering and Revolution," filmed February 2002, accessed October 3, 2011. http://www.ted.com/talks/robert _full_on_engineering_and_evolution.html.

TWO The Feint of Heart

19 *As trivial as it might seem:* Heart Rhythm Society, "Syncope," accessed October 2, 2011. http://www.hrsonline.org/patientinfo/symptomsdiagnosis/fainting/.

19 *emergency rooms handle more:* "National Hospital Ambulatory Medical Care Survey: 2008 Emergency Department Summary Tables," *National Health Statistics Ambulatory Medical Survey* 7 (2008): pp. 11, 18.

19 *About a third of all adults:* Blair P. Grubb, *The Fainting Phenomenon: Understanding Why People Faint and What to Do About It,* Malden: Blackwell-Futura, 2007: p. 3.

20 *writers from Shakespeare:* Kenneth W. Heaton, "Faints, Fits, and Fatalities from Emotion in Shakespeare's Characters: Survey of the Canon," *BMJ* 333 (2006): pp. 1335–38.

20 *Yet this kind of fainting:* Army Casualty Program, "Army Regulation 600-8-1," last modified April 30, 2007, accessed September 20, 2011. http://www.apd.army.mil /pdffiles/r600_8_1.pdf.

20 *And every obstetrician knows:* Edward T. Crosby, and Stephen H. Halpern, "Epidural for Labour, and Fainting Fathers," *Canadian Journal of Anesthesia* 36 (1989): p. 482.

21 the *"clot production" hypothesis:* Paolo Alboni, Marco Alboni, and Giorgio Beterorelle, "The Origin of Vasovagal Syncope: To Protect the Heart or to Escape Predation?" *Clinical Autonomic Research* 18 (2008): pp. 170–78.

21 *A survey of any veterinarian's:* Wendy Ware, "Syncope," Waltham/OSU Symposium: Small Animal Cardiology 2002, accessed February 20, 2009. http://www .vin.com/proceedings/Proceedings.plx?CID=WALTHAMOSU2002&PID=2992.

22 *Wildlife veterinarians:* Personal correspondence between authors and wildlife veterinarians.

22 *"fainted so completely that":* George L. Engel and John Romano, "Studies of Syncope: IV. Biologic Interpretation of Vasodepressor Syncope," *Psychosomatic Medicine* 29 (1947): p. 288.

22 *"not only tremble and turn":* Ibid.

23 *Woodchucks, rabbits, fawns, and monkeys:* Norbert E. Smith and Robert A. Woodruff, "Fear Bradycardia in Free-Ranging Woodchucks, *Marmota monax,*" *Journal of Mammalogy* 61 (1980): p. 750.

23 *Willow grouse, caimans, cats, squirrels:* Ibid.

23 *Once I started looking into:* Nadine K. Jacobsen, "Alarm Bradycardia in White-Tailed Deer Fawns (*Odocoileus virginianus*)," *Journal of Mammalogy* 60 (1979): p. 343.

23 *One noticeable difference between animal:* J. Gert van Dijk, "Fainting in Animals," *Clinical Autonomic Research* 13 (2003): p. 247–55.

24 *One study demonstrated that inexperienced:* Alan B. Sargeant and Lester E. Eberhardt, "Death Feigning by Ducks in Response to Predation by Red Foxes (*Vulpes fulva*)," *American Midland Naturalist* 94 (1975): pp. 108–19.

24 *In 1941, twenty-one-year-old:* UCSB Department of History, "Nina Morecki: My Life, 1922–1945," accessed August 25, 2011. http://www.history.ucsb.edu/projects /holocaust/NinasStory/letter02.htm.

24 *This strategy has been described:* Anatoly Kuznetsov, *Babi Yar: A Document in the Form of a Novel,* New York: Farrar, Straus and Giroux, 1970; Mark Obmascik, "Columbine—Tragedy and Recovery: Through the Eyes of Survivors," *Denver Post,* June 13, 1999, accessed September 12, 2011. http://extras.denverpost.com /news/shot0613a.htm.

24 *A vagal state can make:* Tim Caro, *Antipredator Defenses in Birds and Mammals,* Chicago: University of Chicago Press, 2005.

24 *Rape-prevention educators:* Illinois State Police, "Sexual Assault Information," accessed September 6, 2011. http://www.isp.state.il.us/crime/assault.cfm.

24 *They suggest that when fighting back:* David H. Barlow, *Anxiety and Its Disorders: The Nature and Treatment of Anxiety and Panic,* New York: Guilford, 2001: p. 4; Gallup, Gordon G., Jr., "Tonic Immobility," in *Comparative Psychology: A Handbook,* edited by Gary Greenberg, 780. London: Routledge, 1998.

24 *Many narratives taken:* Karen Human Rights Group, "Torture of Karen Women by SLORC: An Independent Report by the Karen Human Rights Group, February 16, 1993," accessed September 30, 2011. http://www.khrg.org/khrg93/93_02_16b.html; Inquirer Wire Service, "Klaus Barbie: Women Testify of Torture at His Hand," *Philadelphia Inquirer,* March 23, 1987, accessed September 30, 2011. http://writing .upenn.edu/~afilreis/Holocaust/barbie.html; Human Rights Watch, "Egypt: Impunity for Torture Fuels Days of Rage," January 31, 2011, accessed September 30, 2011. http://www.hrw.org/news/2011/01/31/egypt-impunity-torture-fuels-days-rage.

25 *Female robberflies:* Göran Arnqvist and Locke Rowe, *Sexual Conflict,* Princeton: Princeton University Press, 2005.

25 *Some experts believe:* David A. Ball, "The crucifixion revisited," *Journal of the Mississippi State Medical Association* 49 (2008): pp. 67–73.

25 *Canadian scientists studying white-tailed deer:* Aaron N. Moen, M. A. DellaFera, A. L. Hiller, and B. A. Buxton, "Heart Rates of White-Tailed Deer Fawns in Response to Recorded Wolf Howls," *Canadian Journal of Zoology* 56 (1978): pp. 1207–10.

26 *On the night of January:* I. Yoles, M. Hod, B. Kaplan, and J. Ovadia, "Fetal 'Fright-Bradycardia' Brought On by Air-Raid Alarm in Israel," *International Journal of Gynecology Obstetrics* 40 (1993): p. 157.

26 *In the maternity ward:* Ibid., pp. 157–60.

26 *Hiding in the face of:* Caro, *Antipredator Defenses.*

26 *But when oscars get stressed:* Stéphan G. Reebs, "Fishes Feigning Death," howfish behave.ca, 2007, accessed September 12, 2011. http://www.howfishbehave.ca/pdf /Feigning%20death.pdf.

26 *The fish heart has:* Karel Liem, William E. Bemis, Warren F. Walker Jr., and Lance

Grande, *Functional Anatomy of the Vertebrate: An Evolutionary Perspective,* 3rd ed., Belmont, CA: Brooks/Cole, 2001.

26 *Called ampullary organs:* David Hudson Evans and James B. Clairborne, *The Physiology of Fish,* Zug, Switzerland: CRC Press, 2005.

28 *The Volvo car company:* Tom Scocca, "Volvo Drivers Will No Longer Be Electronically Protected from Ax Murderers Lurking in the Back Seat," *Slate.com,* July 22, 2010, accessed October 2, 2011. http://www.slate.com/content/slate/blogs /scocca/2010/07/22/volvo_drivers_will_no_longer_be_electronically_pro- tected_from_ax_murderers_lurking_in_the_back_seat.html.

29 *The same goes for a peculiar:* Caro, *Antipredator Defenses.*

29 *Wildlife biologists call these:* Ibid.

THREE Jews, Jaguars, and Jurassic Cancer

31 *Five times the number:* Centers for Disease Control and Prevention, "Achievements in Public Health, 1900–1999: Decline in Deaths from Heart Disease and Stroke—United States, 1900–1999," *MMWR Weekly* 48 (August 6, 1999): pp. 649–56.

31 *Every two years, starting in 1948:* "Framingham Heart Study," accessed October 7, 2011. http://www.framinghamheartstudy.org/.

32 *In 2012 he began enrolling:* Morris Animal Foundation, "Helping Dogs Enjoy a Healthier Tomorrow," accessed September 28, 2011. http://www.morrisanimalfoundation .org/our-research/major-health-campaigns/clhp.html.

32 *completed in 2005:* Kerstin Lindblad-Toh, Claire M. Wade, Tarjei S. Mikkelsen, Elinor K. Karlsson, David B. Jaffe, Michael Kamal, Michele Clamp, et al., "Genome Sequence, Comparative Analysis and Haplotype Structure of the Domestic Dog," *Nature* 438 (2005): pp. 803–19.

33 *As I patted:* Linda Hettich interview, Anaheim, CA, June 12, 2010.

35 *Smoking, sun exposure, excess alcohol:* National Toxicology Program, U.S. Department of Health and Human Services, "Substances Listed in the Twelfth Report on Carcinogens," *Report on Carcinogens, Twelfth Edition* (2011): pp. 15–16, accessed October 7, 2011. http://ntp.niehs.nih.gov/ntp/roc/twelfth/ListedSubstancesKnown .pdf.

35 *There's also a catalog:* Kathleen Sebelius, U.S. Department of Health and Human Services Secretary, *12th Report on Carcinogens,* Washington, DC: U.S. DHHS (June 10, 2011), accessed October 7, 2011. http://ntp.niehs.nih.gov/ntp/roc/twelfth/roc12 .pdf; National Toxicology Program, "Substances Listed," pp. 15–16.

36 *"The desire to explain sickness":* Charles E. Rosenberg, "Disease and Social Order in America: Perceptions and Expectations," in "AIDS: The Public Context of an Epidemic," *Milbank Quarterly* 64 (1986): p. 50.

36 *Intriguingly, many canine cancers:* David J. Waters, and Kathleen Wildasin, "Cancer Clues from Pet Dogs: Studies of Pet Dogs with Cancer Can Offer Unique Help in the Fight Against Human Malignancies While Also Improving Care for Man's Best Friend," *Scientific American* (December 2006): pp. 94–101.

37 *leukemia or lymphoma:* American Association of Feline Practitioners, "Feline Leukemia Virus," accessed December 19, 2011. http://www.vet.cornell.edu/fhc /brochures/felv.html; PETMD, "Lymphoma in Cats," accessed December 19, 2011. http://www.petmd.com/cat/conditions/cancer/c_ct_lymphoma#.Tu_RQ1Yw28B.

37 *And when a cat's owner:* Giovanni P. Burrai, Sulma I. Mohammed, Margaret A. Miller, Vincenzo Marras, Salvatore Pirino, Maria F. Addis, and Sergio Uzzau, "Spontaneous Feline Mammary Intraepithelial Lesions as a Model for Human Estrogen Receptor and Progesterone Receptor-Negative Breast Lesions," *BMC Cancer* 10 (2010): p. 156.

37 *Rabbit hysterectomies:* Daniel D. Smeak and Barbara A. Lightner, "Rabbit Ovariohysterectomy," Veterinary Educational Videos Collection from Dr. Banga's websites, accessed April 1, 2012. http//video.google.com/videoplay?docid= 5953436041779809619.

37 *Parakeets are prone:* M. L. Petrak and C. E. Gilmore, "Neoplasms," in *Diseases of Cage and Aviary Birds,* ed. Margaret Petrak, pp. 606–37. Philadelphia: Lea & Febiger, 1982.

37 *Zoo veterinarians have reported:* Luigi L. Capasso, "Antiquity of Cancer," *International Journal of Cancer* 113 (2005): pp. 2–13; , S. V. Machotka and G. D. Whitney, "Neoplasms in Snakes: Report of a Probable Mesothelioma in a Rattlesnake and a Thorough Tabulation of Earlier Cases," in *The Comparative Pathology of Zoo Animals,* eds. R. J. Montali and G. Migaki, pp. 593–602. Washington, DC: Smithsonian Institution Press, 1980.

37 *Equine sunburn:* University of Minnesota Equine Genetics and Genomics Laboratory, "Gray Horse Melanoma," accessed October 7, 2011. http://www.cvm.umn .edu/equinegenetics/ghmelanoma/home.html.

37 *may connect more to a genetic issue:* Gerli Rosengren Pielberg, Anna Golovko, Elisabeth Sundström, Ino Curik, Johan Lennartsson, Monika H. Seltenhammer, Thomas Druml, et al., "A *Cis*-Acting Regulatory Mutation Causes Premature Hair Graying and Susceptibility to Melanoma in the Horse," *Nature Genetics* 40 (2008): pp. 1004–09; S. Rieder, C. Stricker, H. Joerg, R. Dummer, and G. Stranzinger, "A Comparative Genetic Approach for the Investigation of Ageing Grey Horse Melanoma," *Journal of Animal Breeding and Genetics* 117 (2000): pp. 73–82; Kerstin Lindblad-Toh telephone interview, July 28, 2010.

37 *Her cancer grew under her horn:* Olsen Ebright, "Rhinoceros Fights Cancer at LA Zoo," *NBC Los Angeles,* November 17, 2009, accessed October 14, 2011. http://www. nbclosangeles.com/news/local/Los-Angeles-Zoo-Randa-Skin-Cancer-70212192 .html.

37 *Cattle also develop squamous cell:* W. C. Russell, J. S. Brinks, and R. A. Kainer, "Incidence and Heritability of Ocular Squamous Cell Tumors in Hereford Cattle," *Journal of Animal Science* 43 (1976): pp. 1156–62.

37 *Strike-branding livestock:* I. Yeruham, S. Perl, and A. Nyska, "Skin Tumours in Cattle and Sheep After Freeze- or Heat-Branding," *Journal of Comparative Pathology* 114 (1996): pp. 101–06.

38 *Osteosarcoma, the cancer:* Stephen J. Withrow and Chand Khanna, "Bridging the Gap Between Experimental Animals and Humans in Osteosarcoma," *Cancer Treatment and Research* 152 (2010): pp. 439–46.

38 *Sadly, a killer whale:* M. Yonezawa, H. Nakamine, T. Tanaka, and T. Miyaji, "Hodgkin's Disease in a Killer Whale (*Orcinus orca*)," *Journal of Comparative Pathology* 100 (1989): pp. 203–07.

38 *And the neuroendocrine cancer:* G. Minkus, U. Jütting, M. Aubele, K. Rodenacker, P. Gais, W. Breuer, and W. Hermanns, "Canine Neuroendocrine Tumors of the Pancreas: A Study Using Image Analysis Techniques for the Discrimination of the Metastatic Versus Nonmetastatic Tumors," *Veterinary Pathology* 37 (1997): pp.

138–145; G. A. Andrews, N. C. Myers III, and C. Chard-Bergstrom, "Immunohistochemistry of Pancreatic Islet Cell Tumors in the Ferret (*Mustela putorius furo*)," *Veterinary Pathology* 34 (1997): pp. 387–93.

38 *Wild sea turtles around:* Denise McAloose and Alisa L. Newton, "Wildlife Cancer: A Conservation Perspective," *Nature Reviews: Cancer* 9 (2009): p. 521.

38 *Genital cancers have become rampant:* Ibid.

38 *Tasmanian devils, found only:* R. Loh, J. Bergfeld, D. Hayes, A. O'Hara, S. Pyecroft, S. Raidal, and R. Sharpe, "The Pathology of Devil Facial Tumor Disease (DFTD) in Tasmanian Devils (*Sarcophilus harrisii*)," *Veterinary Pathology* 43 (2006): pp. 890–95.

38 *Attwater's prairie chickens:* McAloose and Newton, "Wildlife Cancer," pp. 517–26.

38 *Western barred bandicoots:* Ibid.

38 *The disease can even be destructive:* The Huntington Library, Art Collection, and Botanical Gardens, "Do Plants Get Cancer? The Effects of Infecting Sunflower Seedlings with *Agrobacterium tumefaciens*," accessed October 7, 2011. http://www .huntington.org/uploadedFiles/Files/PDFs/GIB-DoPlantsGetCancer.pdf; John H. Doonan and Robert Sablowski, "Walls Around Tumours—Why Plants Do Not Develop Cancer," *Nature* 10 (2010): pp. 794–802.

38 *More than 3,500 years:* James S. Olson, *Bathsheba's Breast: Women, Cancer, and History.* Baltimore: Johns Hopkins University Press, 2002.

39 *"Among ancients, breast cancer":* Ibid.

39 *They've examined Bronze Age:* Mel Greaves, *Cancer: The Evolutionary Legacy,* Oxford: Oxford University Press, 2000; Capasso, "Antiquity of Cancer," pp. 2–13.

39 *In 1997, amateur fossil hunters:* Kathy A. Svitil, "Killer Cancer in the Cretaceous," *Discover Magazine,* November 3, 2003, accessed May 24, 2010. http://discovermagazine .com/2003/nov/killer-cancer1102.

39 *"As the tumor grew":* Ibid.

40 *Other paleo-oncologists:* B. M. Rothschild, D. H. Tanke, M. Helbling, and L. D. Martin, "Epidemiologic Study of Tumors in Dinosaurs," *Naturwissenschaften* 90 (2003): pp. 495–500.

40 *At the University of Pittsburgh:* University of Pittsburgh Schools of the Health Sciences Media Relations, "Study of Dinosaurs and Other Fossil Part of Plan by Pitt Medical School to Graduate Better Doctors Through Unique Collaboration with Carnegie Museum of Natural History," last updated February 28, 2006, accessed March 2, 2012. http://www.upmc.com/MediaRelations/NewsReleases/2006/Pages /StudyFossils.aspx.

40 *And evidence of probable metastatic cancer:* Bruce M. Rothschild, Brian J. Witzke, and Israel Hershkovitz, "Metastatic Cancer in the Jurassic," *Lancet* 354 (1999): p. 398.

40 *About sixty-five million years ago:* G. V. R. Prasad and H. Cappetta, "Late Cretaceous Selachians from India and the Age of the Deccan Traps," *Palaeontology* 36 (1993): pp. 231–48.

40 *Ionizing radiation, toxic volcanic spew:* Tom Simkin, "Distant Effects of Volcanism—How Big and How Often?" *Science* 264 (1994): pp. 913–14.

40 *In fact, cycads and conifers:* Rothschild et al., "Epidemiologic Study," pp. 495–500; Dolores R. Piperno and Hans-Dieter Sues, "Dinosaurs Dined on Grass," *Science* 310 (2005): pp. 1126–28.

41 *"becomes a statistical inevitability":* Greaves, *Cancer.*

41 *Genomics researchers:* John D Nagy, Erin M. Victor, and Jenese H. Cropper, "Why

Don't All Whales Have Cancer? A Novel Hypothesis Resolving Peto's Paradox," *Integrative and Comparative Biology* 47 (2007): pp. 317–28.

41 *larger species, overall:* R. Peto, F. J. C. Roe, P. N. Lee, L. Levy, and J. Clack, "Cancer and Ageing in Mice and Men," *British Journal of Cancer* 32 (1975): pp. 411–26.

43 *In one Swedish study:* Patricio Rivera, "Biochemical Markers and Genetic Risk Factors in Canine Tumors," doctoral thesis, Swedish University of Agricultural Sciences, Uppsala, 2010.

43 *Zoo veterinarians report:* Linda Munson and Anneke Moresco, "Comparative Pathology of Mammary Gland Cancers in Domestic and Wild Animals," *Breast Disease* 28 (2007): pp. 7–21.

44 *Here I should pause:* Christie Wilcox, "Ocean of Pseudoscience: Sharks DO get cancer!" *Science Blogs,* September 6, 2010, accessed October 13, 2011. http://scienceblogs.com/observations/2010/09/ocean_of_pseudoscience_sharks.php.

44 *Professional lactators:* Munson and Moresco, "Comparative Pathology," pp. 7–21.

44 *Some wild bats:* Xiaoping Zhang, Cheng Zhu, Haiyan Lin, Qing Yang, Qizhi Ou, Yuchun Li, Zhong Chen, et al. "Wild Fulvous Fruit Bats (*Rousettus leschenaulti*) Exhibit Human-Like Menstrual Cycle," *Biology of Reproduction* 77 (2007): pp. 358–64.

45 *In fact, worldwide, some 20 percent:* World Health Organization, "Viral Cancers," *Initiative for Vaccine Research,* accessed October 7, 2011. http://www.who.int/vaccine_research/diseases/viral_cancers/en/index1.html.

45 *Across Africa's "lymphoma belt":* S. H. Swerdlow, E. Campo, N. L. Harris, E. S. Jaffe, S. A. Pileri, H. Stein, J. Thiele, et al., *World Health Organization Classification of Tumours of Haematopoietic and Lymphoid Tissues,* Lyon: IARC Press, 2008; Arnaud Chene, Daria Donati, Jackson Orem, Anders Bjorkman, E. R. Mbidde, Fred Kironde, Mats Wahlgren, et al., "Endemic Burkitt's Lymphoma as a Polymicrobial Disease: New Insights on the Interaction Between Plasmodium Falciparum and Epstein-Barr Virus," *Seminars in Cancer Biology* 19 (2009): pp. 411–420.

45 *According to the WHO:* World Health Organization, "Viral Cancers."

45 *In 1982, dead beluga whales:* Daniel Martineau, Karin Lemberger, André Dallaire, Phillippe Labelle, Thomas P. Lipscomb, Pascal Michel, and Igor Mikaelian, "Cancer in Wildlife, a Case Study: Beluga from the St. Lawrence Estuary, Québec, Canada," *Environmental Health Perspectives* 110 (2002): pp. 285–92.

46 *Animals even forewarn us:* Peter M. Rabinowitz, Matthew L. Scotch, and Lisa A. Conti, "Animals as Sentinels: Using Comparative Medicine to Move Beyond the Laboratory," *Institute for Laboratory Animal Research Journal* 51 (2010): pp. 262–67.

46 *Although PCB production and DDT use:* Gina M. Ylitalo, John E. Stein, Tom Hom, Lyndal L. Johnson, Karen L. Tilbury, Alisa J. Hall, Teri Rowles, et al., "The Role of Organochlorides in Cancer-Associated Mortality in California Sea Lions," *Marine Pollution Bulletin* 50 (2005): pp. 30–39; Ingfei Chen, "Cancer Kills Many Sea Lions, and Its Cause Remains a Mystery," *New York Times,* March 4, 2010, accessed March 8, 2010. http://www.nytimes.com/2010/03/05/science/05sfsealion.html.

47 *One study of nose and sinus:* Peter M. Rabinowitz and Lisa A. Conti, *Human-Animal Medicine: Clinical Approaches to Zoonoses, Toxicants and Other Shared Health Risks,* Maryland Heights, MO: Saunders, 2010: p. 60.

47 *Bladder cancer and lymphoma:* Ibid.

47 *And military dogs:* Ibid.

48 *Cats have served:* Ibid.

48 *Where we don't overlap:* Melissa Paoloni and Chand Khanna, "Translation of New

Cancer Treatments from Pet Dogs to Humans," *Nature Reviews: Cancer* 8 (2008): pp. 147–56.

48 *But beyond their possible use:* Ibid.; Chand Khanna, Kerstin Lindblad-Toh, David Vail, Cheryl London, Philip Bergman, Lisa Barber, Matthew Breen, et al., "The Dog as a Cancer Model," letter to the editor, *Nature Biotechnology* 24 (2006): pp. 1065–66; Melissa Paoloni telephone interview, May 19, 2010, and Philip Bergman interview, Anaheim, CA, June 10, 2010.

48 *Currently, the vast majority:* Ira Gordon, Melissa Paoloni, Christina Mazcko, and Chand Khanna, "The Comparative Oncology Trials Consortium: Using Spontaneously Occurring Cancers in Dogs to Inform the Cancer Drug Development Pathway," *PLoS Medicine* 6 (2009): p. e1000161.

48 *Dog cancer cells:* Ibid.

49 *This novel approach:* George S. Mack, "Cancer Researchers Usher in Dog Days of Medicine," *Nature Medicine* 11 (2005): p. 1018; Gordon et al., "The Comparative Oncology Trials"; Paoloni interview; National Cancer Institute, "Comparative Oncology Program," accessed October 7, 2011. https://ccrod.cancer.gov/confluence/display/CCRCOPWeb/Home.

49 *For example, the limb-sparing technique:* Withrow and Khanna, "Bridging the Gap," pp. 439–46.; Steve Withrow telephone interview, May 17, 2010.

50 *Kerstin Lindblad-Toh:* Lindblad-Toh interview; Lindblad-Toh et al., "Genome Sequence," pp. 803–19.

50 *German shepherds:* Paoloni and Khanna, "Translation of New Cancer Treatments," pp. 147–56.

50 *But noticing where cancer isn't:* Ibid.

51 *In many ways, the dinner crowd:* Philip Bergman interview, Orlando, FL, January 17, 2010; Bergman interview, June 10, 2010.

51 *"This is the* Princeton *Club":* Bergman interview, January 17, 2010.

52 *"Do dogs," he asked, "get melanoma?":* Bergman interview, January 17, 2010; Bergman interview, June 10, 2010; Jedd Wolchok telephone interview, June 29, 2010.

52 *"the diseases are essentially one and the same":* Bergman interview, January 17, 2010; Bergman interview, June 10, 2010; Wolchok interview.

52 *xenogeneic:* Bergman interview, January 17, 2010.

53 *The therapy:* Philip J. Bergman, Joanne McKnight, Andrew Novosad, Sarah Charney, John Farrelly, Diane Craft, Michelle Wulderk, et al., "Long-Term Survival of Dogs with Advanced Malignant Melanoma After DNA Vaccination with Xenogeneic Human Tyrosinase: A Phase I Trial," *Clinical Cancer Research* 9 (2003): pp. 1284–90.

53 *In 2009, Merial released:* Merial Limited, "Canine Oral Melanoma and ONCEPT Canine Melanoma Vaccine, DNA," *Merial Limited Media Information,* January 17, 2010.

53 *From a* human *melanoma cell:* Wolchok interview.

54 *For the time being, mice:* Ibid.

54 *"Almost without fail":* Bergman interview, January 17, 2010.

FOUR Roar-gasm

55 *Lancelot was having:* Authors' tour of UC Davis horse barn, Davis, CA, February 12, 2011; Janet Roser telephone interview, August 30, 2011.

56 *"Most people think of stallions":* Sandy Sargent, "Breeding Horses: Why Won't

My Stallion Breed to My Mare," allexperts.com, July 19, 2009, accessed February 18, 2011. http://en.allexperts.com/q/Breeding-Horses-3331/2009/7/won-t-stallion -breed.htm.

56 *Even when copulating:* Katherine A. Houpt, *Domestic Animal Behavior for Veterinarians and Animal Scientists,* 5th ed., Ames, IA: Wiley-Blackwell, 2011: pp. 117–21.

56 *At the other end:* Ibid., p. 119.

57 *Their lower status and forced celibacy:* Ibid., pp. 91–93; Edward O. Price, "Sexual Behavior of Large Domestic Farm Animals: An Overview," *Journal of Animal Science* 61 (1985): pp. 62–72.

57 *"Pain, fear, and confusion":* Jessica Jahiel, "Young Stallion Won't Breed," Jessica Jahiel's Horse-Sense, accessed February 18, 2011. http://www.horse-sense.org/archives /2001027.php.

57 *sex comes in many forms:* Marlene Zuk, *Sexual Selections: What We Can and Can't Learn About Sex from Animals,* Berkeley: University of California Press, 2003; Tim Birkhead, *Promiscuity: An Evolutionary History of Sperm Competition,* Cambridge, MA: Harvard University Press, 2002; Olivia Judson, *Dr. Tatiana's Sex Advice to All Creation: The Definitive Guide to the Evolutionary Biology of Sex,* New York: Henry Holt, 2002.

58 *The earliest single-celled:* Matt Ridley, *The Red Queen: Sex and the Evolution of Human Nature,* New York: Harper Perennial, 1993.

59 *The oldest penis:* David J. Siveter, Mark D. Sutton, Derek E. G. Briggs, and Derek J. Siveter, "An Ostracode Crustacean with Soft Parts from the Lower Silurian," *Science* 302 (2003): pp. 1749–51.

59 *Before it was found:* Jason A. Dunlop, Lyall I. Anderson, Hans Kerp, and Hagen Hass, "Palaeontology: Preserved Organs of Devonian Harvestmen," *Nature* 425 (2003): p. 916.

59 *Paleontologists have speculated:* Discovery Channel Videos, "Tyrannosaurus Sex: Titanosaur Mating," *Discovery Channel,* accessed October 7, 2011. http://dsc .discovery.com/videos/tyrannosaurus-sex-titanosaur-mating.html.

59 *Not all internal:* Birkhead, *Promiscuity,* p. 95.

60 *Spiny anteaters sport:* Nora Schultz, "Exhibitionist Spiny Anteater Reveals Bizarre Penis," *New Scientist,* October 26, 2007, accessed February 8, 2011. http://www .newscientist.com/article/dn12838-exhibitionist-spiny-anteater.

60 *the phalluses of Argentine lake ducks:* Kevin G. McCracken, "The 20-cm Spiny Penis of the Argentine Lake Duck (*Oxyura vittata*)," *The Auk* 117 (2000): pp. 820–25.

60 *Despite a thirty-three-inch:* Birkhead, *Promiscuity,* p. 99.

60 *That title goes to:* Christopher J. Neufeld and A. Richard Palmer, "Precisely Proportioned: Intertidal Barnacles Alter Penis Form to Suit Coastal Wave Action," *Proceedings of the Royal Society B* 275 (2008): pp. 1081–87.

60 *Several species of marine:* Birkhead, *Promiscuity,* p. 98.

60 *Some snakes and lizards:* Ibid.

60 *As for insects:* David Grimaldi and Michael S. Engel, *Evolution of the Insects,* New York: Cambridge University Press, 2005: p. 135.

60 *"It is generally assumed":* Birkhead, *Promiscuity,* p. 95.

60 *Although barnacles:* Ibid.

61 *The sexcapades of krill:* So Kawaguchi, Robbie Kilpatrick, Lisa Roberts, Robert A. King, and Stephen Nicol. "Ocean-Bottom Krill Sex," *Journal of Plankton Research* 33 (2011): pp. 1134–38.

61 *Since arising more than 200 million:* Diane A. Kelly, "Penises as Variable-Volume Hydrostatic Skeletons," *Annals of the New York Academy of Sciences* 1101 (2007): pp. 453–63.

61 *An actual penis bone:* D. A. Kelly, "Anatomy of the Baculum-Corpus Cavernosum Interface in the Norway Rat (*Rattus norvegicus*) and Implications for Force Transfer During Copulation," *Journal of Morphology* 244 (2000): pp. 69–77; correspondence with Diane A. Kelly.

61 *A rope of thick tissue:* Birkhead, *Promiscuity*, p. 97.

61 *But humans, along with armadillos:* Kelly, "Penises," pp. 453–63; Kelly, "The Functional Morphology of Penile Erection: Tissue Designs for Increasing and Maintaining Stiffness," *Integrative and Comparative Biology* 42 (2002): pp. 216–21; Kelly, "Expansion of the Tunica Albuginae During Penile Inflation in the Nine-Banded Armadillo (*Dasypus novemcinctus*)," *Journal of Experimental Biology* 202 (1999): pp. 253–65.

61 *As Diane A. Kelly:* Kelly telephone interview; Kelly, "Penises," pp. 453–63; Kelly, "Functional Morphology," pp. 216–21; Kelly, "Expansion," pp. 253–65.

61 *It starts with the deceptively inert:* Ion G. Motofei and David L. Rowland, "Neurophysiology of the Ejaculatory Process: Developing Perspectives," *BJU International* 96 (2005): pp. 1333–38; Jeffrey P. Wolters and Wayne J. G. Hellstrom, "Current Concepts in Ejaculatory Dysfunction," *Reviews in Urology* 8 (2006): pp. S18–25.

62 *The command to relax:* Motofei and Rowland, "Neurophysiology," pp. 1333–38; Wolters and Hellstrom, "Current Concepts," pp. S18–25.

62 *Next comes a key chemical reaction:* Motofei and Rowland, "Neurophysiology," pp. 1333–38; Wolters and Hellstrom, "Current Concepts," pp. S18–25.

62 *To protect the organ from rupturing:* Kelly, "Penises," pp. 453–63.

62 *(Kelly says it's a trick shared by pufferfish):* Ibid.

63 *A study on certain fish:* R. Brian Langerhans, Craig A. Layman, Thomas J. DeWitt, and David B. Wake, "Male Genital Size Reflects a Tradeoff Between Attracting Mates and Avoiding Predators in Two Live-Bearing Fish Species," *Proceedings of the National Academy of Sciences* 102 (2005): pp. 7618–23.

63 *"point of no return":* W. P. de Silva, "ABC of Sexual Health: Sexual Variations," *BMJ* 318 (1999): pp. 654–56.

63 *But all male mammals:* Kelly, "Penises," pp. 453–63.

63 *And the ejaculation of a male:* Phillip Jobling, "Autonomic Control of the Urogenital Tract," *Autonomic Neuroscience* 165 (2011): pp. 113–126.

63 *Electroencephalograms:* Harvey D. Cohen, Raymond C. Rosen, and Leonide Goldstein, "Electroencephalographic Laterality Changes During Human Sexual Orgasm," *Archives of Sexual Behavior* 5 (1976): pp. 189–99.

63 *Many men describe:* James G. Pfaus and Boris B. Gorzalka, "Opioids and Sexual Behavior," *Neuroscience & Biobehavioral Reviews* 11 (1987): pp. 1–34; James G. Pfaus and Lisa A. Scepkowski, "The Biologic Basis for Libido," *Current Sexual Health Reports* 2 (2005): pp. 95–100.

64 *If you are an ER doc:* Kenia P. Nunes, Marta N. Cordeiro, Michael Richardson, Marcia N. Borges, Simone O. F. Diniz, Valbert N. Cardoso, Rita Tostes, Maria Elena De Lima, et al., "Nitric Oxide–Induced Vasorelaxation in Response to PnTx2–6 Toxin from *Phoneutria nigriventer* Spider in Rat Cavernosal Tissue," *The Journal of Sexual Medicine* 7 (2010): pp. 3879–88.

65 *When a randy stallion:* Houpt, *Domestic Animal Behavior*, p. 114; Roser interview.

65 *Male horses:* Houpt, *Domestic Animal Behavior*, p. 10; L. E. L. Rasmussen, "Source and Cyclic Release Pattern of (Z)-8-Dodecenyl Acetate, the Pre-ovulatory Pheromone of the Female Asian Elephant," *Chemical Senses* 26 (2001): p. 63.

66 *Also called the facial nerve:* Edwin Gilland and Robert Baker, "Evolutionary Patterns of Cranial Nerve Efferent Nuclei in Vertebrates," *Brain, Behavioral Evolution* 66 (2005): pp. 234–54.

66 *Male porcupines:* Uldis Roze, *The North American Porcupine*, 2nd edition. Ithaca, NY: Comstock Publishing, 2009: pp. 135–43, 231.

66 *Male goats:* Edward O. Price, Valerie M. Smith, and Larry S. Katz, "Stimulus Condition Influencing Self-Enurination, Genital Grooming and Flehmen in Male Goats," *Applied Animal Behaviour Science* 16 (1986): pp. 371–81.

66 *Elk bucks:* Dale E. Toweill, Jack Ward Thomas, and Daniel P. Metz, *Elk of North America: Ecology and Management*, Mechanicsburg, PA: Stackpole Books, 1982.

66 *Courting female crayfish:* Fiona C. Berry and Thomas Breithaupt, "To Signal or Not to Signal? Chemical Communication by Urine-Borne Signals Mirrors Sexual Conflict in Crayfish," *BMC Biology* 8 (2010): p. 25.

66 *The urine of male swordtail fish:* Gil G. Rosenthal, Jessica N. Fitzsimmons, Kristina U. Woods, Gabriele Gerlach, and Heidi S. Fisher, "Tactical Release of a Sexually-Selected Pheromone in a Swordtail Fish," *PLoS One* 6 (2011): p. e16994.

67 *For example, the reddening:* C. Bielert and L. A. Van der Walt, "Male Chacma Baboon (*Papio ursinus*) Sexual Arousal: Mediation by Visual Cues from Female Conspecifics," *Psychoneuroendocrinology* 7 (1986): pp. 31–48; Craig Bielert, Letizia Girolami, and Connie Anderson, "Male Chacma Baboon (*Papio ursinus*) Sexual Arousal: Studies with Adolescent and Adult Females as Visual Stimuli," *Developmental Psychobiology* 19 (1986): pp. 369–83.

67 *Blindfolded bulls:* E. B. Hale, "Visual Stimuli and Reproductive Behavior in Bulls," *Journal of Animal Science* 25 (1966): pp. 36–44.

67 *Researchers in Morocco:* Adeline Loyau and Frederic Lacroix, "Watching Sexy Displays Improved Hatching Success and Offspring Growth Through Maternal Allocation," *Proceedings of the Royal Society of London B* 277 (2010): pp. 3453–60.

68 *Similarly, pig breeders:* Price, "Sexual Behavior," p. 66.

68 *"female Kob antelope whistle":* Bruce Bagemihl, *Biological Exuberance: Animal Homosexuality and Natural Diversity*, New York: St. Martin's, 1999.

68 *One fascinating study revealed that female Barbary:* Dana Pfefferle, Katrin Brauch, Michael Heistermann, J. Keith Hodges, and Julia Fischer, "Female Barbary Macaque (*Macaca sylvanus*) Copulation Calls Do Not Reveal the Fertile Phase but Influence Mating Outcome," *Proceedings of the Royal Society of London B* 275 (2008): pp. 571–78.

68 *Bulls have been found:* Houpt, *Domestic Animal Behavior*, p. 100.

69 *But this physiology sets the stage:* Wolters and Hellstrom, "Current Concepts," pp. S18–25; Arthur L. Burnett telephone interview, April 5, 2011; Jacob Rajfer telephone interview, April 29, 2011.

69 *Some five hundred years ago:* I. Goldstein, "Male Sexual Circuitry. Working Group for the Study of Central Mechanisms in Erectile Dysfunction," *Scientific American* 283 (2000): pp. 70–75.

69 *Worldwide, one in ten men:* Minnesota Men's Health Center, P.A., "Facts About Erectile Dysfunction," accessed October 8, 2011. http://www.mmhc-online.com/articles/impotency.html.

69 *According to Arthur L. Burnett:* Burnett interview.

70 *Ring-tailed lemurs:* Lisa Gould telephone interview, April 5, 2011.

71 *The mere presence of a dominant:* Price, "Sexual Behavior," pp. 62–72; Houpt, *Domestic Animal Behavior,* p. 110.

71 *The control of mating:* Nicholas E. Collias, "Aggressive Behavior Among Vertebrate Animals," *Physiological Zoology* 17 (1944): pp. 83–123; Houpt, *Domestic Animal Behavior,* pp. 90–93.

71 *"Some men under stress have difficulty":* Rajfer interview.

72 *According to Arthur L. Burnett:* Burnett interview.

72 *"an expeditious partner who mounts":* Lawrence K. Hong, "Survival of the Fastest: On the Origin of Premature Ejaculation," *Journal of Sex Research* 20 (1984): p. 113.

72 *Human males take:* Chris G. McMahon, Stanley E. Althof, Marcel D. Waldinger, Hartmut Porst, John Dean, Ira D. Sharlip, et al., "An Evidence-Based Definition of Lifelong Premature Ejaculation: Report of the International Society for Sexual Medicine (ISSM) Ad Hoc Committee for the Definition of Premature Ejaculation," *The Journal of Sexual Medicine* 5 (2008): pp. 1590–1606.

72 *Small marine iguanas:* Martin Wikelski and Silke Baurle, "Pre-Copulatory Ejaculation Solves Time Constraints During Copulations in Marine Iguanas," *Proceedings of the Royal Society of London B* 263 (1996): pp. 439–44.

73 *Jacob Rajfer, the UCLA urologist:* Rajfer interview.

74 *As hilariously—and exhaustively:* Mary Roach, *Bonk: The Curious Coupling of Science and Sex,* New York: Norton, 2008; Zuk, *Sexual Selections;* Birkhead, *Promiscuity;* Judson, *Dr. Tatiana's Sex Advice;* Sarah Blaffer Hrdy, *Mother Nature: Maternal Instincts and How They Shape the Human Species.* New York: Ballantine, 1999.

74 *Orangutans self-stimulate:* Judson, *Dr. Tatiana's Sex Advice,* p. 246; Naturhistorisk Museum, "Homosexuality in the Animal Kingdom," accessed October 8, 2011. http://www.nhm.uio.no/besok-oss/utstillinger/skiftende/againstnature/gayanimals.html.

74 *Daddy longlegs spin:* Ed Nieuwenhuys, "Daddy-longlegs, Vibrating or Cellar Spiders," accessed October 14, 2011. http://ednieuw.home.xs4all.nl/Spiders/Pholcidae/Pholcidae.htm.

74 *Livestock farmers and large animal veterinarians:* Houpt, *Domestic Animal Behavior,* pp. 102, 119, 129.

74 *Bats and hedgehogs:* Min Tan, Gareth Jones, Guangjian Zhu, Jianping Ye, Tiyu Hong, Shanyi Zhou, Shuyi Zhang, et al., "Fellatio by Fruit Bats Prolongs Copulation Time," *PLoS One* 4 (2009): p. e7595.

74 *Male-male and female-female:* Price, "Sexual Behavior," p. 64.

75 *Bagemihl includes:* Bagemihl, *Biological Exuberance,* pp. 263–65.

75 *Roughgarden details:* Joan Roughgarden, *Evolution's Rainbow: Diversity, Gender, and Sexuality in Nature and People,* Berkeley: University of California Press, 2004.

75 *Marlene Zuk and Nathan W. Bailey:* Nathan W. Bailey and Marlene Zuk, "Same-Sex Sexual Behavior and Evolution," *Trends in Ecology and Evolution* 24 (2009): pp. 439–46.

75 *"The capacity for behavioral plasticity":* Bagemihl, *Biological Exuberance,* p. 251.

76 *"near elimination of the idea":* Birkhead, *Promiscuity,* pp. 38–39.

76 *"Using information about animal behavior":* Zuk, *Sexual Selections,* pp. 177–78.

76 *Normal reproduction:* Birkhead, *Promiscuity.*

76 *New York City's bedbug:* Göran Arnqvist and Locke Rowe, *Sexual Conflict: Monographs in Behavior and Ecology,* Princeton, NJ: Princeton University Press, 2005.

76 *An animal form of necrophilia:* C. W. Moeliker, "The First Case of Homosexual Necrophilia in the Mallard *Anas platyrhynchos* (Aves: Anatidae)," *Deinsea* 8 (2001): pp. 243–47; Irene Garcia, "Beastly Behavior," *Los Angeles Times,* February 12, 1998, accessed December 20, 2011. http://articles.latimes.com/1998/feb/12 /entertainment/ca-18150.

76 *Sex with relatives and immature:* Carol M. Berman, "Kinship: Family Ties and Social Behavior," in *Primates in Perspective,* 2nd ed., eds. Christina J. Campbell, Agustin Fuentes, Katherine C. MacKinnon, Simon K. Bearder, and Rebecca M. Strumpf, p. 583. New York: Oxford University Press, 2011; Raymond Obstfeld, *Kinky Cats, Immortal Amoebas, and Nine-Armed Octopuses: Weird, Wild, and Wonderful Behaviors in the Animal World,* New York: HarperCollins, 1997: pp. 43–47; Ridley, *The Red Queen,* pp. 282–84; Judson, *Dr. Tatiana's Sex Advice,* pp. 169–86.

77 *"Breeding males are usually highly motivated":* Birkhead, *Promiscuity.*

77 *"Even in nonhumans, sex can":* Zuk, *Sexual Selections.*

77 *"an accidental physiological side effect":* Anders Ågmo, *Functional and Dysfunctional Sexual Behavior: A Synthesis of Neuroscience and Comparative Psychology,* Waltham, MA: Academic Press, 2007. Kindle edition: iii.

78 *"the mating face":* Houpt, *Domestic Animal Behavior,* p. 8.

78 *We're said to have what's called:* Boguslaw Pawlowski, "Loss of Oestrus and Concealed Ovulation in Human Evolution: The Case Against the Sexual-Selection Hypothesis," *Current Anthropology* 40 (1999): pp. 257–76.

78 *Women have been:* Geoffrey Miller, Joshua M. Tybur, and Brent D. Jordan, "Ovulatory Cycle Effects on Tip Earnings by Lap Dancers: Economic Evidence for Human Estrus?" *Evolution and Human Behavior* 27 (2007): pp. 375–81; Debra Lieberman, Elizabeth G. Pillsworth, and Martie G. Haselton, "Kin Affiliation Across the Ovulatory Cycle: Females Avoid Fathers When Fertile," *Psychological Science* (2010): doi: 10.1177/0956797610390385; Martie G. Haselton, Mina Mortezaie, Elizabeth G. Pillsworth, April Bleske-Rechek, and David A. Frederick, "Ovulatory Shifts in Human Female Ornamentation: Near Ovulation, Women Dress to Impress," *Hormones and Behavior* 51 (2007): pp. 40–45.

78 *Men perceive ovulating:* Miller, Tybur, and Jordan, "Ovulatory Cycle Effects," pp. 375–81.

78 *College-aged women:* Lieberman, Pillsworth, and Haselton, "Kin Affiliation."

78 *Physically, female orgasm:* Barry R. Komisaruk, Carlos Beyer-Flores, and Beverly Whipple, *The Science of Orgasm,* Baltimore: Johns Hopkins University Press, 2006.

78 *In developing fetuses:* Kenneth V. Kardong, *Vertebrates: Comparative Anatomy, Function, Evolution,* 4th ed., New York: Tata McGraw-Hill, 2006: pp. 556, 565; Balcombe, Jonathan, *Pleasure Kingdom: Animals and the Nature of Feeling Good,* Hampshire, UK: Palgrave Macmillan, 1997.

79 *A quick comparative survey:* Stefan Anitei, "The Largest Clitoris in the World," *Softpedia,* January 26, 2007, accessed October 14, 2011. http://news.softpedia.com/news /The-Largest-Clitoris-in-the-World-45527.shtml; Balcombe, *Pleasure Kingdom.*

79 *An estimated 40 percent:* Jan Shifren, Brigitta Monz, Patricia A. Russo, Anthony Segreti, and Catherine B. Johannes, "Sexual Problems and Distress in United States Women: Prevalence and Correlates," *Obstetrics & Gynecology* 112 (2008): pp. 970–78.

79 *They affect as many:* J. A. Simon, "Low Sexual Desire—Is It All in Her Head? Patho-physiology, Diagnosis and Treatment of Hypoactive Sexual Desire Disorder," *Post-graduate Medicine* 122 (2010): pp. 128–36; S. Mimoun, "Hypoactive Sexual Desire Disorder, HSDD," *Gynécologie Obstétrique Fertilité* 39 (2011): pp. 28–31; Anita H. Clayton, "The Pathophysiology of Hypoactive Sexual Desire Disorder in Women," *International Journal of Gynecology and Obstetrics* 110 (2010): pp. 7–11.

80 *Low desire and HSDD:* Clayton, "The Pathophysiology," pp. 7–11; Santiago Pala-cios, "Hypoactive Sexual Desire Disorder and Current Pharmacotherapeutic Options in Women," *Women's Health* 7 (2011): pp. 95–107.

80 *Doctors treat HSDD:* Clayton, "The Pathophysiology," pp. 7–11; Palacios, "Hypoac-tive Sexual Desire Disorder," pp. 95–107.

80 *"Cases of dissatisfaction by both partners":* Ralph Myerson, "Hypoactive Sexual Desire Disorder," *Healthline: Connect to Better Health,* accessed October 8, 2011. http://www.healthline.com/galecontent/hypoactive-sexual-desire-disorder.

80 *I asked Dr. Janet Roser:* Roser interview.

80 *Female rats scratch, bite, and vocalize:* James Pfaus telephone interview, February 23, 2011.

81 *Entomologists Randy Thornhill:* Randy Thornhill and John Alcock, *The Evolution of Insect Mating Systems,* Cambridge: Harvard University Press, 1983: p. 469.

81 *James Pfaus, a Concordia University:* Pfaus interview.

82 *Lordosis is a very specific:* Donald Pfaff, *Man and Woman: An Inside Story,* Oxford: Oxford University Press, 2011: p. 78; Donald W. Pfaff, *Drive: Neurobiological and Molecular Mechanisms of Sexual Motivation,* Cambridge, MA: MIT Press, 1999: pp. 76–79.

82 *According to Donald Pfaff:* D. W. Pfaff, L. M. Kow, M. D. Loose, and L. M. Flanagan-Kato, "Reverse Engineering the Lordosis Behavior Circuit," *Hormones and Behav-ior* 54 (2008): pp. 347–54; Pfaff, *Drive,* pp. 76–79.

82 *"ascend[s] the spinal cord":* Pfaff, *Man and Woman,* p. 78.

82 *Like some erections:* Pfaff, *Man and Woman,* p. 78; Pfaff et al., "Reverse Engineer-ing," pp. 347–54.

82 *Receptive female elephant seals:* William F. Perrin, Bernd Wursig, and J. G. M. Thew-issen, *Encyclopedia of Marine Mammals,* Waltham, MA: Academic Press, 2002: p. 394.

82 *"large number of mechanisms for hormone":* Pfaff, *Man and Woman,* p. 78.

83 *"basic, reductionistic":* Pfaff, *Drive,* pp. 76–79.

83 *"The most elementary functions":* Pfaff, *Man and Woman,* p. 57.

84 *"an adequate period of sexual foreplay":* Houpt, *Domestic Animal Behavior,* p. 117.

84 *Dogs, too, engage:* Ibid., pp. 125–27.

84 *Hypersexual behavior occurs:* Ibid., pp. 99, 117.

84 *bellow "like a bull":* Ibid., p. 99.

85 *"assumes a stretching posture":* Masaki Sakai and Mikihiko Kumashiro, "Copula-tion in the Cricket Is Performed by Chain Reaction," *Zoological Science* 21 (2004): p. 716.

86 *The "shudder":* Bagemihl, *Biological Exuberance,* p. 208.

86 *"violet flannel, then the sharpness":* Molly Peacock, "Have You Ever Faked an Orgasm?" in *Cornucopia: New & Selected Poems,* New York: Norton, 2002.

86 *Sexual desire in:* Dreborg et al., "Evolution of Vertebrate Opiod Receptors," pp. 15487–92.

FIVE Zoophoria

88 *In Tasmania:* Jason Dicker, "The Poppy Industry in Tasmania," Chemistry and Physics in Tasmanian Agriculture: A Resource for Science Students and Teachers, accessed July 14, 2010. http://www.launc.tased.edu.au/online/sciences/agsci /alkalo/popindus.htm.

88 *Ignoring security cameras:* Damien Brown, "Tassie Wallabies Hopping High," *Mercury,* June 25, 2009, accessed July 14, 2010. http://www.themercury.com.au/article /2009/06/25/80825_tasmania-news.html.

88 *Even the mug shot:* Ibid.

89 *The medical community's:* National Institutes of Health, "Addiction and the Criminal Justice System," *NIH Fact Sheets,* accessed October 7, 2011. http://report.nih .gov/NIHfactsheets/ViewFactSheet.aspx?csid=22.

89 *Addicts belong to:* K. H. Berge, M. D. Seppala, and A. M. Schipper, "Chemical Dependency and the Physician," *Mayo Clinic Proceedings* 84 (2009): pp. 625–31.

90 *No one issued Flying While Intoxicated:* Emily Beeler telephone interview, October 12, 2011.

90 *The Bohemian waxwings:* Ronald K. Siegel, *Intoxication: Life in Pursuit of Artificial Paradise,* New York: Pocket Books, 1989.

90 *When a horse named Fat Boy:* Luke Salkeld, "Pictured: Fat Boy, the Pony Who Got Drunk on Fermented Apples and Fell into a Swimming Pool," *MailOnline,* October 16, 2008, accessed July 15, 2010. http://www.dailymail.co.uk/news/article-1077831 /Pictured-Fat-Boy-pony-gotdrunk-fermented-apples-fell-swimming-pool.html.

90 *Bighorn sheep:* Siegel, *Intoxication,* pp. 51–52.

90 *In opium-producing regions of Asia:* Ibid., p. 130.

90 *The pen-tailed tree shrew:* Frank Wiens, Annette Zitzmann, Marc-André Lachance, Michel Yegles, Fritz Pragst, Friedrich M. Wurst, Dietrich von Holst, et al., "Chronic Intake of Fermented Floral Nectar by Wild Treeshrews," *Proceedings of the National Academy of Sciences* 105 (2008): pp. 10426–31.

90 *When cattle and horses:* M. H. Ralphs, D. Graham, M. L. Galyean, and L. F. James, "Creating Aversions to Locoweed in Naïve and Familiar Cattle," *Journal of Range Management* 50 (1997): pp. 361–66; Michael H. Ralphs, David Graham, and Lynn F. James, "Social Facilitation Influences Cattle to Graze Locoweed," *Journal of Range Management* 47 (1994): pp. 123–26; United States Department of Agriculture, Agricultural Research Service, "Locoweed (*astragalus* and *Oxytropis* spp.)." Last modified February 7, 2006, accessed March 9, 2010. http://www.ars.usda.gov/services /docs.htm?docid=9948&pf=1&cg_id=0.

91 *A friendly cocker spaniel:* Laura Mirsch, "The Dog Who Loved to Suck on Toads," *NPR,* October 24, 2006, accessed July 14, 2010. http://www.npr.org/templates/story /story.php?storyId=6376594; United States Department of Agriculture, "Locoweed."

91 *In Australia's Northern Territory:* "Dogs Getting High Licking Hallucinogenic Toads!" *StrangeZoo.com,* accessed July 14, 2010. http://www.strangezoo.com /content/item/105766.html.

91 *In colonial New England:* Iain Gately, "Drunk as a Skunk . . . or a Wild Monkey . . . or a Pig," *Proof Blog, New York Times,* January 24, 2009, accessed January 27, 2009. http://proof.blogs.nytimes.com/2009/01/24/drunk-as-a-skunk-or-a-wild-monkey-or-a-pig/.

91 *"they were filled with the husks"*: Ibid.

92 *Apparently the trick:* Ibid.

92 *Darwin also detailed:* Charles Darwin, *The Descent of Man,* in *From So Simple a Beginning: The Four Great Books of Charles Darwin,* ed. Edward O. Wilson. New York: Norton, 2006: pp. 783–1248.

92 *You can see modern-day:* BBC Worldwide, "Alcoholic Vervet Monkeys! Weird Nature—BBC Animals," video, 2009, retrieved October 9, 2011, http://www.you tube.com/watch?v=pSm7BcQHWXk&feature=related.

93 *They lose neuromuscular control:* Toni S. Shippenberg and George F. Koob, "Recent Advances in Animal Models of Drug Addiction," in *Neuropsychopharmacology: The Fifth Generation of Progress,* ed. K. L. Davis, D. Charney, J. T. Coyle, and C. Nemeroff, Philadelphia: Lippincott, Williams and Wilkins, 2002: pp. 1381–97; J. Wolfgramm, G. Galli, F. Thimm, and A. Heyne, "Animal Models of Addiction: Models for Therapeutic Strategies?" *Journal of Neural Transmission* 107 (2000): pp. 649–68.

93 *Bees "dance" more vigorously:* Andrew B. Barron, Ryszard Maleszka, Paul G. Helliwell, and Gene E. Robinson, "Effects of Cocaine on Honey Bee Dance Behaviour," *Journal of Experimental Biology* 212 (2009): pp. 163–68.

93 *Immature zebrafish hang out:* S. Bretaud, Q. Li, B. L. Lockwood, K. Kobayashi, E. Lin, and S. Guo, "A Choice Behavior for Morphine Reveals Experience-Dependent Drug Preference and Underlying Neural Substrates in Developing Larval Zebrafish," *Neuroscience* 146 (2007): pp. 1109–16.

93 *Methamphetamine juices snail:* Kathryn Knight, "Meth(amphetamine) May Stop Snails from Forgetting," *Journal of Experimental Biology* 213 (2010), i, accessed May 31, 2010. doi: 10.1242/jeb.046664.

93 *Spiders on a range of drugs:* "Spiders on Speed Get Weaving," *New Scientist,* April 29, 1995, accessed October 9, 2011. http://www.newscientist.com/article /mg14619750.500-spiders-on-speed-get-weaving.html.

93 *Alcohol can make:* Hyun-Gwan Lee, Young-Cho Kim, Jennifer S. Dunning, and Kyung-An Han, "Recurring Ethanol Exposure Induces Disinhibited Courtship in *Drosophila*," *PLoS One* (2008): p. e1391.

93 *Even humble* Caenorhabditis elegans: Andrew G. Davies, Jonathan T. Pierce-Shimomura, Hongkyun Kim, Miri K. VanHoven, Tod R. Thiele, Antonello Bonci, Cornelia I. Bargmann, et al., "A Central Role of the BK Potassium Channel in Behavioral Responses to Ethanol in *C. elegans*," *Cell* 115: pp. 656–66.

93 *That we can see parallel:* T. Sudhaharan and A. Ram Reddy, "Opiate Analgesics' Dual Role in Firefly Luciferase Activity," *Biochemistry* 37 (1998): pp. 4451–58; K. L. Machin, "Fish, Amphibian, and Reptile Analgesia," *Veterinary Clinics of North American Exotic Animal Practice* 4 (2001): pp. 19–22.

93 *Receptors for opiates:* Susanne Dreborg, Görel Sundström, Tomas A. Larsson, and Dan Larhammar, "Evolution of Vertebrate Opioid Receptors," *Proceedings of the National Academy of Sciences* 105 (2008): pp. 15487–92; Janicke Nordgreen, Joseph P. Garner, Andrew Michael Janczak, Brigit Ranheim, William M. Muir, and Tor Einar Horsberg, "Thermonociception in Fish: Effects of Two Different Doses of Morphine on Thermal Threshold and Post-Test Behavior in Goldfish (*Carassius auratus*)," *Applied Animal Behaviour Science* 119 (2009): pp. 101–07; N. A. Zabala, A. Miralto, H. Maldonado, J. A. Nunez, K. Jaffe, and L. de C. Calderon, "Opiate Receptor in Praying Mantis: Effect of Morphine and Naloxone," *Pharmacology Biochemistry & Behavior* 20 (1984): pp. 683–87; V. E. Dyakonova, F. W. Schurmann,

and D. A. Sakharov, "Effects of Serotonergic and Opioidergic Drugs on Escape Behaviors and Social Status of Male Crickets," *Naturwissenschaften* 86 (1999): pp. 435–37.

94 *Receptors for cannabinoids:* John McPartland, Vincenzo Di Marzo, Luciano De Petrocellis, Alison Mercer, and Michelle Glass, "Cannabinoid Receptors Are Absent in Insects," *Journal of Comparative Neurology* 436 (2001): pp. 423–29; Osceola Whitney, Ken Soderstrom, and Frank Johnson, "CB1 Cannabinoid Receptor Activation Inhibits a Neural Correlate of Song Recognition in an Auditory/Perceptual Region of the Zebra Finch Telencephalon," *Journal of Neurobiology* 56 (2003): pp. 266–74; E. Cottone, A. Guastalla, K. Mackie, and M. F. Franzoni, "Endocannabinoids Affect the Reproductive Functions in Teleosts and Amphibians," *Molecular and Cellular Endocrinology* 286S (2008): pp. S41–S45.

94 *"how the human mind, especially emotions":* Jaak Panksepp, "Science of the Brain as a Gateway to Understanding Play: An Interview with Jaak Panksepp," *American Journal of Play* 3 (2010): p. 250.

95 *Rat tickling came in the mid-1900s:* Ibid., p. 266

96 *Most animals don't vocalize:* Franklin D. McMillan, *Mental Health and Well-Being in Animals,* Hoboken, NJ: Blackwell, 2005: pp. 6–7.

97 *And in some cases tragically:* K. J. S. Anand and P. R. Hickey, "Pain and Its Effects in the Human Neonate and Fetus," *The New England Journal of Medicine* 317 (1987): pp. 1321–29.

97 *In the early 1900s:* Jill R. Lawson, "Standards of Practice and the Pain of Premature Infants," *Recovered Science,* accessed December 18, 2011. http://www.recoveredscience.com/ROP_preemiepain.htm.

98 *how an animal experiences:* Joseph LeDoux, "Rethinking the Emotional Brain," *Neuron* 73 (2012): pp. 653–76.

98 *"Emotions . . . shaped by natural selection":* Randolph M. Nesse and Kent C. Berridge, "Psychoactive Drug Use in Evolutionary Perspective," *Science* 278 (1997): pp. 63–66, accessed February 16, 2010. doi: 0.1126/science.278.5335.63.

98 *"Love joins hate":* E. O. Wilson, *Sociobiology,* Cambridge, MA: Harvard University Press, 1975.

99 *Indeed, when these activities:* Brian Knutson, Scott Rick, G. Elliott Wimmer, Drazen Prelec, and George Loewenstein, "Neural Predictors of Purchases," *Neuron* 53 (2007): pp. 147–56; Ethan S. Bromberg-Martin and Okihide Hikosaka, "Midbrain Dopamine Neurons Signal Preference for Advance Information About Upcoming Rewards," *Neuron* 63 (2009): pp. 119–26.

99 *"from slugs to primates":* Nesse and Berridge, "Psychoactive Drug Use," pp. 63–66.

99 *Opioid receptors and pathways:* Dreborg et al., "Evolution of Vertebrate Opioid Receptors," pp. 15487–92.

100 *Researchers working with Panksepp:* Panksepp, "Science of the Brain," p. 253.

100 *"neurochemical jungle of the human brain":* Shaun Gallagher, "How to Undress the Affective Mind: An Interview with Jaak Panksepp," *Journal of Consciousness Studies* 15 (2008): pp. 89–119.

100 *"Drugs of abuse":* Nesse and Berridge, "Psychoactive Drug Use," pp. 63–66.

101 *"You can't become addicted to a drug":* David Sack telephone interview, July 28, 2010.

103 *"Every mammal has a system in the brain":* Jaak Panksepp, "Evolutionary Substrates of Addiction: The Neurochemistries of Pleasure Seeking and Social Bonding in the

Mammalian Brain," in *Substance and Abuse Emotion*, ed. Jon D. Kassel, Washington, DC: American Psychological Association, 2010, pp. 137–67.

104 *As Gary Wilson:* Gary Wilson interview, Moorpark, CA, May 24, 2011.

105 *David J. Linden:* David J. Linden, *The Compass of Pleasure*, Viking: 2011 (location 113 in ebook).

107 *Extensive study of the effect:* Craig J. Slawecki, Michelle Betancourt, Maury Cole, and Cindy L. Ehlers, "Periadolescent Alcohol Exposure Has Lasting Effects on Adult Neurophysiological Function in Rats," *Developmental Brain Research* 128 (2001): pp. 63–72; Linda Patia Spear, "The Adolescent Brain and the College Drinker: Biological Basis of Propensity to Use and Misuse Alcohol," *Journal of Studies on Alcohol* 14 (2002): pp. 71–81; Melanie L. Schwandt, Stephen G. Lindell, Scott Chen, J. Dee Higley, Stephen J. Suomi, Markus Heilig, and Christina S. Barr, "Alcohol Response and Consumption in Adolescent Rhesus Macaques: Life History and Genetic Influences," *Alcohol* 44 (2010): pp. 67–90.

SIX Scared to Death

110 *On the day of the earthquake:* Jonathan Leor, W. Kenneth Poole, and Robert A. Kloner, "Sudden Cardiac Death Triggered by an Earthquake," *New England Journal of Medicine* 334 (1996): pp. 413–19.

111 *Admissions to hospitals for chest pain:* Laura S. Gold, Leslee B. Kane, Nona Soto-odehnia, and Thomas Rea, "Disaster Events and the Risk of Sudden Cardiac Death: A Washington State Investigation," *Prehospital and Disaster Medicine* 22 (2007): pp. 313–17.

111 *Statisticians combing through the numbers:* S. R. Meisel, K. I. Dayan, H. Pauzner, I. Chetboun, Y. Arbel, D. David, and I. Kutz, "Effect of Iraqi Missile War on Incidence of Acute Myocardial Infarction and Sudden Death in Israeli Civilians," *Lancet* 338 (1991): pp. 660–61.

112 *The number of life-threatening:* Omar L. Shedd, Samuel F. Sears, Jr., Jane L. Harvill, Aysha Arshad, Jamie B. Conti, Jonathan S. Steinberg, and Anne B. Curtis, "The World Trade Center Attack: Increased Frequency of Defibrillator Shocks for Ventricular Arrhythmias in Patients Living Remotely from New York City," *Journal of the American College of Cardiology* 44 (2004): pp. 1265–67.

112 *Take the 1998 soccer World Cup:* Paul Oberjuerge, "Argentina Beats Courageous England 4–3 in Penalty Kicks," *Soccer-Times.com*, June 30, 1998, accessed December 8, 2010. http://www.soccertimes.com/worldcup/1998/games/jun30a.htm.

113 *And that day heart attacks:* Douglas Carroll, Shah Ebrahim, Kate Tilling, John Macleod, and George Davey Smith, "Admissions for Myocardial Infarction and World Cup Football Database Survey," *BMJ* 325 (2002): pp. 21–8.

113 *Interestingly, soccer matches:* L. Toubiana, T. Hanslik, and L. Letrilliart, "French Cardiovascular Mortality Did Not Increase During 1996 European Football Championship," *BMJ* 322 (2001): p. 1306.

113 *Richard Williams, a sportswriter:* Richard Williams, "Down with the Penalty Shootout and Let the 'Games Won' Column Decide," *Sports Blog, Guardian*, October 24, 2006, accessed October 5, 2011. http://www.guardian.co.uk/football/2006/oct/24/sport.comment3.

114 *Takotsubo cardiomyopathy:* K. Tsuchihashi, K. Ueshima, T. Uchida, N. Oh-mura,

K. Kimura, M. Owa, M. Yoshiyama, et al., "Transient Left Ventricular Apical Ballooning Without Coronary Artery Stenosis: A Novel Heart Syndrome Mimicking Acute Myocardial Infarction," *Journal of the American College of Cardiology* 38 (2001): pp. 11–18; Yoshiteru Abe, Makoto Kondo, Ryota Matsuoka, Makoto Araki, Kiyoshi Dohyama, and Hitoshi Tanio, "Assessment of Clinical Features in Transient Left Ventricular Apical Ballooning," *Journal of the American College of Cardiology* 41 (2003): pp. 737–42; Kevin A. Bybee and Abhiram Prasad, "Stress-Related Cardiomyopathy Syndromes," *Circulation* 118 (2008): pp. 397–409; Scott W. Sharkey, Denise C. Windenburg, John R. Lesser, Martin S. Maron, Robert G. Hauser, Jennifer N. Lesser, Tammy S. Haas, et al., "Natural History and Expansive Clinical Profile of Stress (Tako-Tsubo) Cardiomyopathy," *Journal of the American College of Cardiology* 55 (2010): pp. 333–41.

114 *Every year your heart:* Matthew J. Loe and William D. Edwards, "A Light-Hearted Look at a Lion-Hearted Organ (Or, a Perspective from Three Standard Deviations Beyond the Norm) Part 1 (of Two Parts)," *Cardiovascular Pathology* 13 (2004): pp. 282–92.

115 *Tragically, however, this steadfast:* National Institutes of Health, "Researchers Develop Innovative Imaging System to Study Sudden Cardiac Death," *NIH News— National Heart, Lung and Blood Institute,* October 30, 2009, accessed October 14, 2011. http://www.nih.gov/news/health/oct2009/nhlbi-30.htm.

117 *At forty-one:* Dan Mulcahy interview, Tulsa, OK, October 27, 2009.

117 *The term describes a syndrome:* Jessica Paterson, "Capture Myopathy," in *Zoo Animal and Wildlife Immobilization and Anesthesia,* edited by Gary West, Darryl Heard, and Nigel Caulkett, Ames, IA: Blackwell, 2007: 115, pp. 115–21.

118 *Veterinarians divide:* Ibid.

118 *Fishermen trawling:* G. D. Stentiford and D. M. Neil, "A Rapid Onset, Post-Capture Muscle Necrosis in the Norway Lobster, *Nephrops norvegicus* (L.), from the West Coast of Scotland," *Journal of Fish Diseases* 23 (2000): pp. 251–63.

118 *stress before being slaughtered:* Purdue University Animal Services, "Meat Quality and Safety," accessed October 14, 2011. http://ag.ansc.purdue.edu/meat_quality/mqf_stress.html.

118 *Special care is taken:* Mitchell Bush and Valerius de Vos, "Observations on Field Immobilization of Free-Ranging Giraffe (*Giraffa camelopardalis*) Using Carfentanil and Xylazine," *Journal of Zoo Animal Medicine* 18 (1987): pp. 135–40; H. Ebedes, J. Van Rooyen, and J. G. Du Toit, "Capturing Wild Animals," in *The Capture and Care Manual: Capture, Care, Accommodation and Transportation of Wild African Animals,* edited by Andrew A. McKenzie, Pretoria: South African Veterinary Foundation, 1993, pp. 382–440.

118 *Deer, elk, and reindeer:* "Why Deer Die," Deerfarmer.com: Deer & Elk Farmers' Information Network, July 25, 2003, accessed October 5, 2011. http://www.deer -library.com/artman/publish/article_98.shtml.

119 *yearly helicopter roundups:* Scott Sonner, "34 Wild Horses Died in Recent Nevada Roundup, Bureau of Land Management Says," *L.A. Unleashed* (blog), *Los Angeles Times,* August 5, 2010, accessed March 3, 2012. http://latimesblogs.latimes .com/unleashed/2010/08/thirtyfour-wild-horses-died-in-recent-nevada-roundup- bureau-of-land-management-says.html.

119 *Military physicians:* J. A. Howenstine, "Exertion-Induced Myoglobinuria and Hemoglobinuria," *JAMA* 173 (1960): pp. 495–99; J. Greenberg and L. Arneson, "Exertional Rhabdomyolysis with Myoglobinuria in a Large Group of Military

Trainees," *Neurology* 17 (1967): pp. 216–22; P. F. Smith, "Exertional Rhabdomyolysis in Naval Officer Candidates," *Archives of Internal Medicine* 121 (1968): pp. 313–19; S. A. Geller, "Extreme Exertion Rhabdomyolysis: a Histopathologic Study of 31 Cases," *Human Pathology* (1973): pp. 241–50.

119 *Extreme athletes:* Mark Morehouse, "12 Football Players Hospitalized with Exertional Condition," *Gazette,* January 25, 2011, accessed October 5, 2011. http://the gazette.com/2011/01/25/ui-release-12-football-players-in-hospital-with-undisclosed -illness/.

120 *Examples of animals dying:* Paterson, "Capture Myopathy."

120 *Bighorn sheep:* Bureau of Land Management, "Status of the Science: On Questions That Relate to BLM Plan Amendment Decisions and Peninsular Ranges Bighorn Sheep," last modified March 14, 2001, accessed October 5, 2011. http://www.blm .gov/pgdata/etc/medialib//blm/ca/pdf/pdfs/palmsprings_pdfs.Par.95932cf3.File. pdf/Stat_of_Sci.pdf.

120 *Pet rabbits:* Department of Health and Human Services, "Rabbits," accessed October 5, 2011. http://ori.hhs.gov/education/products/ncstate/rabbit.htm.

120 *Fireworks blasts:* Blue Cross, "Fireworks and Animals: How to Keep Your Pets Safe," accessed November 26, 2009. http://www.bluecross.org.uk/2154–88390/fireworks -and-animals.html; Maggie Page, "Fireworks and Animals: A Survey of Scottish Vets in 2001," accessed November 26, 2009. http://www.angelfire.com/co3/NCFS/ survey/sspca/scottishspca.html; Don Jordan, "Rare Bird, Spooked by Fireworks, Thrashes Itself to Death," *Palm Beach Post News,* January 1, 2009, accessed November 26, 2009. http://www.palmbeachpost.com/localnews/content/local_news/epaper /2009/01/01/0101deadbird.html.

120 *In Copenhagen in the mid-1990s:* Associated Press, " 'Killer' Opera: Wagner Fatal to Zoo's Okapi," *The Spokesman-Review,* August 10, 1994, accessed March 3, 2012. http://news.google.com/newspapers?nid=1314&dat=19940810&id=joxAAAAIBA J&sjid=5AkEAAAAIBAJ&pg=3036,5879969.

120 *Loud, frightening noises:* World Health Organization: Regional Office for Europe, "Health Effects of Noise," accessed October 5, 2011. http://www.euro.who.int/en/ what-we-do/health-topics/environment-and-health/noise/facts-and-figures/ health-effects-of-noise.

120 *One study published:* Wen Qi Gan, Hugh W. Davies, and Paul A. Demers, "Exposure to Occupational Noise and Cardiovascular Disease in the United States: The National Health and Nutrition Examination Survey 1999–2004," *Occupational and Environmental Medicine* (2010): doi:10.1136/oem.2010.055269, accessed October 6, 2011. http://oem.bmj.com/content/early/2010/09/06/oem.2010.055269 .abstract.

121 *Dalmatians born with long QT syndrome:* W. R. Hudson and R. J. Ruben, "Hereditary Deafness in the Dalmatian Dog," *Archives of Otolaryngology* 75 (1962): p. 213; Thomas N. James, "Congenital Deafness and Cardiac Arrhythmias," *American Journal of Cardiology* 19 (1967): pp. 627–43.

121 *Four captive zebras:* Darah Hansen, "Investigators Probe Death of Four Zebras at Greater Vancouver Zoo," *Vancouver Sun,* April 20, 2009, accessed March 3, 2012. http://forum.skyscraperpage.com/showthread.php?t=168150.

121 *Some bird-watchers:* Jacquie Clark and Nigel Clark, "Cramp in Captured Waders: Suggestions for New Operating Procedures in Hot Conditions and a Possible Field Treatment," *IWSG Bulletin* (2002): 49.

122 *Circumstances that are not:* Alain Ghysen, "The Origin and Evolution of the

Nervous System," *International Journal of Developmental Biology* 47 (2003): pp. 555–62.

122 *It's possible that our imaginative:* Martin A. Samuels, "Neurally Induced Cardiac Damage. Definition of the Problem," *Neurologic Clinics* 11 (1993): p. 273.

122 *"We know that stress":* Carolyn Susman, "What Ken Lay's Death Can Teach Us About Heart Health," *Palm Beach Post,* July 7, 2006, accessed October 4, 2011. http://findarticles.com/p/news-articles/palm-beach-post/mi_8163/is_20060707/ken-lays-death-teach-heart/ai_n51923077/.

122 *Indeed, studies have demonstrated:* Joel E. Dimsdale, "Psychological Stress and Cardiovascular Disease," *Journal of the American College of Cardiology* 51 (2008): pp. 1237–46.

123 *"a unifying hypothesis . . . to explain":* M. A. Samuels, "Neurally Induced Cardiac Damage. Definition of the Problem," *Neurologic Clinics* 11 (1993): p. 273.

123 *Voodoo curses and overly ominous thoughts:* Helen Pilcher, "The Science of Voodoo: When Mind Attacks Body," *New Scientist,* May 13, 2009, accessed May 14, 2009. http://www.newscientist.com/article/mg20227081.100-the-science-of-voodoo-when-mind-attacks-body.html.

123 *"Surgeons are wary":* Brian Reid, "The Nocebo Effect: Placebo's Evil Twin," *Washington Post,* April 30, 2002, accessed November 26, 2009. http://www.washingtonpost.com/ac2/wp-dyn/A2709-2002Apr29.

124 *Arthur Barsky, a psychiatrist:* Ibid.

124 *Sudden unexpected nocturnal:* Ronald G. Munger and Elizabeth A. Booton, "Bangungut in Manila: Sudden and Unexplained Death in Sleep of Adult Filipinos," *International Journal of Epidemiology* 27 (1998): pp. 677–84.

124 *Potato leaves and tubers:* Anna Swiedrych, Katarzyna Lorenc-Kukula, Aleksandra Skirycz, and Jan Szopa, "The Catecholamine Biosynthesis Route in Potato Is Affected by Stress," *Plant Physiology and Biochemistry* 42 (2004): pp. 593–600; Jan Szopa, Grzegorz Wilczynski, Oliver Fiehn, Andreas Wenczel, and Lothar Willmitzer, "Identification and Quantification of Catecholamines in Potato Plants (*Solanum tuberosum*) by GC-MS," *Phytochemistry* 58 (2001): pp. 315–20.

125 *The evolutionary medicine expert:* Randolph M. Nesse, "The Smoke Detector Principle: Natural Selection and the Regulation of Defensive Responses," *Annals of the New York Academy of Sciences* 935 (2001): pp. 75–85.

125 *"Being killed greatly decreases":* S. L. Lima and L. M. Dill, "Behavioral Decisions Made Under the Risk of Predation: A Review and Prospectus," *Canadian Journal of Zoology* 68 (1990): pp. 619–40.

127 *In human medical circles:* Wanda K. Mohr, Theodore A. Petti, and Brian D. Mohr, "Adverse Effects Associated with Physical Restraint," *Canadian Journal of Psychiatry* 48 (2003): pp. 330–37.

128 *Sudden infant death syndrome:* Centers for Disease Control and Prevention, "Sudden Infant Death Syndrome—United States, 1983–1994," *Morbidity and Mortality Weekly Report* 45 (1996): pp. 859–63; M. Willinger, L. S. James, and C. Catz, "Defining the Sudden Infant Death Syndrome (SIDS): Deliberations of an Expert Panel Convened by the National Institute of Child Health and Human Development," *Pediatric Pathology* 11 (1991): pp. 677–84; Roger W. Byard and Henry F. Krous, "Sudden Infant Death Syndrome: Overview and Update," *Pediatric and Developmental Pathology* 6 (2003): 112–27.

128 *"the sudden death of an infant":* National SIDS Resource Center, "What Is SIDS?," accessed October 5, 2011. http://sids-network.org/sidsfact.htm.

128 *Theories abound:* Centers for Disease Control and Prevention, "Sudden Infant Death Syndrome," pp. 859–63; Willinger, James, and Catz, "Defining the Sudden Infant Death Syndrome," pp. 677–84; Byard and Krous, "Sudden Infant Death Syndrome," pp. 112–27.

129 *These reflexes likely share:* B. Kaada, "Electrocardiac Responses Associated with the Fear Paralysis Reflex in Infant Rabbits and Rats: Relation to Sudden Infant Death," *Functional Neurology* 4 (1989): pp. 327–40.

129 *The heart rates of animals:* E. J. Richardson, M. J. Shumaker, and E. R. Harvey, "The Effects of Stimulus Presentation During Cataleptic, Restrained, and Free Swimming States on Avoidance Conditioning of Goldfish (*Carassius auratus*)," *Psychological Record* 27 (1997): pp. 63–75; P. A. Whitman, J. A. Marshall, and E. C. Keller, Jr., "Tonic Immobility in the Smooth Dogfish Shark, *Mustelus canis* (Pisces, Carcharhinidae)," *Copeia* (1986): pp. 829–32; L. Lefebvre and M. Sabourin, "Effects of Spaced and Massed Repeated Elicitation on Tonic Immobility in the Goldfish (*Carassius auratus*)," *Behavioral Biology* 21 (1997): pp. 300–5; A. Kahn, E. Rebuffat, and M. Scottiaux, "Effects of Body Movement Restraint on Cardiac Response to Auditory Stimulation in Sleeping Infants," *Acta Paediatrica* 81 (1992): 959–61; Laura Sebastiani, Domenico Salamone, Pasquale Silvestri, Alfredo Simoni, and Brunello Ghelarducci, "Development of Fear-Related Heart Rate Responses in Neonatal Rabbits," *Journal of the Autonomic Nervous System* 50 (1994): pp. 231–38.

129 *Birger Kaada connected:* Birger Kaada, "Why Is There an Increased Risk for Sudden Infant Death in Prone Sleeping? Fear Paralysis and Atrial Stretch Reflexes Implicated?" *Acta Paediatrica* 83 (1994): pp. 548–57.

130 *And, interestingly, swaddling:* Patricia Franco, Sonia Scaillet, José Groswaasser, and André Kahn, "Increased Cardiac Autonomic Responses to Auditory Challenges in Swaddled Infants," *Sleep* 27 (2004): pp. 1527–32.

SEVEN Fat Planet

132 *But there I was:* American Association of Zoo Veterinarians Annual Conference with the Nutrition Advisory Group, Tulsa, OK, October 2009.

133 *Exact numbers are hard to pin down:* I. M. Bland, A. Guthrie-Jones, R. D. Taylor, and J. Hill. "Dog Obesity: Veterinary Practices' and Owners' Opinions on Cause and Management," *Preventive Veterinary Medicine* 94 (2010): pp. 310–15; Alexander J. German, "The Growing Problem of Obesity in Dogs and Cats," *Journal of Nutrition* 136 (2006): pp. 19405–65; Elizabeth M. Lund, P. Jane Armstrong, Claudia A. Kirk, and Jeffrey S. Klausner, "Prevalence and Risk Factors for Obesity in Adult Dogs from Private US Veterinary Practice," *International Journal of Applied Research in Veterinary Medicine* 4 (2006): pp. 177–86.

133 *But studies in both the United States and Australia:* Bland et al., "Dog Obesity"; German, "The Growing Problem," pp. 19405–65; Lund et al., "Prevalence and Risk Factors," pp. 177–86.

133 *close to a jaw-dropping 70 percent:* Cynthia L. Ogden and Margaret D. Carroll, "Prevalence of Overweight, Obesity, and Extreme Obesity Among Adults: United States, Trends 1960–1962 Through 2007–2008," National Center for Health Statis-

tics, June 2010, accessed October 12, 2011. http://www.cdc.gov/nchs/data/hestat/obesity_adult_07_08/obesity_adult_07_08.pdf.

134   *With our pets' excess pounds:* Lund et al., "Prevalence and Risk Factors"; C. A. Wyse, K. A. McNie, V. J. Tannahil, S. Love, and J. K. Murray, "Prevalence of Obesity in Riding Horses in Scotland," *Veterinary Record* 162 (2008): pp. 590–91.

134   *Some dogs are put on:* Rob Stein, "Something for the Dog That Eats Everything: A Diet Pill," *Washington Post,* January 6, 2007, accessed October 12, 2011. http://www.washingtonpost.com/wp-dyn/content/article/2007/01/05/AR2007010501753.html.

134   *Liposuction has been the treatment:* P. Bottcher, S. Kluter, D. Krastel, and V. Grevel, "Liposuction—Removal of Giant Lipomas for Weight Loss in a Dog with Severe Hip Osteoarthritis," *Journal of Small Animal Practice* 48 (2006): pp. 46–48.

134   *Companion felines have been placed:* Jessica Tremayne, "Tell Clients to Bite into 'Catkins' Diet to Battle Obesity, Expert Advises," *DVM Newsmagazine,* August 1, 2004, accessed March 3, 2012. http://veterinarynews.dvm360.com/dvm/article/articleDetail.jsp?id=110710.

134   *Veterinarians increasingly treat "portly ponies":* Caroline McGregor-Argo, "Appraising the Portly Pony: Body Condition and Adiposity," *Veterinary Journal* 179 (2009): pp. 158–60.

134   *If you've ever tallied:* Jennifer Watts interview, Tulsa, OK, October 27, 2009; CBS News, "When Lions Get Love Handles: Zoo Nutritionists Are Rethinking Ways of Feeding Animals in Order to Avoid Obesity," March 17, 2008, accessed January 30, 2010. http://www.cbsnews.com/stories/2008/03/17/tech/main3944935.shtml.

134   *In Indianapolis, zookeepers:* Ibid.

134   *In Toledo, plump giraffes:* Ibid.

135   *City rats crawling:* Yann C. Klimentidis, T. Mark Beasley, Hui-Yi Lin, Giulianna Murati, Gregory E. Glass, Marcus Guyton, Wendy Newton, et al., "Canaries in the Coal Mine: A Cross Species Analysis of the Plurality of Obesity Epidemics," *Proceedings of the Royal Society B* (2010): pp. 2, 3–5. doi:10.1098/rspb.2010.1980.

136   *Big felines, like lions:* Joanne D. Altman, Kathy L. Gross, and Stephen R. Lowry, "Nutritional and Behavioral Effects of Gorge and Fast Feeding in Captive Lions," *Journal of Applied Animal Welfare Science* 8 (2005): pp. 47–57.

136   *"We're all hardwired":* Mark Edwards interview, San Luis Obispo, CA, February 5, 2010.

136   *In fact, when presented with unlimited:* Katherine A. Houpt, *Domestic Animal Behavior for Veterinarians and Animal Scientists,* 5th ed., Ames, IA: Wiley-Blackwell, 2011: p. 62.

136   *A seal with the catchy nickname:* Jim Braly, "Swimming in Controversy, Sea Lion C265 Is First to Be Killed," *Oregon-Live,* April 17, 2009, accessed April 27, 2010. http://www.oregonlive.com/news/index.ssf/2009/04/swimming_in_controversy_c265_w.html.

136   *The weight of blue whales:* Dan Salas telephone interview, September 21, 2010.

136   *In the Colorado Rockies:* Arpat Ozgul, Dylan Z. Childs, Madan K. Oli, Kenneth B. Armitage, Daniel T. Blumstein, Lucretia E. Olsen, Shripad Tuljapurkar, et al., "Coupled Dynamics of Body Mass and Population Growth in Response Environmental Change," *Nature* 466 (2010): pp. 482–85.

136   *"As the snow has melted":* Dan Blumstein interview, Los Angeles, CA. February 29, 2012.

136   *A study Blumstein copublished:* Ibid.

137   *And if this doesn't seem like a lot:* Cynthia L. Ogden, Cheryl D. Fryar, Margaret

D. Carroll, and Katherine M. Flegal, "Mean Body Weight, Height, and Body Mass Index, United States 1960–2002," *Centers for Disease Control and Prevention Advance Data from Vital and Health Statistics* 347, October 27, 2004, accessed October 13, 2011. http://www.cdc.gov/nchs/data/ad/ad347.pdf.

137 *Slovakians living at the base:* Eugene K. Balon, "Fish Gluttons: The Natural Ability of Some Fishes to Become Obese When Food Is in Extreme Abundance," *Hydrobiologia* 52 (1977): pp. 239–41.

137 *"Obesity is a disease of the environment":* "Dr. Richard Jackson of the Obesity Epidemic," video, Media Policy Center, accessed October 13, 2011. http://dhc .mediapolicycenter.org/video/health/dr-richardjackson-obesity-epidemic.

137 *"One of the problems":* Ibid.

138 *Excess sugar, fat, and salt:* David Kessler, *The End of Overeating: Taking Control of the Insatiable American Appetite,* Emmaus, PA: Rodale, 2009.

139 *A survey of almost 300,000:* Medscape News Cardiology, Cardiologist Lifestyle Report 2012," accessed March 1, 2012. http://www.medscape.com/features/slide show/lifestyle/2012/cardiology.

139 *The evolutionary biologist Peter Gluckman:* Peter Gluckman, and Mark Hanson, *Mismatch: The Timebomb of Lifestyle Disease,* New York: Oxford University Press, 2006: pp. 161–62.

139 *In the dry western United States:* Peter Nonacs interview, Los Angeles, April 13, 2010.

140 *These sandy-blond rodents:* Ibid.

140 *"You don't have to eat a lot of meat":* Ibid.

140 *Evolutionary biologists think the desire:* Ibid.

141 *Even insects' body fat:* Caroline M. Pond, *The Fats of Life,* Cambridge: Cambridge University Press, 1998.

142 *"It's what the animal":* Mads Bertelsen interview, Tulsa, OK, October 27, 2009.

143 *The gorge-and-fast regimen:* Altman, Gross, and Lowry, "Nutritional and Behavioral Effects," pp. 47–57.

143 *Environmental enrichment:* Jill Mellen and Marty Sevenich MacPhee, "Philosophy of Environmental Enrichment: Past, Present and Future," *Zoo Biology* 20 (2001): pp. 211–26.

143 *Settings that allowed:* Ibid.; Ruth C. Newberry, "Environmental Enrichment: Increasing the Biological Relevance of Captive Environments," *Applied Animal Behaviour Science* 44 (1995): pp. 229–43.

143 *At the Smithsonian National Zoo:* Smithsonian National Zoological Park, "Conservation & Science: Zoo Animal Enrichment," accessed October 12, 2011. http:// nationalzoo.si.edu/SCBI/AnimalEnrichment/default.cfm.

144 *Nutritionists provide smaller:* Newberry, "Environmental Enrichment."

145 *Every autumn:* Jennifer Watts, telephone interview by Kathryn Bowers, April 19, 2010.

145 *Every day, as it has:* Volodymyr Dvornyk, Oxana Vinogradova, and Eviatar Nevo, "Origin and Evolution of Circadian Clock Genes in Prokaryotes," *Proceedings of the National Academy of Sciences* 100 (2003): pp. 2495–500.

145 *The cells of all:* Jay C. Dunlap, "Salad Days in the Rhythms Trade," *Genetics* 178 (2008): pp. 1–13; John S. O'Neill and Akhilesh B. Reddy, "Circadian Clocks in Human Red Blood Cells," *Nature* 469 (2011): pp. 498–503; John S. O'Neill, Gerben van Ooijen, Laura E. Dixon, Carl Troein, Florence Corellou, François-Yves Bouget, Akhilesh B. Reddy, et al., "Circadian Rhythms Persist Without Transcription in a Eukaryote," *Nature* 469 (2011): pp. 554–58; Judit Kovac, Jana Husse, and Henrik

Oster, "A Time to Fast, a Time to Feast: The Crosstalk Between Metabolism and the Circadian Clock," *Molecules and Cells* 28 (2009): pp. 75–80.

146 *So-called higher creatures:* Dunlap, "Salad Days"; O'Neill and Reddy, "Circadian Clocks"; O'Neill et al., "Circadian Rhythms"; Kovac, Husse, and Oster, "A Time to Fast."

146 *Several studies have linked:* L. C. Antunes, R. Levandovski, G. Dantas, W. Caumo, and M. P. Hidalgo, "Obesity and Shift Work: Chronobiological Aspects," *Nutrition Research Reviews* 23 (2010): pp. 155–68; L. Di Lorenzo, G. De Pergola, C. Zocchetti, N. L'Abbate, A. Basso, N. Pannacciulli, M. Cignarelli, et al., "Effect of Shift Work on Body Mass Index: Results of a Study Performed in 319 Glucose-Tolerant Men Working in a Southern Italian Industry," *International Journal of Obesity* 27 (2003): pp. 1353–58; Yolande Esquirol, Vanina Bongard, Laurence Mabile, Bernard Jonnier, Jean-Marc Soulat, and Bertrand Perret, "Shift Work and Metabolic Syndrome: Respective Impacts of Job Strain, Physical Activity, and Dietary Rhythms," *Chronobiology International* 26 (2009): pp. 544–59.

146 *A rodent study:* Laura K. Fonken, Joanna L. Workman, James C. Walton, Zachary M. Weil, John S. Morris, Abraham Haim, and Randy J. Nelson, "Light at Night Increases Body Mass by Shifting the Time of Food Intake," *Proceedings of the National Academy of Sciences* 107 (2010): pp. 18664–69.

146 *"subjected to dim lighting":* Naheeda Portocarero, "Background: Get the Light Right," *World Poultry,* accessed March 1, 2011. http://worldpoultry.net/background /get-the-light-right-8556.html.

146 *Studies have shown that disrupting:* John Pavlus, "Daylight Savings Time: The Extra Hour of Sunshine Comes at a Steep Price," *Scientific American* (September 2010): p. 69.

147 *Latitude does seem:* William Galster and Peter Morrison, "Carbohydrate Reserves of Wild Rodents from Different Latitudes," *Comparative Biochemistry and Physiology Part A: Physiology* 50 (1975): pp. 153–57.

147 *The guts of some small songbirds:* Franz Bairlein, "How to Get Fat: Nutritional Mechanisms of Seasonal Fat Accumulation in Migratory Songbirds," *Naturwissenschaften* 89 (2002): pp. 1–10.

148 *When they've fattened:* Herbert Biebach, "Phenotypic Organ Flexibility in Garden Warblers *Sylvia borin* During Long-Distance Migration," *Journal of Avian Biology* 29 (1998): pp. 529–35; Scott R. McWilliams and William H. Karasov, "Migration Takes Gut: Digestive Physiology of Migratory Birds and Its Ecological Significance," in *Birds of Two Worlds: The Ecology and Evolution of Migration,* ed. Peter P. Marra and Russell Greenberg, pp. 67–78. Baltimore: Johns Hopkins University Press, 2005; Theunis Piersma and Ake Lindstrom, "Rapid Reversible Changes in Organ Size as a Component of Adaptive Behavior, *Trends in Ecology and Evolution* 12 (1997): pp. 134–38.

148 *observed in fish:* John Sweetman, Arkadios Dimitroglou, Simon Davies, and Silvia Torrecillas, "Nutrient Uptake: Gut Morphology a Key to Efficient Nutrition," *International Aquafeed* (January–February 2008): pp. 26–30.

148 *frogs:* Elizabeth Pennesi, "The Dynamic Gut," *Science* 307 (2005): pp. 1896–99.

148 *squirrels, voles, and mice:* Terry L. Derting and Becke A. Bogue, "Responses of the Gut to Moderate Energy Demands in a Small Herbivore (*Microtus pennsylvanicus*)," *Journal of Mammalogy* 74 (1993): pp. 59–68.

148 *Jared Diamond, a UCLA physiologist:* Pennesi, "The Dynamic Gut."

148 *Deep inside every animal colon:* Ruth E. Ley, Micah Hamady, Catherine Lozupone,

Peter J. Turnbaugh, Rob Roy Ramey, J. Stephen Bircher, Michael L. Schlegel, et al., "Evolution of Mammals and Their Gut Microbes," *Science* 320 (2008): pp. 1647–51.

149 *It turns out that within our microbiomes:* Peter J. Turnbaugh, Ruth E. Ley, Michael A. Mahowald, Vincent Magrini, Elaine R. Mardis, and Jeffrey I. Gordon, "An Obesity-Associated Gut Microbiome with Increased Capacity for Energy Harvest," *Nature* 444 (2006): pp. 1027–31.

149 *Obese humans had:* Ibid.

150 *"The bacteria in obese mice":* Ibid.; Matej Bajzer and Randy J. Seeley, "Obesity and Gut Flora," *Nature* 444 (2006): p. 1009.

150 *"Feed the gut bugs":* Watts interview.

152 *The Harvard medical sociologist:* Nicholas A. Christakis and James Fowler, "The Spread of Obesity in a Large Social Network over 32 Years," *New England Journal of Medicine* 357: pp. 370–79.

152 *"It has been proven that animals":* Nikhil V. Dhurandhar, "Infectobesity: Obesity of Infectious Origin," *Journal of Nutrition* 131 (2001): pp. 2794S–97S; Robin Marantz Henig, "Fat Factors," *New York Times,* August 13, 2006, accessed February 26, 2010. http://www.nytimes.com/2006/08/13/magazine/13obesity.html; Nikhil V. Dhurandhar, "Chronic Nutritional Diseases of Infectious Origin: An Assessment of a Nascent Field," *Journal of Nutrition* 131 (2001): pp. 2787S–88S.

152 *James Marden is an entomologist and professor:* James Marden telephone interview, September 1, 2011.

153 *Fat was collecting:* Rudolph J. Schilder and James H. Marden, "Metabolic Syndrome and Obesity in an Insect," *Proceedings of the National Academy of Sciences* 103 (2006): pp. 18805–09; Rudolph J. Schilder, and James H. Marden, "Metabolic Syndrome in Insects Triggered by Gut Microbes," *Journal of Diabetes Science and Technology* 1 (2007): pp. 794–96.

153 *Dragonfly blood:* Marden interview.

153 *Metabolic syndrome:* National Diabetes Information Clearinghouse, "Insulin Resistance and Pre-diabetes," accessed October 13, 2011. http://diabetes.niddk.nih .gov/DM/pubs/insulinresistance/#metabolicsyndrome.

153 *When Marden looked:* Schilder and Marden, "Metabolic Syndrome and Obesity"; Marden interview.

153 *What the parasites caused:* Schilder and Marden, "Metabolic Syndrome and Obesity."

154 *By measuring the way:* Marden interview.

154 *"specific components of their metabolism":* Ibid.

154 *The gregarine infections:* Schilder and Marden, "Metabolic Syndrome in Insects"; Schilder and Marden, "Metabolic Syndrome and Obesity."

154 *Intriguingly, the parasites:* Marden interview; Schilder and Marden, "Metabolic Syndrome and Obesity."

154 *But although I had never:* Justus F. Mueller, "Further Studies on Parasitic Obesity in Mice, Deer Mice, and Hamsters," *Journal of Parasitology* 51 (1965): pp. 523–31.

155 *In 2005, the Australians:* NobelPrize.org, "The Nobel Prize in Physiology or Medicine 2005: Barry J. Marshall, J. Robin Warren," Nobel Prize press release, October 3, 2005, accessed October 1, 2011. http://www.nobelprize.org/nobel_prizes/medicine /laureates/2005/press.html.

155 *The road to the Nobel:* Melissa Sweet, "Smug as a Bug," *Sydney Morning Herald,* August 2, 1997, accessed October 1, 2011. http://www.vianet.net.au/~bjmrshll /features2.html.

155 *"really wasn't much response":* Marden interview.
156 *"Metabolic disease isn't some strange":* Penn State Science, "Dragonfly's Metabolic Disease Provides Clues About Human Obesity," November 20, 2006, accessed October 13, 2011. http://science.psu.edu/news-and-events/2006-news/Marden11-2006 .htm/.
156 *Watts had decided:* Watts interview.
157 *The fruit zookeepers buy:* Edwards interview.

EIGHT Grooming Gone Wild

159 *"I am so worried":* "Need Help with Feather Picking in Baby," African Grey Forum, board post dated Feb. 17, 2009, by andrea1981, accessed July 3, 2009. http://www. africangreyforum.com/forum/f38/need-help-with-feather-picking-in-baby; "Sydney Is the Resident Nudist Here," African Grey Forum, board post dated April 25, 2008, by Lisa B., accessed July 3, 2009. http://www.africangreyforum.com/forum /showthread.php/389-ok-so-who-s-grey-has-plucking-or-picking-issues; "Quaker Feather Plucking," New York Bird Club, accessed July 3, 2009. http://www.lucie-dove.websitetoolbox.com/post?id=1091055; "Feather Plucking: Help My Bird Has a Feather Plucking Problem," QuakerParrot Forum, accessed July 3, 2009. http:// www.quakerparrots.com/forum/indexphp?act=idx; Theresa Jordan, "Quaker Mutilation Syndrome (QMS): Part I," *Winged Wisdom Pet Bird Magazine,* January 1998, accessed July 3, 2009. http://www.birdsnways.com/wisdom/ww19eiv.htm; "My Baby Is Plucking," Quaker Parrot Forum, accessed July 3, 2009. http://www. quakerparrots.com/forum/index.php?showtopic=49091.
160 *"Lately she has been pulling":* "Feather Plucking."
160 *Its name says it all:* E. David Klonsky and Jennifer J. Muehlenkamp, "Self-Injury: A Research Review for the Practitioner," *Journal of Clinical Psychology: In Session* 63 (2007): pp. 1045–56; E. David Klonsky, "The Function of Deliberate Self-Injury: A Review of the Evidence," *Clinical Psychology Review* 27 (2007): pp. 226–39; E. David Klonsky, "The Functions of Self-Injury in Young Adults Who Cut Themselves: Clarifying the Evidence for Affect Regulation," *Psychiatry Research* 166 (2009): pp. 260–68; Nicola Madge, Anthea Hewitt, Keith Hawton, Erik Jan de Wilde, Paul Corcoran, Sandor Fakete, Kees van Heeringen, et al., "Deliberate Self-Harm Within an International Community Sample of Young People: Comparative Findings from the Child & Adolescent Self-harm in Europe (CASE) Study," *Journal of Child Psychology and Psychiatry* 49 (2008): pp. 667–77; Keith Hawton, Karen Rodham, Emma Evans, and Rosamund Weatherall, "Deliberate Self Harm in Adolescents: Self Report Survey in Schools in England," *BMJ* 325 (2002): pp. 1207–11; Marilee Strong, *A Bright Red Scream: Self-Mutilation and the Language of Pain,* London: Penguin (Non-Classics): 1999; Steven Levenkron, *Cutting: Understanding and Overcoming Self-Mutilation,* New York: Norton, 1998; Mary E. Williams, *Self-Mutilation (Opposing Viewpoints),* Farmington Hills: Greenhaven, 2007.
160 *Psychiatrists call cutters "self-injurers":* Klonsky and Muehlenkamp, "Self-Injury"; Klonsky, "The Function of Deliberate Self-Injury"; Madge et al., "Deliberate Self-Harm"; Hawton et al., "Deliberate Self Harm"; Strong, *A Bright Red Scream;* Levenkron, *Cutting;* Williams, *Self-Mutilation.*
161 *I, for one, was shocked:* BBC News, "The Panorama Interview," November 2005,

accessed October 2, 2011. http://www.bbc.co.uk/news/special/politics97/diana/panorama.html; Andrew Morton, *Diana: Her True Story in Her Own Words,* New York: Pocket Books, 1992.

161 *While Angelina Jolie:* "Angelina Jolie Talks Self-Harm," video, 2010, retrieved October 2, 2011, from http://www.youtube.com/watch?v=IW1Ay4u5JDE; Jolie, *20/20* interview, video, 2010, retrieved October 3, 2011, from http://www.youtube.com/watch?v=rfzPhag_09E&feature=related.

161 *Christina Ricci:* David Lipsky, "Nice and Naughty," *Rolling Stone* 827 (1999): pp. 46–52.

161 *Johnny Depp:* Chris Heath, "Johnny Depp—Portrait of the Oddest as a Young Man," *Details* (May 1993): pp. 159–69, 174.

161 *Colin Farrell:* Chris Heath, "Colin Farrell—The Wild One," *GQ Magazine* (2004): pp. 233–39, 302–3.

161 *"I began cutting my arms":* "Self Inflicted Injury," Cornell Blog: An Unofficial Blog About Cornell University, accessed October 9, 2011. http://cornell.elliottback.com/self-inflicted-injury/.

162 *And they confirm:* Klonsky and Muehlenkamp, "Self-Injury."

162 *Psychiatrists have linked:* Ibid.; Klonsky, "The Function of Deliberate Self-Injury"; Klonsky, "The Functions of Self-Injury"; Madge et al., "Deliberate Self-Harm"; Hawton et al., "Deliberate Self Harm"; Strong, *A Bright Red Scream;* Levenkron, *Cutting;* Williams, *Self-Mutilation.*

162 *In the fourth version:* American Psychiatric Association, *DSM-IV: Diagnostic and Statistical Manual of Mental Disorders,* 4th Ed., Arlington: American Psychiatric Publishing, 1994.

162 *But it turns out:* Klonsky and Muehlenkamp, "Self-Injury," p. 1047; Lorrie Ann Dellinger-Ness and Leonard Handler, "Self-Injurious Behavior in Human and Non-human Primates," *Clinical Psychology Review* 26 (2006): pp. 503–14.

162 *Some begin the behavior:* Klonsky and Muehlenkamp, "Self-Injury," p. 1046.

163 *It's a common diagnosis:* L. S. Sawyer, A. A. Moon-Fanelli, and N. H. Dodman, "Psychogenic Alopecia in Cats: 11 Cases (1993–1996)," *Journal of the American Veterinary Medical Association* 214 (1999): pp. 71–74.

163 *The diagnosis of acral lick dermatitis:* Anita Patel, "Acral Lick Dermatitis," *UK Vet* 15 (2010): pp. 1–4; Mark Patterson, "Behavioural Genetics: A Question of Grooming," *Nature Reviews: Genetics* 3 (2002): p. 89; A. Luescher, "Compulsive Behavior in Companion Animals," *Recent Advances in Companion Animal Behavior Problems,* ed. K. A. Houpt, Ithaca: International Veterinary Information Service, 2000.

163 *"Flank biters":* Katherine A. Houpt, *Domestic Animal Behavior for Veterinarians and Animal Scientists,* 5th ed., Ames, IA: Wiley-Blackwell, 2011: pp. 121–22.

164 *Nicholas Dodman, a veterinarian:* N. H. Dodman, E. K. Karlsson, A. A. Moon-Fanelli, M. Galdzicka, M. Perloski, L. Shuster, K. Lindblad-Toh, et al., "A Canine Chromosome 7 Locus Confers Compulsive Disorder Susceptibility," *Molecular Psychiatry* 15 (2010): pp. 8–10.

164 *Whether OCD in humans and CCD in dogs:* N. H. Dodman, A. A. Moon-Fanelli, P. A. Mertens, S. Pflueger, and D. J. Stein, "Veterinary Models of OCD," In *Obsessive Compulsive Disorders,* edited by E. Hollander and D. J. Stein. New York: Marcel Dekker, 1997 pp. 99–141; A. A. Moon-Fanelli and N. H. Dodman, "Description and Development of Compulsive Tail Chasing in Terriers and Response to Clomipramine Treatment," *Journal of the American Veterinary Medical Association* 212 (1998): pp. 1252–57.

164 *In contrast, with animals:* Karen L. Overall and Arthur E. Dunham, "Clinical Features and Outcome in Dogs and Cats with Obsessive-Compulsive Disorder: 126 Cases (1989–2000)," *Journal of the American Veterinary Medical Association* 221 (2002): pp. 1445–52; Dellinger-Ness and Handler, "Self-Injurious Behavior."

164 *compulsive vocalization:* Dan J. Stein, Nicholas H. Dodman, Peter Borchelt, and Eric Hollander, "Behavioral Disorders in Veterinary Practice: Relevance to Psychiatry," *Comprehensive Psychiatry* 35 (1994): pp. 275–85; Nicholas H. Dodman, Louis Shuster, Gary J. Patronek, and Linda Kinney, "Pharmacologic Treatment of Equine Self-Mutilation Syndrome," *International Journal of Applied Research in Veterinary Medicine* 2 (2004): pp. 90–98.

164 *They call it, simply:* Alice Moon-Fanelli "Feline Compulsive Behavior," accessed October 9, 2011. http://www.tufts.edu/vet/vet_common/pdf/petinfo/dvm/case _march2005.pdf; Houpt, *Domestic Animal Behavior,* p. 167.

165 *Some chimps pick parasites:* Christophe Boesch, "Innovation in Wild Chimpanzees," *International Journal of Primatology* 16 (1995): pp. 1–16.

165 *Japanese macaques:* Ichirou Tanaka, "Matrilineal Distribution of Louse Egg-Handling Techniques During Grooming in Free-Ranging Japanese Macaques," *American Journal of Physical Anthropology* 98 (1995): pp. 197–201; Ichirou Tanaka, "Social Diffusion of Modified Louse Egg-Handling Techniques During Grooming in Free-Ranging Japanese Macaques," *Animal Behaviour* 56 (1998): pp. 1229–36.

165 *plays a vital role in the social structure:* Megan L. Van Wolkenten, Jason M. Davis, May Lee Gong, and Frans B. M. de Waal, "Coping with Acute Crowding by *Cebus Apella,*" *International Journal of Primatology* 27 (2006): pp. 1241–56.

165 *Some groups of chimps:* Kristin E. Bonnie and Frans B. M. de Waal, "Affiliation Promotes the Transmission of a Social Custom: Handclasp Grooming Among Captive Chimpanzees," *Primates* 47 (2006): pp. 27–34.

165 *When lower-ranking bonnet macaques:* Joseph H. Manson, C. David Navarrete, Joan B. Silk, and Susan Perry, "Time-Matched Grooming in Female Primates? New Analyses from Two Species," *Animal Behaviour* 67 (2004): pp. 493–500.

165 *A tropical reef dweller:* Karen L. Cheney, Redouan Bshary, and Alexandra S. Grutter, "Cleaner Fish Cause Predators to Reduce Aggression Toward Bystanders at Cleaning Stations," *Behavioral Ecology* 19 (2008): pp. 1063–67.

165 *Scientists have found:* Ibid.

166 *Cats and rabbits may spend:* Houpt, *Domestic Animal Behavior,* p. 57.

166 *Sea lions and seals:* Hilary N. Feldman and Kristie M. Parrott, "Grooming in a Captive Guadalupe Fur Seal," *Marine Mammal Science* 12 (1996): pp. 147–53.

166 *Birds roll in dirt:* Peter Cotgreave and Dale H. Clayton, "Comparative Analysis of Time Spent Grooming by Birds in Relation to Parasite Load," *Behaviour* 131 (1994): pp. 171–87.

166 *Snakes, lacking napkins:* Daniel S. Cunningham and Gordon M. Burghardt, "A Comparative Study of Facial Grooming After Prey Ingestion in Colubrid Snakes," *Ethology* 105 (1999): pp. 913–36.

166 *Grooming actually alters the neurochemistry:* Allan V. Kalueff and Justin L. La Porte, *Neurobiology of Grooming Behavior,* New York: Cambridge University Press, 2010.

166 *Even simply petting:* Karen Allen, "Are Pets a Healthy Pleasure? The Influence of Pets on Blood Pressure," *Current Directions in Psychological Science* 12 (2003): pp. 236–39; Sandra B. Barker, "Therapeutic Aspects of the Human-Companion Ani-

mal Interaction," *Psychiatric Times* 16 (1999), accessed October 10, 2011. http://www.psychiatrictimes.com/display/article/10168/54671?pageNumber=1.

168 *It turns out that both pain and grooming:* Kalueff and La Porte, *Neurobiology of Grooming Behavior;* G. C. Davis, "Endorphins and Pain," *Psychiatric Clinics of North America* 6 (1983): pp. 473–87.

168 *Researchers in Massachusetts:* Melinda A. Novak, "Self-Injurious Behavior in Rhesus Monkeys: New Insights into Its Etiology, Physiology, and Treatment," *American Journal of Primatology* 59 (2003): pp. 3–19.

170 *Call a veterinarian to treat a flank biter:* Sue M. McDonnell, "Practical Review of Self-Mutilation in Horses," *Animal Reproduction Science* 107 (2008): pp. 219–28; Houpt, *Domestic Animal Behavior,* pp. 121–22; Nicholas H. Dodman, Jo Anne Normile, Nicole Cottam, Maria Guzman, and Louis Shuster, "Prevalence of Compulsive Behaviors in Formerly Feral Horses," *International Journal of Applied Research in Veterinary Medicine* 3 (2005): pp. 20–24.

170 *Isolation can also provoke it:* I. H. Jones and B. M. Barraclough, "Auto-mutilation in Animals and Its Relevance to Self-Injury in Man," *Acta Psychiatrica Scandinavica* 58 (1978): pp. 40–47.

170 *Birds—even ones:* Franklin D. McMillan, *Mental Health and Well-Being in Animals,* Hoboken: Blackwell, 2005: pp. 289.

170 *Many stallions stop:* McDonnell, "Practical Review," pp. 219–28; Houpt, *Domestic Animal Behavior,* p. 121–22.

170 *Animals of many kinds:* McDonnell, "Practical Review," pp. 219–28.

170 *As mentioned earlier, environmental enrichment:* Robert J. Young, *Environmental Enrichment for Captive Animals,* Hoboken: Universities Federation for Animal Welfare and Blackwell, 2003; Ruth C. Newberry, "Environmental Enrichment: Increasing the Biological Relevance of Captive Environments," *Applied Animal Behaviour Science* 44 (1995): pp. 229–43.

171 *In 1985, the USDA:* Jodie A. Kulpa-Eddy, Sylvia Taylor, and Kristina M. Adams, "USDA Perspective on Environmental Enrichment for Animals," *Institute for Laboratory Animal Research Journal* 46 (2005): pp. 83–94.

171 *When the coyote handler:* Hilda Tresz, Linda Ambrose, Holly Halsch, and Annette Hearsh, "Providing Enrichment at No Cost," *The Shape of Enrichment: A Quarterly Source of Ideas for Environmental and Behavioral Enrichment* 6 (1997): pp. 1–4.

171 *Trainers give horses:* McDonnell, "Practical Review," pp. 219–28.

173 *Some therapists counsel:* Deb Martinsen, "Ways to Help Yourself Right Now," American Self-Harm Information Clearinghouse, accessed December 20, 2011. http://www.selfinjury.org/docs/selfhelp.htm.

174 *A survey comparing:* John P. Robinson and Steven Martin, "What Do Happy People Do?" *Social Indicators Research* 89 (2008): pp. 565–71.

NINE Fear of Feeding

177 *strikes 1 in 200:* H. W. Hoek, "Incidence, Prevalence and Mortality of Anorexia Nervosa and Other Eating Disorders," *Current Opinion in Psychiatry* 19 (2006): pp. 389–94.

177 *It's surprisingly lethal:* Joanna Steinglass, Anne Marie Albano, H. Blair Simpson, Kenneth Carpenter, Janet Schebendach, and Evelyn Attia, "Fear of Food as a Treat-

ment Target: Exposure and Response Prevention for Anorexia Nervosa in an Open Series," *International Journal of Eating Disorders* (2011), accessed March 3, 2012. doi: 10.1002/eat.20936.

177 *Bulimia nervosa:* James I. Hudson, Eva Hiripi, Harrison G. Pope, Jr., and Ronald C. Kessler, "The Prevalence and Correlates of Eating Disorders in the National Comorbidity Survey Replication," *Biological Psychiatry* 61 (2007): pp. 348–58.

177 *the World Health Organization has:* W. Stewart Agras, *The Oxford Handbook of Eating Disorders,* New York: Oxford University Press, 2010.

177 *In the two decades:* Ibid.

177 *Because disordered eating:* Ibid.

177 *Anxiety disorders are frequently:* Walter H. Kaye, Cynthia M. Bulik, Laura Thornton, Nicole Barbarich, Kim Masters, and Price Foundation Collaborative Group, "Comorbidity of Anxiety Disorders with Anorexia and Bulimia Nervosa," *The American Journal of Psychiatry* 161 (2004): pp. 2215–21.

177 *They report enjoying:* Agras, *The Oxford Handbook.*

181 *scientists at Yale built:* Dror Hawlena and Oswald J. Schmitz, "Herbivore Physiological Response to Predation Risk and Implications for Ecosystem Nutrient Dynamics," *Proceedings of the National Academy of Sciences* 107 (2010): pp. 15503–7; Emma Marris, "How Stress Shapes Ecosystems," *Nature News,* September 21, 2010, accessed August 25, 2011. http://www.nature.com/news/2010/100921/full/news.2010.479.html.

181 *When stressed out:* Dror Hawlena, telephone interview, September 29, 2010.

181 *The threat of predation:* Dror Hawlena and Oswald J. Schmitz, "Physiological Stress as a Fundamental Mechanism Linking Predation to Ecosystem Functioning," *American Naturalist* 176 (2010): pp. 537–56.

182 *Psychiatrists studying eating disorders:* Marian L. Fitzgibbon and Lisa R. Blackman, "Binge Eating Disorder and Bulimia Nervosa: Differences in the Quality and Quantity of Binge Eating Episodes," *International Journal of Eating Disorders* 27 (2000): pp. 238–43.

182 *In a study of gerbils:* Tim Caro, *Antipredator Defenses in Birds and Mammals,* Chicago: University of Chicago Press, 2005.

182 *Another study, on rodents:* Ibid.

182 *Scorpions have shown a similar aversion:* Ibid.

182 *It's known that light:* Masaki Yamatsuji, Tatsuhisa Yamashita, Ichiro Arii, Chiaki Taga, Noaki Tatara, and Kenji Fukui, "Season Variations in Eating Disorder Subtypes in Japan," *International Journal of Eating Disorders* 33 (2003): pp. 71–77.

183 *"with its large carnivores gone":* David Baron, *The Beast in the Garden: A Modern Parable of Man and Nature,* New York: Norton, 2004: p. 19.

183 *For fifty years:* Scott Creel, John Winnie Jr., Bruce Maxwell, Ken Hamlin, and Michael Creel, "Elk Alter Habitat Selection as an Antipredator Response to Wolves," *Ecology* 86 (2005): pp. 3387–97; John W. Laundre, Lucina Hernandez, and Kelly B. Altendorf, "Wolves, Elk, and Bison: Reestablishing the 'landscape of fear' in Yellowstone National Park, U.S.A.," *Canadian Journal of Zoology* 79 (2001): pp. 1401–9; Geoffrey C. Trussell, Patrick J. Ewanchuk, and Mark D. Bertness, "Trait-Mediated Effects in Rocky Intertidal Food Chains: Predator Risk Cues Alter Prey Feeding Rates," *Ecology* 84 (2003): pp. 629–40; Aaron J. Wirsing and Willilam J. Ripple, "Frontiers in Ecology and the Environment: A Comparison of Shark and Wolf Research Reveals Similar Behavioral Responses by Prey," *Frontiers in Ecology and the Environment* (2010). doi: 10.1980/090226.

184 *beyond squirrels pushing nuts:* Stephen B. Vander Wall, *Food Hoarding in Animals,* Chicago: University of Chicago Press, 1990.

185 *Some moles create worm farms:* Ibid.

186 *For example, food hoarding is often:* Mark D. Simms, Howard Dubowitz, and Moira A. Szilagyi, "Health Care Needs of Children in the Foster Care System," *Pediatrics* 105 (2000): pp. 909–18.

186 *Compulsive hoarding:* Alberto Pertusa, Miguel A. Fullana, Satwant Singh, Pino Alonso, Jose M. Mechon, and David Mataix-Cols. "Compulsive Hoarding: OCD Symptom, Distinct Clinical Syndrome, or Both?" *American Journal of Psychiatry* 165 (2008): pp. 1289–98.

186 *OCD is linked:* Walter H. Kaye, Cynthia M. Bulik, Laura Thornton, Nicole Barbarich, Kim Masters, and Price Foundation Collaborative Group, "Comorbidity of Anxiety Disorders with Anorexia and Bulimia Nervosa," *American Journal of Psychiatry* 161 (2004): pp. 2215–21.

187 *"the affected animals restrict":* Janet Treasure and John B. Owen, "Intriguing Links Between Animal Behavior and Anorexia Nervosa," *International Journal of Eating Disorders* 21 (1997): p. 307.

187 *"spend more time on nonnutritive":* Ibid.

187 *"pigs, especially those":* Ibid.

187 *"led to the uncovering of recessive":* Ibid., p. 308.

187 *"an analogous genetic basis":* Ibid.

187 *Studies of twins and generations:* Ibid., pp. 307–11.

188 *"People with anorexia nervosa":* Michael Strober interview, Los Angeles, CA, February 2, 2010.

188 *It strikes most often during:* Treasure and Owen, "Intriguing Links," pp. 307–11.

188 *Weaning is a vulnerable:* Ibid.; S. C. Kyriakis, and G. Andersson, "Wasting Pig Syndrome (WPS) in Weaners—Treatment with Amperozide," *Journal of Veterinary Pharmacology and Therapeutics* 12 (1989): pp. 232–36.

188 *Farmers keep an eye out:* Treasure and Owen, "Intriguing Links," p. 308.

189 *Connecting fearful states to eating:* Treasure and Owen, "Intriguing Links," pp. 307–11; "Thin Sow Syndrome," ThePigSite.com, accessed September 10, 2010. http://www.thepigsite.com/pighealth/article/212/thin-sow-syndrome.

189 *"There is no treatment":* "Diseases: Thin Sow Syndrome," PigProgress.Net, accessed December 19, 2011. http://www.pigprogress.net/diseases/thin-sow-syndrome-d89.html.

189 *Farmers advise making sure:* "Thin Sow Syndrome"; "Diseases: Thin Sow Syndrome."

189 *Similarly, rodent researchers found that warmer:* Robert A. Boakes, "Self-Starvation in the Rat: Running Versus Eating," *Spanish Journal of Psychology* 10 (2007): p. 256.

189 *Pig farmers also recommend:* "Thin Sow Syndrome"; Treasure and Owen, "Intriguing Links," p. 308.

190 *Some eating disorders, say psychiatrists:* Christian S. Crandall, "Social Cognition of Binge Eating," *Journal of Personality and Social Psychology* 55 (1988): pp. 588–98.

190 *Today's aspiring bulimics and anorexics:* Beverly Gonzalez, Emilia Huerta-Sanchez, Angela Ortiz-Nieves, Terannie Vazquez-Alvarez, and Christopher Kribs-Zaleta, "Am I Too Fat? Bulimia as an Epidemic," *Journal of Mathematical Psychology* 47 (2003): pp. 515–26; "Tips and Advice." Thinspiration, accessed September 14, 2010. http://mytaintedlife.wetpaint.com/page/Tips+and+Advice.

190 *Images of skeletal celebrities:* "Tips and Advice," Thinspiration.

190 *"the voluntary, retrograde movement":* Kristen E. Lukas, Gloria Hamor, Mollie A. Bloomsmith, Charles L. Horton, and Terry L. Maple, "Removing Milk from Captive Gorilla Diets: The Impact on Regurgitation and Reingestion (R/R) and Other Behaviors," *Zoo Biology* 18 (1999): p. 516.

190 *An affected gorilla:* Ibid., pp. 515–28.

191 *"might be socially enhanced":* Ibid., p. 526.

191 *R and R is widely believed not:* Ibid., p. 516.

192 *The black vultures in McKinney:* Sheryl Smith-Rodgers, "Scary Scavengers," *Texas Parks and Wildlife,* October 2005, accessed November 9, 2010. http://www.tpwmagazine.com/archive/2005/oct/legend/.

192 *Some caterpillars, too:* Jacqualine Bonnie Grant, "Diversification of Gut Morphology in Caterpillars Is Associated with Defensive Behavior," *Journal of Experimental Biology* 209 (2006): pp. 3018–24.

192 *some animals defecate:* Caro, *Antipredator Defenses.*

TEN The Koala and the Clap

194 *When monster wildfires scorched:* Fox News, "Scorched Koala Rescued from Australia's Wildfire Wasteland," February 10, 2009, accessed August 25, 2011. http://www.foxnews.com/story/0,2933,490566,00.html.

194 *But six months later:* ABC News, "Sam the Bushfire Koala Dies," August 7, 2009, accessed August 25, 2011. http://www.abc.net.au/news/2009-08-06/sam-the-bushfire-koala-dies/1381672.

195 *Technically, the disease:* Robin M. Bush and Karin D. E. Everett, "Molecular Evolution of the Chlamydiaceae," *International Journal of Systematic and Evolutionary Microbiology* 51 (2001): pp. 203–20; L. Pospisil and J. Canderle, "*Chlamydia (Chlamydiophila) pneumoniae* in Animals: A Review," *Veterinary Medicine—Czech* 49 (2004): pp. 129–34.

195 *An international survey of physicians:* Dag Album and Steinar Westin, "Do Diseases Have a Prestige Hierarchy? A Survey Among Physicians and Medical Students," *Social Science and Medicine* 66 (2008): p. 182.

196 *Among biologists, the handful of professional:* Rob Knell, telephone interview, October 21, 2009.

196 *HIV/AIDS is the world's:* World Health Organization, "Global Health Risks: Mortality and Burden of Disease Attributable to Selected Major Risks," 2009, accessed September 30, 2011. http://www.who.int/healthinfo/global_burden_disease/GlobalHealthRisks_report_full.pdf.

196 *Consider the following:* Ann B. Lockhart, Peter H. Thrall, and Janis Antonovics, "Sexually Transmitted Diseases in Animals: Ecological and Evolutionary Implications," *Biological Reviews of the Cambridge Philosophical Society* 71 (1996): pp. 415–71.

196 *Sexually spread brucellosis, leptospirosis, and trichomoniasis:* G. Smith and A. P. Dobson, "Sexually Transmitted Diseases in Animals," *Parasitology Today* 8 (1992): pp. 159–66.

196 *Pig litters can be decimated:* Ibid., p. 161.

196 *Venereal diseases in farmed geese:* Ibid.

196 *Contagious equine metritis so predictably:* APHIS Veterinary Services, "Contagious Equine Metritis," last modified June 2005, accessed August 25, 2011. http://

www.aphis.usda.gov/publications/animal_health/content/printable_version/fs_ ahcem.pdf.

196 *Dog STDs can cause abortions:* Smith and Dobson, "Sexually Transmitted Diseases," p. 161.

196 *Dungeness crabs, for example, are vulnerable:* Ibid., p. 163.

197 *Two-dot ladybugs:* Knell interview.

197 *A postcoital housefly that lands:* Lockhart, Thrall, and Antonovics, "Sexually Transmitted Diseases," p. 422.

197 *Astonishingly, some of the diseases:* Ibid., p. 432; Robert J. Knell and K. Mary Webberley, "Sexually Transmitted Diseases of Insects: Distribution, Evolution, Ecology and Host Behaviour," *Biological Review* 79 (2004): pp. 557–81.

197 *Indeed, STDs have been found thriving:* Lockhart, Thrall, and Antonovics, "Sexually Transmitted Diseases," pp. 418, 423.

197 *For example, rabbit syphilis:* Smith and Dobson, "Sexually Transmitted Diseases," p. 163.

197 *These nasty bacteria cause spontaneous:* University of Wisconsin–Madison School of Veterinary Medicine, "Brucellosis," accessed October 5, 2010. http://www .vetmed.wisc.edu/pbs/zoonoses/brucellosis/brucellosisindex.html.

197 *Cattle, pigs, and dogs transmit it:* J. D. Oriel and A. H. S. Hayward, "Sexually Transmitted Diseases in Animals," *British Journal of Venereal Diseases* 50 (1974): p. 412.

198 *brucellosis is a major public health concern:* Centers for Disease Control and Prevention, "Brucellosis," accessed September 15, 2011. http://www.cdc.gov/ncidod/ dbmd/diseaseinfo/brucellosis_g.htm.

198 *(In developed countries, it has become mercifully rare):* Ibid.

198 *zookeepers in Japan:* International Society for Infectious Diseases, "Brucellosis, Zoo Animals, Human—Japan," last modified June 25, 2001, accessed August 25, 2010. http://www.promedmail.org/pls/otn/f?p=2400:1001:16761574736063971049 ::::F2400_P1001_BACK_PAGE,F2400_P1001_ARCHIVE_NUMBER,F2400_ P1001_USE_ARCHIVE:1202,20010625.1203,Y.

198 *And although they're rare:* Ibid.

199 *Nowadays, "trich" is:* Centers for Disease Control and Prevention, "Diseases Characterized by Vaginal Discharge," *Sexually Transmitted Diseases Treatment Guidelines, 2010,* accessed September 15, 2011. http://www.cdc.gov/std/treatment/2010/ vaginal-discharge.htm.

199 *But contemporary* T. vag: Jane M. Carlton, Robert P. Hirt, Joana C. Silva, Arthur L. Delcher, Michael Schatz, Qi Zhao, Jennifer R. Wortman, et al., "Draft Genome Sequence of the Sexually Transmitted Pathogen *Trichomonas vaginalis*," *Science* 315 (2007): pp. 207–12.

199 *Ancient, ancestral* T. vaginalis *resided*: Ibid.

199 T. tenax, *for example, thrives:* Ibid.

199 T. foetus *causes chronic diarrhea:* H. D. Stockdale, M. D. Givens, C. C. Dykstra, and B. L. Blagburn, "*Tritrichomonas foetus* Infections in Surveyed Pet Cats," *Veterinary Parasitology* 160 (2009): pp. 13–17; Lynette B. Corbeil, "Use of an Animal Model of Trichomoniasis as a Basis for Understanding This Disease in Women," *Clinical Infectious Diseases* 21 (1999): pp. S158–61.

199 T. gallinae *(or its close cousin):* Ewan D. S. Wolff, Steven W. Salisbury, John R. Horner, and David J. Varricchio, "Common Avian Infection Plagued the Tyrant Dinosaurs," *PLoS One* 4 (2009): p. e7288.

199 *Recent research on Sue:* Ibid.

200 *For example, several hundred years ago:* Kristin N. Harper, Paolo S. Ocampo, Bret M. Steiner, Robert W. George, Michael S. Silverman, Shelly Bolotin, Allan Pillay, et al., "On the Origin of the Treponematoses: A Phylogenetic Approach," *PLoS Neglected Tropical Disease* 2 (2008): p. e148.

200 *Before it discovered its current preference:* Ibid.

200 *Sex and mother's milk:* Beatrice H. Hahn, George M. Shaw, Kevin M. De Cock, and Paul M. Sharp, "AIDS as a Zoonosis: Scientific and Public Health Implications," *Science* 28 (2000): pp. 607–14; A. M. Amedee, N. Lacour, and M. Ratterree, "Mother-to-infant transmission of SIV via breast-feeding in rhesus macaques," *Journal of Medical Primatology* 32 (2003): pp. 187–93.

201 *The theory is that, by eating the meat:* Martine Peeters, Valerie Courgnaud, Bernadette Abela, Philippe Auzel, Xavier Pourrut, Frederic Bilollet-Ruche, Severin Loul, et al., "Risk to Human Health from a Plethora of Simian Immunodeficiency Viruses in Primate Bushmeat," *Emerging Infectious Diseases* 8 (2002): pp. 451–57.

202 *Hydrophobia, or fear of water:* Centers for Disease Control and Prevention, "Rabies," accessed September 15, 2011. http://www.cdc.gov/rabies/.

202 *Or take* Toxoplasma gondii: Ajai Vyas, Seon-Kyeong Kim, Nicholas Giacomini, John C. Boothroyd, and Robert M. Sapolsky, "Behavioral Changes Induced by *Toxoplasma* Infection of Rodents Are Highly Specific to Aversion of Cat Odors," *Proceedings of the National Academy of Sciences* 104 (2007): pp. 6442–47.

203 *Humans are "dead-end" hosts for toxo:* Ibid.; J. P. Dubey, "*Toxoplasma gondii,*" in *Medical Microbiology,* 4th ed., ed. S. Baron, chapter 84. Galveston: University of Texas Medical Branch at Galveston, 1996.

203 *Exposure to toxo:* Vyas et al., "Behavioral Changes," p. 6446.

203 *other parasites have been shown:* Frederic Libersat, Antonia Delago, and Ram Gal, "Manipulation of Host Behavior by Parasitic Insects and Insect Parasites," *Annual Review of Entomology* 54 (2009): pp. 189–207; Amir H. Grosman, Arne Janssen, Elaine F. de Brito, Eduardo G. Cordeiro, Felipe Colares, Juliana Oliveira Fonseca, Eraldo R. Lima, et al., "Parasitoid Increases Survival of Its Pupae by Inducing Hosts to Fight Predators," *PLoS One* 3 (2008): p. e2276.

204 *Male* Gryllodes sigillatus *crickets:* Marlene Zuk, and Leigh W. Simmons, "Reproductive Strategies of the Crickets (Orthoptera: Gryllidae)," in *The Evolution of Mating Systems in Insects and Arachnids,* ed. Jae C. Choe and Bernard J. Crespi, Cambridge: Cambridge University Press, 1997, pp. 89–109.

204 *When infected with the sexually transmitted:* Knell and Webberley, "Sexually Transmitted Diseases of Insects," p. 574.

204 *Male swamp milkweed beetles infected:* Ibid., pp. 573–74.

204 *The white campion flower:* Peter H. Thrall, Arjen Biere, and Janis Antonovics, "Plant Life-History and Disease Suspectibility: The Occurrence of *Ustilago violacea* on Different Species Within the Caryophyllaceae," *Journal of Ecology* 81 (1993): pp. 489–90.

204 *A Duke University botanical disease ecologist:* Lockhart, Thrall, and Antonovics, "Sexually Transmitted Diseases," p. 423.

205 *A similar "strategy":* Smith and Dobson, "Sexually Transmitted Diseases," pp. 159–60.

205 *Intriguingly, scientists and veterinarians report anecdotally:* Knell interview.

206 *(The increasing incidence of STDs in people over fifty):* Centers for Disease Control and Prevention, "Persons Aged 50 and Older: Prevention Challenges," accessed September 29, 2011. http://www.cdc.gov/hiv/topics/over50/challenges.htm.

206 *An STD of deer:* Colorado Division of Wildlife, "Wildlife Research Report—Mammals—July 2005," accessed October 11, 2011. http://wildlife.state.co.us/Site

CollectionDocuments/DOW/Research/Mammals/Publications/2004–2005
WILDLIFERESEARCHREPORT.pdf.

206 *When* Brucella abortus *causes a cow:* Oriel and Hayward, "Sexually Transmitted Diseases in Animals," p. 414.

207 *It's called cloacal pecking:* B. C. Sheldon, "Sexually Transmitted Disease in Birds: Occurrence and Evolutionary Significance," *Philosophical Transactions of the Royal Society of London B* 339 (1993): pp. 493, 496; N. B. Davies, "Polyandry, Cloaca-Pecking and Sperm Competition in Dunnocks," *Nature* 302 (1983): pp. 334–36.

207 *Cloacal pecking may aid:* Ibid.

207 *Rats that are prevented:* Sheldon, "Sexually Transmitted Disease in Birds," p. 493.

207 *Many birds preen:* Ibid.

207 *In humans, genital scrubbing:* Allan M. Brandt, *No Magic Bullet: A Social History of Venereal Disease in the United States Since 1880,* New York: Oxford University Press, 1987.

207 *A study of Cape ground squirrels:* J. Waterman, "The Adaptive Function of Masturbation in a Promiscuous African Ground Squirrel," *PLoS One* 5 (2010): p. e13060.

207 *A recent study showed that simply:* Mark Schaller, Gregory E. Miller, Will M. Gervais, Sarah Yager, and Edith Chen, "Mere Visual Perception of Other People's Disease Symptoms Facilitates a More Aggressive Immune Response," *Psychological Science* 21 (2010): 649–52.

207 *For example, in males:* Matt Ridley, *The Red Queen: Sex and the Evolution of Human Nature,* New York: Harper Perennial, 1993.

208 *David Strachan was pondering:* David P. Strachan, "Hay Fever, Hygiene and Household Size," *British Medical Journal* 299 (1989): pp. 1259–60.

208 *A few years later, a German scientist:* PBS, "Hygiene Hypothesis," accessed October 4, 2011. http://www.pbs.org/wgbh/evolution/library/10/4/l_104_07.html.

208 *Most animals have multiple sexual partners:* Ridley, *The Red Queen.*

210 *"There is no imperative":* Janis Antonovics telephone interview, September 30, 2009.

210 *Timms, along with his colleagues at the Queensland:* Peter Timms telephone interview, October 5, 2009.

211 *This is why, although HIV:* Randy Dotinga, "Genetic HIV Resistance Deciphered," *Wired.com,* January 7, 2005, accessed November 9, 2010. http://www.wired.com/medtech/health/news/2005/01/66198#ixzz13JfSSBIj.

211 *A dramatic recent example:* Mark Schoofs, "A Doctor, a Mutation and a Potential Cure for AIDS," *Wall Street Journal,* November 7, 2008, accessed October 11, 2011. http://online.wsj.com/article/SB122602394113507555.html.

ELEVEN Leaving the Nest

212 *You will not spot female:* Tim Tinker telephone interview, July 28, 2011.

213 *Parental provisioning:* T. H. Clutton-Brock, *The Evolution of Parental Care,* Princeton: Princeton University Press, 1991.

213 *In other animals:* Kate E. Evans and Stephen Harris, "Adolescence in Male African Elephants, *Loxodonta africana,* and the Importance of Sociality," *Animal Behaviour* 76 (2008): pp. 779–87; "Life Cycle of a Housefly," accessed October 10, 2011. http://www.vtaide.com/png/housefly.htm.

213 *For zebra finches:* Tim Ruploh e-mail correspondence, August 5, 2011.

213 *In vervet monkeys:* Lynn Fairbanks interview, Los Angeles, CA, May 3, 2011.

213 *Even lowly, single-celled:* Marine Biological Laboratory, *The Biological Bulletin,* vols. 11–12. Charleston: Nabu Press, 2010: p. 234.

214 *"Adolescent medicine":* Society for Adolescent Health and Medicine, "Overview," accessed October 12, 2011. http://www.adolescenthealth.org/Overview/2264.htm.

214 *once children have survived infancy:* Centers for Disease Control and Prevention, "Worktable 310: Deaths by Single Years of Age, Race, and Sex, United States, 2007," last modified April 22, 2010, accessed October 14, 2011. http://www.cdc.gov/nchs/data/dvs/MortFinal2007_Worktable310.pdf.

214 *The Centers for Disease Control:* Arialdi M. Minino, "Mortality Among Teenagers Aged 12–19 Years: United States, 1999–2006," *NCHS Data Brief* 37 (May 2010), accessed October 14, 2011. http://www.cdc.gov/nchs/data/databriefs/db37.pdf.

214 *At about age twenty-five:* Melonie Heron, "Deaths: Leading Causes for 2007," *National Vital Statistics Reports* 59 (2011), accessed October 14, 2011. http://www.cdc.gov/nchs/data/nvsr/nvsr59/nvsr59_08.pdf.

214 *Young [animals] suffer:* Tim Caro, *Antipredator Defenses in Birds and Mammals,* Chicago: University of Chicago Press, 2005: p. 15.

215 *Since they can't run as fast:* Maritxell Genovart, Nieves Negre, Giacomo Tavecchia, Ana Bistuer, Luís Parpal, and Daniel Oro, "The Young, the Weak and the Sick: Evidence of Natural Selection by Predation," *PLoS One* 5 (2010): p. e9774; Sarah M. Durant, Marcella Kelly, and Tim M. Caro, "Factors Affecting Life and Death in Serengeti Cheetahs: Environment, Age, and Sociality," *Behavioral Ecology* 15 (2004): pp. 11–22; Caro, *Antipredator Defenses,* p. 15.

215 *What kills adolescents disproportionately:* Margie Peden, Kayode Oyegbite, Joan Ozanne-Smith, Adnan A. Hyder, Christine Branche, AKM Fazlur Rahman, Frederick Rivara, and Kidist Bartolomeos, "World Report on Child Injury Prevention," Geneva: World Health Organization, 2008.

215 *The CDC reports that 35 percent:* Minino, "Mortality Among Teenagers," p. 2.

215 *According to the World Health Organization:* Peden et al., "World Report."

215 *In some parts of the world:* World Health Organization, "Global Health Risks: Mortality and Burden of Disease Attributable to Selected Major Risks," 2009, accessed September 30, 2011. http://www.who.int/healthinfo/global_burden_disease/Global HealthRisks_report_full.pdf.

215 *And a scarlet rectangle:* Chris Megerian, "N.J. Officials Unveil Red License Decals for Young Drivers Under Kyleigh's Law," *New Jersey Real-Time News,* March 24, 2010, accessed October 10, 2011. http://www.nj.com/news/index.ssf/2010/03/nj_officials_decide_how_to_imp.html.

215 *But extensive, new neurological research:* Linda Spear, *The Behavioral Neuroscience of Adolescence,* New York: Norton, 2010; Linda Van Leijenhorst, Kiki Zanole, Catharina S. Van Meel, P. Michael Westenberg, Serge A. R. B. Rombouts, and Eveline A. Crone, "What Motivates the Adolescent? Brain Regions Mediating Reward Sensitivity Across Adolescence," *Cerebral Cortex* 20 (2010): pp. 61–69; Laurence Steinberg, "The Social Neuroscience Perspective on Adolescent Risk-Taking," *Developmental Review* 28 (2008): pp. 78–106; Laurence Steinberg, "Risk Taking in Adolescence: What Changes, and Why?" *Annals of the New York Academy of Sciences* 1021 (2004): pp. 51–58; Stephanie Burnett, Nadege Bault, Girgia Coricelli, and Sarah-Jayne Blakemore, "Adolescents' Heightened Risk-Seeking in a Probabilistic Gambling Task," *Cognitive Development* 25 (2010): pp. 183–96; Linda Patia Spear, "Neurobehavioral Changes in Adolescence," *Current Directions in Psychological*

Science 9 (2000): pp. 111–14; Cheryl L. Sisk, "The Neural Basis of Puberty and Adolescence," *Nature Neuroscience* 7 (2004): pp. 1040–47; Linda Patia Spear, "The Biology of Adolescence," last updated February 2, 2010, accessed October 10, 2011.

216 *Researchers from Rome's Istituto:* Giovanni Laviola, Simone Macrì, Sara Morley-Fletcher, and Walter Adriani, "Risk-Taking Behavior in Adolescent Mice: Psychobiological Determinants an Early Epigenetic Influence," *Neuroscience and Biobehavioral Reviews* 27 (2003): pp. 19–31.

216 *Adolescent rats display:* Kirstie H. Stansfield, Rex M. Philpot, and Cheryl L. Kirstein, "An Animal Model of Sensation Seeking: The Adolescent Rat," *Annals of the New York Academy of Sciences* 1021 (2004): pp. 453–58.

216 *Similarly, when primatologists:* Lynn A. Fairbanks, "Individual Differences in Response to a Stranger: Social Impulsivity as a Dimension of Temperament in Vervet Monkeys (*Cercopithecus aethiops sabaeus*), *Journal of Comparative Psychology* 115 (2001): pp. 22–28; Fairbanks interview.

216 *Preadult zebra finches:* Ruploh e-mail correspondence.

216 *Transitioning sea otters:* Tinker interview; Gena Bentall interview, Moss Landing, CA, August 4, 2011.

217 *"Young animals may approach and inspect":* Caro, *Antipredator Defenses,* p. 20.

217 *For example, instead of hiding:* Clare D. Fitzgibbon, "Anti-predator Strategies of Immature Thomson's Gazelles: Hiding and the Prone Response," *Animal Behaviour* 40 (1990): pp. 846–55.

217 *"Mobbing is a way to impress":* Judy Stamps telephone interview, August 4, 2011.

219 *Looking-away responses:* N. J. Emery, "The Eyes Have It: The Neuroethology, Function and Evolution of Social Gaze," *Neuroscience and Biobehavioral Reviews* 24 (2000): pp. 581–604.

219 *House sparrows:* Carter et al., "Subtle Cues," pp. 1709–15.

219 *Studies of humans:* Emery, "The Eyes Have It," pp. 581–604.

219 *As they test their danger-detection skills:* Caro, *Antipredator Defenses.*

219 *Vervet monkeys make:* Fairbanks interview; Lynn A. Fairbanks, Matthew J. Jorgensen, Adriana Huff, Karin Blau, Yung-Yu Hung, and J. John Mann, "Adolescent Impulsivity Predicts Adult Dominance Attainment in Male Vervet Monkeys," *American Journal of Primatology* 64 (2004): pp. 1–17.

220 *She told me that:* Fairbanks interview.

221 *"the idea that an age-limited increase":* Fairbanks et al., "Adolescent Impulsivity."

221 *"age-specific behavioral characteristics":* Spear, "Neurobehavioral Changes."

221 *"are deeply embedded":* Spear, "The Biology of Adolescence."

223 *Adolescent African elephants:* Kate E. Evans and Stephen Harris, "Adolescence in Male African Elephants, *Loxodonta africana,* and the Importance of Sociality," *Animal Behaviour* 76 (2008): pp. 779–87.

223 *These groups of young male:* Ibid.

223 *Gena Bentall has cataloged:* Bentall interview.

223 *Male wild horses:* Claudia Feh, "Social Organisation of Horses and Other Equids," Havemeyer Equine Behavior Lab, accessed April 15, 2010. http://research.vet.upenn.edu/HavemeyerEquineBehaviorLabHomePage/ReferenceLibraryHavemeyerEquineBehaviorLab/HavemeyerWorkshops/HorseBehaviorandWelfare1316June2002/HorseBehaviorandWelfare2/RelationshipsandCommunicationinSociallyNatura/tabid/3119/Default.aspx.

223 *Female wild horses:* Ibid.

223 *Evans and Harris spotted:* Evans and Harris, "Adolescence."

223 *Notorious periods:* Ibid.

224 *For California condors:* Michael Clark interview, Los Angeles, CA, July 21, 2011.

224 *As Michael Clark:* Ibid.

224 "Lord of the Flies *situation*": Ibid.

225 *Thanks to the Los Angeles Zoo:* Ibid.

225 *As Alan Kazdin, a professor:* Alan Kazdin telephone interview, July 26, 2011.

225 *"Having peers around":* Alan Kazdin and Carlo Rotella, "No Breaks! Risk and the Adolescent Brain," *Slate,* February 4, 2010, accessed October 10, 2011. http://www .slate.com/articles/life/family/2010/02/no_brakes_2.html.

225 *Zebra finches, too:* Ruploh e-mail correspondence.

225 *Ancient adolescents also formed groups:* David J. Varricchio, Paul C. Sereno, Zhao Xijin, Tan Lin, Jeffery A. Wilson, and Gabrielle H. Lyon, "Mud-Trapped Herd Captures Evidence of Distinctive Dinosaur Sociality," *Acta Palaeontologica Polonica* 53 (2008): pp. 567–78.

225 *Pink salmon, too, grow:* Jean-Guy J. Godin, "Behavior of Juvenile Pink Salmon (*Oncorhynchus gorbuscha* Walbaum) Toward Novel Prey: Influence of Ontogeny and Experience, *Environmental Biology of Fishes* 3 (1978): pp. 261–66.

227 *Susan Perry notes in* Manipulative Monkeys: Susan Perry, with Joseph H. Manson, *Manipulative Monkeys: The Capuchins of Lomas Barbudal,* Cambridge: Harvard University Press, 2008: p. 51.

227 *"extremely high social intelligence" and "great interpersonal skills":* Susan Perry telephone interview, May 12, 2011.

227 *She followed one monkey:* Ibid.

228 *"Delinquency and criminal behavior":* Laurence Steinberg, *The 10 Basic Principles of Good Parenting,* New York: Simon & Schuster, 2004; Laurence Steinberg and Kathryn C. Monahan, "Age Differences in Resistance to Peer Influence," *Developmental Psychology* 43 (2007): pp. 1531–43.

228 *In September 2010, six teens:* LGBTQNation, "Two More Gay Teen Suicide Victims—Raymond Chase, Cody Barker—Mark 6 Deaths in September," October 1, 2010, accessed October 10, 2011. http://www.lgbtqnation.com/2010/10/two-more-gay-teen-suicide-victims-raymond-chase-cody-barker-mark-6-deaths-in-september/.

228 *Their deaths were added:* Centers for Disease Control and Prevention, "Suicide Prevention: Youth Suicide," accessed October 14, 2011. http://www.cdc.gov /violenceprevention/pub/youth_suicide.html.

229 *But recent research suggests:* U.S. Department of Health and Human Services, Health Resources and Services Administration, Stop Bullying Now!, "Children Who Bully," accessed October 14, 2011. http://stopbullying.gov/community/tip _sheets/children_who_bully.pdf.

230 *The Oxford zoologist T. H. Clutton-Brock:* T. H. Clutton-Brock and G. A. Parker, "Punishment in Animal Societies," *Nature* 373 (1995): pp. 209–16.

230 *That a propensity to bully:* Martina S. Müller, Elaine T. Porter, Jacquelyn K. Grace, Jill A. Awkerman, Kevin T. Birchler, Alex R. Gunderson, Eric G. Schneider, et al., "Maltreated Nestlings Exhibit Correlated Maltreatment As Adults: Evidence of A 'Cycle of Violence,' in Nazca Boobies (*Sula Granti*)," *The Auk* 128 (2011): pp. 615–19.

231 *The parents of Kloss's gibbons:* Clutton-Brock, *The Evolution of Parental Care.*

231 *Three-toed sloth mothers:* Ibid.

232 *Early exposure to alcohol:* Linda Spear, "Modeling Adolescent Development and Alcohol Use in Animals," *Alcohol Res Health* 24 (2000): pp. 115–23.

233 *"You will be a disgrace":* Charles Darwin, "The Autobiography of Charles Darwin," The Complete Work of Charles Darwin Online, accessed October 13, 2011. http://darwin-online.org.uk/content/frameset?itemID=F1497&viewtype-text&pageseq=1.

233 *"My father, who was the kindest man":* Darwin, "The Autobiography."

TWELVE Zoobiquity

234 *When crows by the hundreds:* Tracey McNamara interview, Pomona, CA, May 2009; George V. Ludwig, Paul P. Calle, Joseph A. Mangiafico, Bonnie L. Raphael, Denise K. Danner, Julie A. Hile, Tracy L. Clippinger, et al., "An Outbreak of West Nile Virus in a New York City Captive Wildlife Population," *American Journal of Tropical Medicine and Hygiene* 67 (2002): pp. 67–75; Robert G. McLean, Sonya R. Ubico, Douglas E. Docherty, Wallace R. Hansen, Louis Sileo, and Tracey S. McNamara, "West Nile Virus Transmission and Ecology in Birds," *Annals of the New York Academy of Sciences* 951 (2001): pp. 54–57; K. E. Steele, M. J. Linn, R. J. Schoepp, N. Komar, T. W. Geisbert, R. M. Manduca, P. P. Calle, et al., "Pathology of Fatal West Nile Virus Infections in Native and Exotic Birds During the 1999 Outbreak in New York City, New York," *Veterinary Pathology* 37 (2000): pp. 208–24; Peter P. Marra, Sean Griffing, Carolee Caffrey, A. Marm Kilpatrick, Robert McLean, Christopher Brand, Emi Saito, et al., "West Nile Virus and Wildlife," *BioScience* 54 (2004): pp. 393–402; Caree Vander Linden, "USAMRIID Supports West Nile Virus Investigations," accessed October 11, 2011. http://ww2.dcmilitary.com/dcmilitary_archives/stories/100500/2027-1.shtml; Rosalie T. Trevejo and Millicent Eidson, "West Nile Virus," *Journal of the American Veterinary Medical Association* 232 (2008): pp. 1302–09.

236 *"If you see encephalitis":* American Museum of Natural History, "West Nile Fever: A Medical Detective Story," accessed October 10, 2011. http://www.amnh.org/sciencebulletins/biobulletin/biobulletin/story1378.html.

236 *"I had barrels full of dead birds":* McNamara interview.

237 *"hair stand on end":* Ibid.

238 *"That's when it clicked":* Ibid.

238 *Within forty-eight hours:* Linden, "USAMRIID."

238 *"best of what science can be":* McNamara interview.

238 *In the years since it first emerged:* James J. Sejvar, "The Long-Term Outcomes of Human West Nile Virus Infection," *Emerging Infections* 44 (2007): pp. 1617–24; Douglas J. Lanska, "West Nile Virus," last modified January 28, 2011, accessed October 13, 2011. http://www.medlink.com/medlinkcontent.asp.

238 *In a report to Congress:* United States General Accounting Office, "West Nile Virus Outbreak: Lessons for Public Health Preparedness," *Report to Congressional Requesters,* September 2000, accessed October 10, 2011. http://www.gao.gov/new.items/he00180.pdf.

239 *"The veterinary medicine community":* Ibid.

239 *Other groups in the United States:* Donald L. Noah, Don L. Noah, and Harvey R. Crowder, "Biological Terrorism Against Animals and Humans: A Brief Review and Primer for Action," *Journal of the American Veterinary Medical Association,* 221 (2002): pp. 40–43; Wildlife Disease News Digest, accessed October 10, 2011. http://wdin.blogspot.com/.

239 *The Canary Database:* Canary Database, "Animals as Sentinels of Human Environmental Health Hazards," accessed October 10, 2011. http://canarydatabase .org/.

239 *The U.S. Agency for International Development:* USAID press release, "USAID Launches Emerging Pandemic Threats Program," October 21, 2009.

240 *Jonna Mazet:* University of California, Davis, "UC Davis Leads Attack on Deadly New Diseases," *UC Davis News and Information,* October 23, 2009, accessed on October 10, 2011. http://www.news.ucdavis.edu/search/news_detail.lasso?id=9259.

240 *The program brings together:* USAID spokesperson, March 19, 2012.

240 *"We don't know what diseases":* Jonna Mazet interviewed on Capital Public Radio, by *Insight* host Jeffrey Callison, October 26, 2009. http://www.facebook.com/ video/video.php?v=162741314486.

240 *"Over $200 billion":* Marguerite Pappaioanou address to the University of California, Davis Wildlife and Aquatic Animal Medicine Symposium, February 12, 2011, Davis, CA.

240 *Recently a third-year veterinary student :* One Health, One Medicine Foundation, "Health Clinics," accessed October 10, 2011. http://www.onehealthonemedicine .org/Health_Clinics.php.

241 *A program at Tufts:* North Grafton, "Dogs and Kids with Common Bond of Heart Disease to Meet at Cummings School," Tufts University Cummings School of Veterinary Medicine, April 22, 2009, accessed October 10, 2011. http://www.tufts.edu/ vet/pr/20090422.html.

241 *Similarly, Winter, the dolphin:* Clearwater Marine Aquarium, "Maja Kazazic," accessed October 10, 2011. http://www.seewinter.com/winter/winters-friends/ maja.

242 *In fact, swine flu:* Matthew Scotch, John S. Brownstein, Sally Vegso, Deron Galusha, and Peter Rabinowitz, "Human vs. Animal Outbreaks of the 2009 Swine-Origin H1N1 Influenza A Epidemic," *EcoHealth* (2011): doi: 10/1007/s10393-011-0706-x.

243 *The E. coli–tainted fresh baby spinach:* Michele T. Jay, Michael Cooley, Diana Carychao, Gerald W. Wiscomb, Richard A. Sweitzer, Leta Crawford-Miksza, Jeff A. Farrar, et al., "*Escherichia coli* O157:H7 in Feral Swine Near Spinach Fields and Cattle, Central California Coast," *Emerging Infectious Diseases* 13 (2007): pp. 1908–11; Michele T. Jay and Gerald W. Wiscomb, "Food Safety Risks and Mitigation Strategies for Feral Swine (*Sus scrofa*) Near Agriculture Fields," in *Proceedings of the Twenty-third Vertebrate Pest Conference,* edited by R. M. Timm and M. B. Madon. University of California, Davis, 2008.

243 *One of the world's:* Laura H. Kahn, "Lessons from the Netherlands," *Bulletin of the Atomic Scientists,* January 10, 2011, accessed October 10, 2011. http://www.the bulletin.org/web-edition/columnists/laura-h-kahn/lessons-the-netherlands.

243 *"Q" stands for "query":* Ibid.

243 *But, like the loose Soviet nukes:* Laura H. Kahn, "An Interview with Laura H. Kahn," *Bulletin of the Atomic Scientists,* last updated October 8, 2011, accessed October 10, 2011. http://www.thebulletin.org/web-edition/columnists/laura-h-kahn/interview.

243 *Five of the six:* Centers for Disease Control and Prevention, "Bioterrorism Agents/ Diseases," accessed October 10, 2011. http://www.bt.cdc.gov/agent/agentlist-cate gory.asp; C. Patrick Ryan, "Zoonoses Likely to Be Used in Bioterrorism," *Public Health Reports* 123 (2008): pp. 276–81.

243 *The sixth agent:* Centers for Disease Control and Prevention, "Bioterrorism Agents/ Diseases."

243 *In March 2007:* U.S. Food and Drug Administration, "Melamine Pet Food Recall—Frequently Asked Questions," accessed October 13, 2011. http://www.fda .gov/animalveterinary/safetyhealth/RecallsWithdrawals/ucm129932.htm.

244 *Animals can also be sentinels:* Melissa Trollinger, "The Link Among Animal Abuse, Child Abuse, and Domestic Violence," Animal Legal and Historical Center, September 2001, accessed October 10, 2011. http://www.animallaw.info/articles/ arus30sepcololaw29.htm.

Index

acral lick dermatitis, 163
addictions, 11, 87–108
  in animals, 89–93, 106
  behavioral, 101–5
  as chronic illness, 105
  criminalization of, 89n
  epigenetics in, 106
  heterogeneity in, 105–6
  multispecies approach to, 89–90, 92,
    93–4, 95, 103
  recovery from, 108
  as self-destructive, 89, 94, 100–1, 107
  survival advantages of, 90, 94, 101,
    103–4
adolescence, 212–33
  addiction in, 107
  bullying in, 228–31
  coed groups in, 225
  death in, 214, 215
  delinquency in, 227–8
  males in, 214, 220–3
  mentoring in, 223–5
  misjudgment in, 215–16
  multispecies approach to, 216, 221–2,
    227–33
  overreaction in, 219
  parenting in, 215, 231–3
  peer groups in, 214, 217–18, 225–8
  risk taking in, 213–17, 221–2, 228
  social adjustment in, 220, 223, 226–7,
    229
  suicide in, 228–9
  thrill seeking in, 214, 218
  as transitional, 213–14, 220, 228
adolescent medicine, 214
adrenaline, 4, 22, 71, 108, 116
aerophobia, 202
Ågmo, Anders, 77
Agras, W. Stewart, 177

Agriculture Department, U.S. (USDA), 237–8
air bags, 125n
alarm bradycardia, 23–6, 28, 29–30, 129
Alcock, John, 81
alcohol intoxication, in animals, 90–2
alcoholism, 11
  brain-disease theory of, 103
alligators, diet of, 145, 146
aluminum contaminants, 46
American Medical Association
  (AMA), 8
American Veterinary Medical Association
  (AVMA), 8
ampullary organs, 27–8
amygdala, 71n, 104n
Ancestor's Tale, The (Dawkins), 15
animal abuse, 244n
animal relocation, capture myopathy in,
  118–19
animal testing and experimentation, 10,
  49, 54
animal training, 104–5, 171n
"anorexia gene," 187–8
anorexia nervosa, 174, 177–8, 186–90, 193
anther smut, 204–5
anthrax, 46, 243
anthropomorphism, 12, 142–3
antibiotics, in livestock, 151
Antipredator Defenses in Birds and
  Mammals (Caro), 217
Antonovics, Janis, 209–10
ants, mismatch in diet of, 140–1
anxiety
  in animals, 11
  behavioral causes of, 102
  in eating disorders, 177, 186, 188, 193
  in humans, 4
  in sexual dysfunction, 71
  as survival mechanism, 98

aorta, ruptured, 6
apes, human similarities to, 13–14, 15, 17, 18
apoptosis, 35, 42
Aristotle, 91–2
Arnqvist, Göran, 25*n*
Ashkenazi Jews, 6, 43, 45, 50*n*
assisted reproduction, 68
ataxic myoglobinuric syndrome, 118*n*
attachment disorders, food hoarding and, 186
Australia, wildfires in, 194
autism, 161*n*
autoimmune disorders, 125, 208
autonomic nervous system, 28
avian influenza, 8, 46, 235, 239, 243
Axhi (grizzly bear), 132–3, 156–8

Babec (gorilla), 171–3
Babi Yar massacre, 24
baboons, grooming in, 165
bachelor groups, 222–3, 225
bacteria
    in gastric ulcers, 155
    intestinal, 148–54
Bacteroidetes, 149–50
Bagemihl, Bruce, 68, 75
Bailey, Nathan W., 75
Balanus glandula, 60
Barker, Cody J., 228
Baron, David, 183
Barsky, Arthur, 124
Bathsheba's Breast (Olson), 39
bats, cancer in, 41, 42
beagles, cancer immunity in, 51
Beast in the Garden, The (Baron), 183
beavers, food storage by, 185
bedbugs, mating in, 76
bees, drugs in brains of, 99
Behavioral Neuroscience of Adolescence, The (Spear), 221
behaviorists, 96
Bekoff, Marc, 97*n*
Benson, Herbert, 123–4
Bentall, Gina, 222–3
Bergman, Philip, 51–4
"Berlin patient," 211*n*
Bernese mountain dogs, cancer in, 50
Bertelsen, Mads, 142–3
bestiality, 76–7
bichon frises, cancer in, 53
Biggie (horse), 85
bighorn sheep

drug-seeking by, 90
    sudden death of, 120
binge-eating, 7, 177, 178, 181–2
binge-purge disorder, *see* bulimia
    nervosa
biohazards, 237–8
Biological Exuberance (Bagemihl), 75
"biology is destiny" theory, 14
biomimicry, 17, 28*n*
birds
    adolescence in, 225
    capture myopathy in, 118
    "cramp" in, 122
    drunkenness in, 90
    fainting in, 22
    feather-picking in, 159–60, 163, 164, 170
    food storage by, 185
    grooming in, 168
    intestines of, 147–8
    mating in, 59*n*, 60, 72
    STDs in, 199, 207
    sudden cardiac death in, 111, 117
    West Nile virus in, 234–8
Birkhead, Tim, 59*n*, 60*n*, 74, 76–7
Birt-Hogg Dubé syndrome, 50
biting, compulsive, 11
bladder cancer, 47, 50
Blind Watchmaker, The (Dawkins), 15
blood pressure
    and heart disease, 31
    lowered, 22, 23, 25*n*
Blumstein, Daniel, 136–7
bobcats, feeding by, 178–9
body mass index (BMI), 146, 150*n*
Bonar, Chris, 44
bonobos, mating in, 72, 74, 79
boobies, bullying in, 230–1
borderline personality disorders, 163*n*
boredom, in self-injury, 169–70, 173, 175
bottleneck populations, 50*n*
botulism, 243
boxers, cancer in, 50
bradycardia, 22–8, 1129
brain
    adolescent, 107, 216, 221, 225
    anorexia and, 189
    behaviors releasing chemicals in, 99
    in drug addiction, 89
    and emotions, 96–100
    encephalitis and, 235–8
    in fear response, 122, 123–5

and grooming, 166
and heart, 5–6, 20–1, 26–7, 30, 113–14,
    117, 122, 125
in infants and children, 106
natural narcotics in, 99
neurochemical codes in, 106–7
and obesity, 146
and pain, 96–7
in sexual dysfunction, 69–72
in sexual stimulation, 64–8, 82–3
brainworms, 203
branding, 37–8
BRCA1 mutation, 6, 43, 50n
"breaking tolerance," 52n
breast cancer, 6, 36
in ancient cultures, 39
breast-feeding and prevention of, 44–5
in dogs, 48
multispecies approach to, 42–6
prevention for, 44
breast-feeding, 44–6
broken-heart syndrome, 114
Bronx zoo, West Nile virus at, 234–5
Brown, Asher, 228
brucella, 197–8, 200
Brucella abortus, 206
bulimia nervosa, 177–8, 182, 183, 192
and R and R, 191
bullying, 228–31
Burnett, Arthur L., 69, 72

C-265 (seal), 136
California condors, mentoring in
    reintroduction of, 224
Canary Database, 239
cancer, 31–54
    animal size and, 41–2
    animal warnings for, 45–7
    benign vs. malignant, 35
    causes of, 34–6, 38, 40, 45
    in dinosaurs, 12–13, 16, 36, 39–40
    guilt and blame in, 33–4, 36
    in human antiquity, 38–9
    immunotherapy for, 52–4
    as intrinsic, 36, 40
    lifestyle, 45
    and marketing, 36
    multispecies approach to, 33, 36–9, 41,
        42–6, 49–54
    natural triggers for, 40
    in wild species, 42
    see also specific cancers and species

"cancer à deux," 45
Cancer: The Evolutionary Legacy
    (Greaves), 41
cane toads, hallucinogens in skin of, 91
canine compulsive disorder (CCD), 164
Canine Lifetime Health Project, 32–3
cannabinoids, 94, 100
    in brain, 99
captivity
    eating disorders of animals in, 191
    psychological state of animals in, 170–1
capture myopathy, 4–5, 117–24, 127–8
    and takotsubo, 5, 17, 114, 116
capture shock syndrome, 118n
carbohydrates, binging on, 181–2
carcass feeding, 141–3
carcinogens, 35, 40
    viruses as, 45
cardiomyopathy, 4–5, 18
Caro, Tim, 217
carp, obesity in, 137
Carroll, Sean B., 15
catecholamines, 22, 116, 119–20, 122, 124,
    168
"Catkins" diet, 134
cats
    cancer in, 37, 48n
    grooming in, 166
    kidney failure in, 243n
    licking in, 163
    mating in, 72n
    overgrooming in, 164
    toxo in, 202–3
cattle
    addiction in, 90
    cancer in, 37
cave paintings, 60n
cell replication, cancer-causing mutations
    in, 34–5, 40–1, 42, 45
Center for Disease Control (CDC), 236–8
Center for Evolution and Cancer, 42
Chase, Raymond, 228
cheetahs, as bottleneck population, 50n
chemical restraint, 126
chewing, 167n
    of cud, 191
chickens, light and weight in, 146
childhood, children
    animal, 220
    multispecies programs for, 241
    neurochemical codes developed in,
        106–7

chimpanzees, 13–14, 15, 18
  fainting in, 22
  grooming in, 165
  mating in, 72
  R and R in, 191
  SIV in, 200
chipmunks, feeding by, 178
chlamydia, in koalas, 6, 194–5, 210–11
Chlamydophila, 195*n*
chordomas, 35
chow chows, cancer in, 50
circadian rhythms, 145–7, 151
Clark, Michael, 224
cleansing
  excessive, 208, 209
  postsex, 208
Clementi, Tyler, 228
clicker training, 104*n*
clitoris, 79
cloaca, 207*n*
cloacal kiss, 72
cloacal pecking, 207
clock genes, 145–6
cloning, 58
clot-production hypothesis, 21*n*
Clutton-Brock, T. H., 230
cocaine, in heart attacks, 114*n*
cocker spaniels
  addiction in, 91
  cancer in, 53
cockroaches, 17
coercive copulation, 58, 76, 81
coitus accelerando, 72
coitus interruptus, 71–2
collagen, 62–3
colon cancer, 48
Colymbosathon ecplecticos, 59
common ancestors, 15–16, 18, 100
commotio cordis, 116*n*
comparative cardiology course, 241
comparative medicine, 8*n*
comparative oncology, 49–51
Comparative Oncology Program (COP),
  49, 51
Comparative Oncology Trials
  Consortium, 49
Compass of Pleasure, The (Linden), 105
competition
  in adolescence, 227
  in feeding, 188–9
  for females, 222
  sperm, 70–3, 76, 208–9

contraception, 209
convenience polyandry, 81
Copenhagen Zoo, carcass feeding at,
  141–3
coronary artery disease, 139
corpus cavernosum, 62
courtship displays, 67–8
cows
  nymphomania in, 84
  STDs in, 206
coyotes, boredom and pacing in, 171
crabs, STDs in, 196–7
"cramp," 122
crickets, STDs in, 204
Crohn's disease, 155
crows, West Nile virus in, 234–8
crucifixion, 25*n*
crypsis, 26–7
cutters, 11, 160–2, 168–9, 173, 174, 175
cystic fibrosis, 50*n*

dachsunds, cancer in, 51
Dalmatians, deafness in, 121
Darwin, Charles, 10, 12, 13, 14, 92, 96
  adolescence of, 232–3
Darwin, Robert, 232–3
Davis, Ron, 8
Dawkins, Richard, 15
DDT, 46
death
  cluster, 45–7
  feigning of, 24
deep homology, 15–16
deer, feeding strategies of, 183
defecation, as survival mechanism, 24, 192
defensive regurgitation, 192
defibrillators, 127
delayed pericute syndrome, 118*n*
depression, 11, 147
  adolescent bullying and, 231, 232
  behavioral causes of, 102
  linked to sexual dysfunction, 79
  after sex, 85
Descartes, René, 96
Descent of Man, The (Darwin), 92
detumescence, 64
devil facial tumor disease, 38
Dhurandhar, Nikhil, 152
diabetes, 7
  type 2, 153, 154
Diamond, Jared, 148
Diana, Princess, 11, 161, 166

diarrhea, 150*n*
diet
    animal, human responsibility for, 134–5,
        143, 150–1, 156–8
    animal weight fluctuations and, 136–7,
        141
    in captivity vs. in the wild, 143–4, 156–8
    carcass feeding, 141–3
    circadian rhythm and, 145–7
    environmental enrichment in, 144
    human, 138–40, 144
    mismatch in, 139
    R and R and milk products in, 191*n*–2*n*
digestion, 147–50
    in survival strategy, 192
Dill, Lawrence, 125
dinosaurs
    diseases of, 12–13, 16, 36, 39–40
    peer groups in, 225
    sex in, 59–60
diseases
    animal warning signs for, 45–51, 239, 243
    as common to humans and animals, 5,
        7, 8; *see also* multispecies medicine;
        zoobiquity
    as natural, 210
    in terrorist threats, 243
    see also specific diseases
"diseases of civilization," 16
dispersal, 219–22
diurnal rhythms, 145
dizziness, 20
DNA
    and BRCA1, 43
    and cancer, 34–6, 40–1, 42, 45, 53
    epigenetic alterations to, 16
    in reproduction, 59
    in viruses, 242
Dobermans, self-injury in, 164
doctors, hierarchy of, 9–10
Dodman, Nicholas, 164
dogs
    addiction in, 91
    adolescence in, 213
    arrhythmia in, 241
    breed variations in, 49–50
    cancer in, 32–3, 36, 42, 47–51, 53
    heart disease in, 241
    kidney failure in, 243*n*
    melanoma in, 51–4
    military, 47–8
    self-injury in, 163, 164

    in shelters, 170
    STDs in, 196
    see also specific breeds
dolphins
    prosthesis for, 241
    R and R in, 191
    STDs in, 196
Dolphin Tale, 241
dopamine, 99, 100, 101, 102, 104, 225
double-muscling, 187
dourine, 205
dragonflies, obesity in, 152–4
driver's ed, 218
drugs
    addiction to, 87–101
    as fast track to reward, 100–1
    receptors for, 93–4
    restricted access to, 101, 107
ducks, 24, 60
dyspareunia, 79

ear-piercing, 20–1, 23
earthquakes, 109–11
eastern equine encephalitis (EEE), 235
eating disorders, 7, 176–93
    multispecies approach to, 178, 180, 184,
        186–9, 193
    as socially contagious, 190–2
    susceptibility to, 177–8
    see also specific disorders
Ebola, 243
E. coli, 243
ecology of fear, 181–3, 192, 193
Edwards, Mark, 136
ejaculation, 63–4, 72, 85
electrocardiogram (EKG), 115
elephants, bachelor groups of, 223
elephant seals, mating in, 82
elk, feeding strategies of, 183–4, 193
Emerging Pandemic Threats program, 239
emotions
    animal experience of, 96–9
    brain in, 96–100
    evolutionary roots of, 11–12, 96–8, 100
    in heart failure, 5–6, 20–1
    as survival mechanisms, 100
    see also specific emotions
encephalitis, 235–6, 238
encounter avoidance, 184
endorphins, in self-injury, 168, 175
engineering, biomimicry in, 17, 28*n*
enhanced vigilance, 184

Enterobius vermicularis (pinworms), 202, 203
entrapment, 21–2, 120–1
    symbolic human, 122
environmental cancers, 35–6, 40, 45–7
environmental enrichment
    in obesity, 143–4
    in stereotypies, 170–1
epigenetics, 16, 106
Epstein-Barr virus, 45
erectile dysfunction (ED), 69–72
erection, penile, 58, 61–5, 69–71, 83, 85
escape, fleeing vs. hiding, 125
estrogen, 44
eternal harvest, 158
Ettinger, Stephen, 241
eugenics, 14
Evans, Kate E., 223
evolution
    adolescence in, 221, 226
    in diet, 139–40, 144
    of emotions, 11–12, 96–8, 100
    and fainting, 21
    feeding strategies and, 180, 182, 183–8
    in parenting, 232
    and sexuality, 58–61
    shared human-animal, 10, 13–17, 242
    of STDs, 199–200, 210–11
evolutionary biology, 14, 16
evolutionary medicine, 98
Evolution's Rainbow (Roughgarden), 75
exotic animals, obesity in, 134
*Expression of the Emotions in Man and
    Animals, The* (Darwin), 96
extreme athletes, rhabdo in, 119
eye gaze
    in adolescence, 219
    in capture myopathy, 4, 23

facial expressions
    in animals, 96
    sexual, 66–7
facial nerve, 66
failed copulation, 70–2
fainting (syncope), 6, 19–30
    as alternative to fight-or-flight, 22–3,
        28, 30
    in animals, 21–6, 30
    as feigned death, 23–4
    prevalence of, 19–20
    as rape prevention strategy, 24–5
    as sign of weakness, 29–30
    survival advantages of, 23–7, 29

    and torture, 25
    in women, 21*n*, 24–5
Fairbanks, Lynn, 220–1
Falkland Islands War, 112
Farrell, Colin, 161
Fat Boy (intoxicated horse), 90
fawns, stillness in, 25
fear
    of eating, *see* anorexia nervosa; bulimia
        nervosa
    in eating disorders, 177, 188
    ecology of, 181–3, 192, 193
    and ethnicity, 124*n*
    fainting and, 22–3
    as heart attack trigger, 110–11, 116, 123–4
    in heart failure, 8
    heart rate and, 122, 125, 129
    in loss of bodily control, 24
    of moving air, 202
    overreaction to, 125
    of predators, 178–81, 183–4
    as protective mechanism, 131
    sexuality and, 71
    and SIDS, 129–30
    as survival mechanism, 98
    of water, 202
fear bradycardia, *see* alarm bradycardia
feather-picking, 11, 159–60, 163, 164
fecal therapy, 150*n*
feeding strategies
    competition in, 188–9
    effect of light on, 182–3
    predation and, 178–81, 183–4
    survival mechanisms in, 181, 182
fight-or-flight response, 21–3, 28, 30
Filipinos, 124*n*
Firmicutes, 149–50
fish
    fainting in, 27
    grooming in, 165–6
Fishbein, Michael, 18
flank-biters, 163–4, 170
flatlining, 115
flavivirus, 237–8
flehmen, 65–6, 84
flexural stiffness, 62–3
food hoarding, 177, 178
    by animals, 184–6
    as OCD, 186
foraging, 171*n*, 174
foreplay, in animals, 84
Fossey, Dian, 14

"founder effect," 50*n*
foxes, hunting by, 24
FRADE (fear/restraint-associated death
    events), 123–8
Framingham Heart Study, 31–2
Fred Hutchinson Cancer Research
    Center, 49
fruit, cultivated, 157–8
Furchgott, Robert, 62*n*

Galen, 39
galls, 38
gambling, as foraging, 103–4
gams, 225
Gaston (alligator), 145, 146
gastric ulcers, bacteria in, 155
Gately, Iain, 92
gays, bullying of, 230
gaze aversion, 219
gazelles, adolescent risk taking by, 217, 221,
    233
General Accounting Office, U.S.
    (Government Accountability Office,
    GAO), 238–9
genes
    in bottleneck populations, 50*n*
    evolutionary development altered by, 16
genetics
    in addiction, 106
    in causing cancer, 35–6
    culture vs., 14–16
    in dog cancers, 32
    in eating disorders, 177, 187–8
    multispecies sharing of, 10, 15, 242
    in parenting, 232
    in resistance to STDs, 211
    sexual reproduction and, 58, 76
    in social agendas, 14
    survival instincts in, 67–8
genital cancers, 38
genital inspection, 206–7
genocide, 14, 24
genome
    canine, 32, 50
    overlap in animal and human, 15
genome mapping, 14
German shepherds, cancer in, 50, 53
gibbons, parenting in, 231
giraffes
    capture myopathy in, 118
    obesity in, 132
Girl, Interrupted, 161

Giuliani, Rudy, 236
Gizmo (monkey), 227–8
Gluckman, Peter, 139
glutamate, 100
Goodall, Jane, 13–14
gorging
    in animals, 135–6
    and fasting pattern, 143*n*
Gorgosaurus, brain tumor of, 12–13, 39–40
gorillas, 13–14
    adolescence in, 226
    heart surgery on, 171–3
    R and R in, 190–1
    SIV in, 200
Gould, Lisa, 70
Gould, Stephen Jay, 14
Grable, Betty, 83
Grandin, Temple, 97*n*
gray horse melanoma, 37
Greaves, Mel, 41
gregarines, 153–4
grizzly bears, obesity in, 132–3, 156–8
grooming, 165–9
    as calming, 165–6
    as distraction, 172–3
    human, 166–7
    and social structure, 165
Gulf War, 26, 111
Guy, Michael, 32

hadrosaurs, tumors in, 40
hallucinogens, 91
hand washing, obsessive, 164
Harris, Stephen, 223
Harrison Narcotics Act (1914), 89*n*
Hawlena, Dror, 181
heart
    and brain, 5–6, 20–1, 26–7, 30
    electrical system of, 115
    of fish, 27–8
    functioning of, 114–15
    plumbing system of, 114
    in response to fear, 109–31
    in response to stress, 22–7
    self-injury and, 168–9
    in SIDS, 129
heart attacks, 5, 6, 109–31, 147
    following disasters, 109–12
    multispecies approach to, 116–17, 123,
        128, 129
    and spectator stress, 112–13
    types of, 114–16

heartbeat detectors, 27–8
heart disease, 31, 153*n*
    multispecies approach to, 241
heart failure, 3–4, 18
    fainting and, 20, 22–3
    fear in, 8
    in gorillas, 171–3
Heart Rhythm, 122
heart sounds, 27*n*
heart spasm, 114*n*
heavy metals, 46
Helicobacter pylori, 155
hemangiosarcoma, 50
hemolymph, 153*n*
hepatitis A, 198–9
hepatitis B and, C, 45, 196
herpes virus
    as causing cancer, 38, 45, 47
    in STDs, 195
"hesitation marks," 162*n*
heterogeneity, 105–6
Hettich, Linda, 33
hibernation, and diet, 141, 157
high heels, lordosis and, 83*n*
Hippocrates, 38–9
HIV/AIDS, 195, 196, 206, 210, 215*n*
    dementia in, 203
    immunity to, 211
    and SIV, 200–1
Hmong, 124*n*
homosexuality
    in animals, 58, 74–5
    bullying and, 230
Hong, Lawrence, 72
hormones, sex, 82
horses
    addiction in, 90
    dispersal in, 223*n*
    isolation and boredom in, 171
    self-injury in, 163–4, 170
    sexuality in, 56–7, 65–6, 72, 77–8, 80, 84, 85
    skin cancer in, 37
    STDs in, 196, 205
hospitals, addiction in personnel of, 88
houbara bustard, sexuality in, 67–8
Houpt, Katherine, 78, 84, 85*n*
Hrdy, Sarah Blaffer, 74
human beings
    animal anti-predation and self-
        protection in, 28–30
    animal nature of, 12, 13–18
    as dominant species, 10, 12, 13, 28–9

as responsible for animal disease, 47
    shared medical conditions between
        animals and, *see* multispecies
        medicine
human violent conflict hypothesis, 21*n*
hunting, capture myopathy and, 118
hydrophobia, 202
hygiene hypothesis, 208
    in genital cleansing, 209
hypercapnia, 129*n*
hypoactive sexual desire disorder (HSDD),
        79–81, 84
Hz-2V virus, 204

ICU psychosis, 126–7
IDENTIFY, 239*n*
Ignarro, Louis, 62*n*
iguanas, mating in, 72–3
immune system, in cancer therapy, 52–4
implant cardioverter defibrillators (ICDs),
        112*n*
impulse-control disorders, 162*n*
incest, 76
infant mortality, 128–31, 214
infants
    animal, 220
    fatness in, 141
    fear response in, 25–6
    perceived as impervious to pain, 97
    sleeping posture in, 128–30
    see also sudden infant death syndrome
infection
    and cancer, 45
    during mating, 73
    in obesity, 152–6
    preventive strategies against, 201–2
    various pathogen pathways in, 198–200
    see also viruses
infectious disease, 234–44
    multispecies approach to, 238–44
infectobesity, 152
infertility, STDs and, 205–6
inflatable penises, 61–3
insects
    body fat in, 141
    cancer in, 38
    feeding by, 179, 181, 182
    food storage by, 185
    STDs in, 197, 203, 204
insulin-resistance syndrome (metabolic
        syndrome), 153, 154
internal fertilization, 59

Internet, 137–8, 197
  eating disorders enabled by, 190
  self-injury enabled by, 174
intestines, 147–51, 157
  bacteria in, 148–51
intimidation, and sudden death, 121
"Intriguing Links Between Animal
    Behavior and Anorexia Nervosa"
    (Treasure and Owen), 187
irritable bowel syndrome, 155
isolation, in self-injury, 169–70, 173, 174,
  175

Jackson, Richard, 137–8, 144
jaguars, cancer in, 43, 45
Jahiel, Jessica, 57
Jim (grizzly bear), 132–3, 156–8
Jolie, Angelina, 111, 161
Journal of Veterinary Internal Medicine, 9
Judson, Olivia, 74
Jurassic cancer, *see* cancer, in dinosaurs

Kaada, Birger, 129
Kaposi's sarcoma, 45
Katz, Andrea, 70
Kelly, Diane A., 61
Kessler, David, 138
Khanna, Chand, 50–1
kidney cancer, 50
King, Brittany, 240–1
koalas, chlamydia in, 6, 194–5, 210–11
krill, reproduction in, 61

lactation, 44–6
Lady (cocker spaniel), 91
Lancelot (horse), 55–6, 67, 85
Last Prom, The, 218
Lawson, Jeffrey, 97*n*
Lawson, Jill, 97*n*
Lay, Kenneth, 122
leaving the nest, 220–2
LeDoux, Joseph, 98, 100
Leonardo da Vinci, 69*n*
Lesch-Nyhan syndrome, 161*n*
leukemia, 6, 37, 45, 211*n*
Lhasa apsos, cancer in, 53
Libella pulchella (dragonfly), 152–4
licking, in cats and dogs, 163
Liem, Karel, 29
light
  in circadian rhythm, 146–7
  in feeding strategies, 182

Lima, Steven, 125
Limax redii, 60
Lindblad-Toh, Kerstin, 50
Linden, David L., 105
lions
  carcass feeding of, 141–3
  gorge-and-fast in, 143*n*
liver cancer, 45
livestock
  antibiotics in, 151
  STDs in, 197–8
lobsters, capture myopathy in, 116
locoweed, 90–1
longitudinal medical studies, 31–2
long QT syndrome, 121, 130
lordosis, 82–4
Los Angeles Zoo, 3, 10, 57, 241
Lost Boys (monkeys), 227–8
Lucas, William, 228
lung cancer, 48
lymphoma, 37, 38, 45, 47, 49, 51
lymphoma belt, 45

macaques
  grooming in, 165
  mating in, 81
mad cow disease, 243
Mahr, Roger, 8
Man and Woman: An Inside Story
    (Pfaff), 83
Manipulative Monkeys (S. Perry), 227
Marden, James, 152–56
mare hotel, 77
marmots
  dietary mismatch in, 140
  obesity in, 136–7
Marshall, Barry, 155
mass murderers, 244*n*
Masson, Jeffrey, 97*n*
masturbation, animal, 72–3, 74, 207
mating face, 78
Mazet, Jonna, 240
McNamara, Tracey, 234–8
medial forebrain pleasure circuit, 105
melamine, 243*n*–4*n*
melanoma, 6, 35, 51–4
Memorial Sloan-Kettering hospital, 52–4
menstruation, 44, 186
Merial drug company, 53
metabolic syndrome (insulin-resistance),
    153, 154
metastasis, 35

mice
    BMI study of, 146
    in cancer research, 48, 54*n*
    feeding studies on, 182
    obesity in, 149–50, 154
    toxo in, 202–3
microbiome, 149–51, 155, 158
Microcosm (Zimmer), 149*n*
migraines, 114*n*
migration, and diet, 141
military basic training, rhabdo in, 119
miscarriage, 206
mismatch, dietary, 139
mobbing, 218
moles, food storage by, 185
monkeypox, 243
monkeys
    adolescence in, 216, 219–21, 227–8, 233
    intoxication in, 92
    peer group threats in, 227
Monroe, Marilyn, 83
moose, STDs in, 198
Morecki, Nina, 24
Morrill Land-Grant Acts, 8
Morris, Desmond, 13
Morris Animal Foundation, 32
mosquitoes, as vector for West Nile virus,
    235–6, 238
Moss Landing, 222–3, 226
moths, STDs in, 204
motor vehicle deaths, 215, 218
Mulcahy, Dan, 116–17, 120
multispecies medicine
    in addictions, 89–90, 92, 93–4, 95, 103
    in adolescence, 216, 221–2, 227–33
    in cancer, 33, 36–9, 41, 42–6, 49–54
    for children, 241
    conference on, 241
    in eating disorders, 178, 180, 184, 186–9,
        193
    in heart ailments, 116–17, 123, 128, 129,
        241
    in infectious disease, 238–44
    and nervous system, 124–5
    in obesity, 133, 137, 144, 148, 151, 156–8
    in self-injury, 160, 162–4, 173, 175
    in sexuality, 57–8, 63–4, 68, 70–2, 75–6,
        79, 81, 83–6
    and STDs, 196, 203, 210–11
Munson, Linda, 43*n*
Murad, Ferid, 62*n*
musth, 223*n*

mutations
    in cells, 34–5, 40–1
    of pathogens, 200–1
myocardial infarction, 114
myopathy, 4

Naked Ape, The (Morris), 13
National Cancer Institute, 49
    Comparative Oncology Program of,
        43, 51
National Center for Emerging and
        Zoonotic Infectious Diseases, 239
National Institutes of Health (NIH), 35, 45
natural disasters, heart attacks in, 110–11
natural selection, *see* evolution
Nature, 4, 15, 150, 151, 230
nature/nurture controversy, 14–16, 106
Nature Reviews Cancer, 51
nausea, 22–3
Nazi concentration camps, 24
"near fainting while conscious" syndrome,
    23–4, 28
necrophilia, 76
necropsy, 18, 42
Neisseria gonorrhoeae, 195*n*
nervous system, multispecies approach to,
    124–5
Nesse, Randolph, 98, 99, 100, 125
neurochemicals, 106–7
neuroendocrine cancer, 38
Newcastle disease, 235
New England Journal of Medicine, 110
night eating, 177, 178
Nightmare on Elm Street, 124*n*
nightmares, 124*n*
nitric oxide, 62
Nobel Prize in Medicine, 62*n*, 155
nocebo effect, 124
noise, in sudden death, 120–1
Nonacs, Peter, 140
nonreceptivity, sexual, 80–1, 83–4
normal sinus rhythm, 115
Northridge earthquake, 109–11
nose and sinus cancers, 47
nucleotides, 34
nymphomania, 84

obesity, 7, 35, 47, 132–58
    antibiotics in, 151
    as a disease of the environment, 137–8,
        141, 144, 156
    epidemic, 133–5

fear of, 177
geographic effect on, 147
infectious, 152–8
intestinal bacteria in, 148–51
multispecies approach to, 133, 137, 144,
    148, 151, 156–8
social factors in, 151–2
warnings about, 137–8
in the wild, 125–6, 141
see also diet
obsessive-compulsive disorder (OCD), 11,
    164, 186
Occupational and Environmental Medicine,
    120
octopamine, 99
octopuses, environmental enrichment for,
    143–4
okapi, capture myopathy in, 120
Olson, James S., 39
Oncept, 53
One Health clinic, 240–1
One Health movement, 8–9
One Medicine, 8*n*
oophorectomies, 48
opioids
    in animals, 100
    in brain, 99, 101
    receptors, 99–100
opium, 88, 90
oral sex, animal, 74
orgasm
    animal, 85–6
    female, 78–9
    male, 63–4
Origin of Species, The (Darwin), 13
oscars (*Astronotus ocellatus*), 27
oscillators, 145–6, 151
Osler, William, 7*n*
osteosarcoma, 36, 38, 41–2, 49
ovarian cysts, in nymphomania, 84
overeating, 7
overgrooming, 164–5, 168, 175
ovulation, 78, 80, 83
Owen, John, 187
Oxford Handbook of Eating Disorders, The
    (Stewart), 177
oxytocin, 99, 100

p38 MAP kinase, 154
Pacific kelp beds, 46–7
pacing behavior, 143
pain

in addiction, 96
in animals, 96–7
brain and, 96–7
consciousness perceived as prerequisite
    to, 96–7
emotional, 96–7
in infants, 97
physical, 96–7
pleasure and, 168–9, 175
in self-injury, 168–9
Panda's Thumb, The (Gould), 14
Panksepp, Jaak, 94–6, 100, 103, 106
Paoloni, Melissa, 50–1
papilloma virus, as cancer cause, 38, 45,
    196
Pappaioanou, Marguerite, 240
parasites, 73, 201, 203
    in obesity, 153–4
parasympathetic nervous system, 23, 76
parenting, of adolescents, 215, 231–3
parrots, 11
pathogens
    sexually beneficial, 208–9
    survival techniques of, 202–6
    various pathways of, 198–200
Peacock, Molly, 85
pedophilia, 76
peer groups, 214, 217–18, 225–8
penalty kick shoot-outs, 113
penguins, gorging in, 136
penises
    animal variations of, 58–61
    human, 61–5
peregrine falcons, reintroduction
    of, 224
perfectionism, 184
perineal elaboration, 67
Perry, Katy, 83
Perry, Susan, 227
pesticides, 47
Peto, Richard, 41
Peto's paradox, 41–2, 45, 48
pets
    fainting in, 21–2
    obesity in, 134–5
    as therapy, 174
    see also specific species
Pfaff, Donald, 82–3
Pfaus, James, 81
phallology, 60*n*–1*n*
phantom mare, 56
pheromones, 66

physical restraint
    animal self-injury in, 163
    fainting and, 21–2
    of humans, 120–31
    sudden death and, 120–3
physician arrogance, 9
Pierce, Franklin, 30
pigs
    anorexia in, 186–8
    intoxication in, 91–2
    STDs in, 196
pikas, sudden death in, 120
placebo effect, 124
plague, 243
Planet of Viruses, A (Zimmer), 149*n*
Plankton Research, 61
plants
    cancer in, 38
    fat in, 141
    STDs in, 197, 204–5
    threat responses of, 124–5
plaque buildup, 116
plasmids, 53*n*
Plato, 13
play
    in animals, 95
    fighting, 230
pleasure
    as reward, 98
    sexual, 77, 85–6
polar bears, obesity in, 132
polychlorinated biphenyls (PCBs), 46
polycyclic aromatic hydrocarbons
    (PAHs), 46
polycystic ovary syndrome (PCOS), 84–5
poppies, 88, 90
pornography, 73–4, 103
predation, predators, 24
    defenses against, 26, 218–19
    eating strategies and, 178–80, 183–4,
        193
    food storage in, 185
    humans as, 180–1
    human self-protection as related to,
        29–30
    recognizing of, 216–17
    signals of unprofitability in, 29
predator inspection, 217
PREDICT, 239*n*, 240
premature ejaculation (PE), 72–3
Presley, Elvis, 65–6
PREVENT, 239*n*

*Proceedings of the National Academy of
    Sciences,* 146, 155
professional lactators, 44, 51
promiscuity
    in animals, 58, 75–6
    and STDs, 204
    see also sex addiction
Promiscuity (Birkhead), 76
prostate cancer, 36
protein
    avoidance of, 181*n*
    drive for, 140
pseudopenis, 79
psychiatric disorders, in animals, 11–12
psychodynamic psychotherapy, 102
psychogenic alopecia, 163
psychogenic erection, 64–5
psychological autopsies, 229
psychological castration, 57, 71
psychotherapy, for self-injury, 174
pudendal nerves, 62
pufferfish, 62–3
"Punishment in Animal Societies"
    (Clutton-Brock), 230
pythons, intestines of, 148
Pyxis MedStation 3500, 87–8, 99, 100, 101

Q fever, 243

rabbits
    feeding by, 179
    STDs in, 197, 198
rabies, 202, 203
Rajfer, Jacob, 71, 73
rape
    in animals, 58, 76, 81
    prevention strategies against, 24–5
    proof of, 30
rats
    addiction research on, 92–3
    adolescent risk taking in, 216, 221
    mating in, 72*n*
    obesity in, 135
    tickling research on, 95
    toxo in, 202–3
Raynaud's syndrome, 114*n*
red (color), 207
Red Asphalt, 218
Red Queen, The, 209
reflexogenic erection, 64–5, 82
regurgitation, as defensive strategy, 24,
    192

regurgitation and reingestion (R and R), 190–2
release-relief loop, 167–9, 173, 174
religion, 15
reptiles
    cancer in, 37
    grooming in, 168
    self-injury in, 163
RESPOND, 240*n*
restlessness, in anorexia, 187
retrievers, cancer in, 33, 36, 37, 50
reverse zoonosis, 242*n*
rewards
    for behaviors, 104–5
    drugs as fast track to, 100–1
    from peers, 225
rhabdomyolysis (rhabdo), 119
rhesus monkeys, in heart and self-injury study, 168–9
rhinoceroses, cancer in, 37
Ricci, Christina, 161
Ridley, Matt, 209
ring-tailed lemurs, mating by, 70
risk
    in adolescence, 213–17, 221–2, 227–8, 232
    from peers, 227–8
    in sex addiction, 102
Roach, Mary, 74
robberflies, 25*n*
roller coasters, 218
Rosenberg, Charles, 36
Roser, Janet, 80
Roughgarden, Joan, 75
rumination disorder, 192
ruptured muscle syndrome, 118*n*
Rwandan genocide, 24

Sack, David, 101
safe sex, 195, 206, 209
St. Lawrence Estuary, 45
St. Louis encephalitis (SLE), 236–8
salmon, adolescence in, 225–6, 233
salukis, tumors in, 50
Sam (koala), 194–5, 210–11
Samuels, Martin A., 123
SARS, 46, 243
sauropods, 59–60
scent, in sexuality, 65–7
Schilder, Rudolf, 154
schizophrenia, 161*n*, 203
Scottish terriers, cancer in, 50

sea lions
    cancer in, 46
    grooming in, 166
seals
    gorging in, 136
    grooming in, 166
sea otters
    adolescent risk taking in, 213, 216, 219, 233
    bachelor groups in, 222–3, 226
"second hits," 43
Secretary, 161
secret eating, 177, 178
selective serotonin reuptake inhibitors (SSRIs), 80
self-fulfilling prophecy, fear in, 124
self-injury, 11, 159–75
    dangers of, 162, 169
    distraction as therapy for, 173
    multispecies approach to, 160, 162–4, 173, 175
    pain in, 168–9
    release and relief in, 167–9, 173, 174
    social preening rituals and, 175
    technology as enabling, 174
    triggers for, 169–70, 173, 174, 175
Selfish Gene, The (Dawkins), 15
self-soothing behavior, 167
semen, 56, 63, 209–10
sensual stimulation, 171*n*
September 11, 2001 terrorist attacks, 112, 239
severe acute respiratory syndrome (SARS), 46, 243
sex addiction, 102, 104
sexual dysfunction
    ED, 69–72
    failed copulation, 70–2
    in horses, 56–7
    human, 57
    survival advantages of, 71–2
    in women, 77–81, 84
sexuality, 55–86
    aural stimulation in, 68
    competition in, 70–3
    and evolution, 58–61
    female, 78–86
    human, 58, 61–5, 76–8
    immorality and illegality in, 76–7
    interspecies, 76–7
    mating receptivity in, 67, 78, 80–4
    multiple animal forms of, 57–8, 74–5

sexuality *(continued)*
    multispecies approach to, 57–8, 63–4, 68, 70–2, 75–6, 79, 81, 83–6
    nonreceptivity in, 80–1, 83–4
    pleasure in, 77, 85–6
    reproductive advantages of, 67–8
    scent in, 65–7
    as separate from reproduction, 77
    sex drive in, 80
    survival advantage of speedy mating in, 72–3
    taboos in, 76–7
    visual stimulation in, 66–8
sexually transmitted diseases (STDs), 6, 194–211
    in animals, 58, 194–8
    multispecies approach to, 196, 203, 210–11
    nonsexual spread of, 197–9
    and sexual attractiveness, 203–5
    stigma against, 195–6, 210
    survival techniques of pathogens in, 202–3
    see also specific diseases
sexual reproduction, 58–61
Sexual Selections (Zuk), 75
Shakespeare (shepherd mix), 241
sharks, 27–8
    attacks by, 213–14
    and cancer, 44*n*
Shayk, Irina, 83
shock paddles, 115
shopping addiction, 103, 105
shrews, alcohol consumption by, 90
Shubin, Neil, 15
Siberian huskies, cancer in, 53
siblings, bullying by, 230
signals of unprofitability, 29
simian immunodeficiency virus (SIV), 200–1
skin cancer, 37–8
Skinner, B. F., 96
slaughter, stress reduction techniques in, 118*n*
sloths, parenting in, 231–2
smallpox, 243*n*
smoking, and heart disease, 31
soccer World Cup (1998), 112–13
social grouping, 171*n*
social phobia, 184
Sociobiology (Wilson), 13–14
souls, 96

Spear, Linda, 221
spectator stress, 112–13
spermatophore, 59*n*
sperm competition, 70–3, 76, 208–9
spiders, erection-causing venom of, 64*n*
Spitzbuben (tamarin), 3–4, 18
sporting events, and heart attacks, 112–13
Sports Illustrated swimsuit edition, 83
Springer spaniels, cancer in, 43
stage fright, 22–3
Stamps, Judy, 217
starvation, and food hoarding, 184–6
Steinberg, Laurence, 228
stereotypies, 11, 164, 167*n*, 170–1, 174
stillness, as survival mechanism, 25–7, 29
stotting, 29
Strachan, David, 208
Streptococcus A, 198
stress
    in eating disorders, 182, 188
    heart and, 23, 114
    in R and R, 192
    in self-injury, 169–70, 173, 174, 175
    and sexual dysfunction, 71
    spectator, 112–13
    in weaning, 188
Strober, Michael, 192*n*
stroke, 153*n*
structure and substrate, 171*n*
submarines, 28
sudden cardiac death (SCD; sudden death), 111, 115–23
sudden infant death syndrome (SIDS), 128–31
sudden unexpected nocturnal death syndrome (SUNDS), 124*n*
Sue (*T. rex*), 199
sugar
    binging on, 181–2
    drive for, 140
suicide
    in animals, 11–112
    bullying and, 228–9, 232
    and cutters, 161–2
suicide codes, 35, 42
superorganisms, 149
suprachiasmatic nucleus (SCN), 146
survival advantages
    of fainting, 23–7
    in reproduction, 67
swaddling, 130–1
swallowers, 160

Swedes, HIV immunity in, 211
swine flu (H1N1), 242
sympathetic nervous system, 23, 71, 78
syncope, *see* fainting
syphilis, 200
    dementia in, 203

Tabin, Cliff, 15
tachycardia, 22, 169
TAG teaching (teaching with acoustical
        guidance), 104*n*
takotsubo cardiomyopathy, 5, 17, 114, 116
tamarins
    gorging in, 135
    heart disease in, 3–4, 18
Tasha (boxer), 32
Tasmania, wallaby opium addiction in, 88
Tasmanian devils, cancer in, 38
tattooing, 38, 175
T cells, 53*n*
technology:
    addiction to, 103, 104
    see also Internet
Tel Aviv
    maternity ward in, 26
    Scud missile attacks on, 111
Tessa (Labrador retriever), 33, 36, 37
testicular cancer, 47
testosterone, 79, 80, 84
    in adolescence, 223
thanatosis, 25*n*
thin sow syndrome, 186–9
Thirteen, 161
Thornhill, Randy, 81
thought leaders, 190
Thrall, Peter, 204
Tiboy (alligator), 145, 146
Timms, Peter, 210
titanosaurs, 59
toad licking, 91
Tolstoy, Leo, 69*n*
torture, 25
Tourette's syndrome, 161*n*, 162*n*, 164
toxins, cancer caused by, 35–6, 40, 45–7
Toxoplasmosis gondii (toxo), 202–3, 212
toys, 171
translational, 54
Treasure, Janet, 187
trehalose, 153*n*
Triangle of Death, 212–13, 216, 222, 233
Trichomonas vaginalis (trich; *T. vag*),
        199–200

trichotillomania, 160
Trollinger, Melissa, 244*n*
tularemia, 243
tumors, 35
type 2 diabetes, 153, 154
tyrosinase, 53*n*, 54*n*

unsafe sex, 7
urination, as survival mechanism, 24
urine
    rust-colored, 119
    in sexuality, 65–7, 84
urophiliacs, 67*n*
U.S. Agency for International
        Development (USAID), 239–40
uterine cancer, 37

vaginismus, 79
vasovagal syncope (VVS), 20–1, 23, 24,
        25*n*, 28
ventrical tachycardia (VT), 115–16, 126–7
ventricular fibrillation (VF), 115–16
vesicles, 98
veterinarians, veterinary medicine
    cooperation between physicians and, 3,
        7, 8–9, 11, 12, 16, 42, 238–44
    fusion of human medicine and, *see*
        multispecies medicine; zoobiquity
    GAO recognition of, 239
    segregation of human medicine from,
        7–9, 51, 54, 237–8
Viagra, 62*n*
Viloria, Joel, 55–6
violence
    in adolescence, 215
    sexual, *see* rape
viral hemorrhagic fevers, 243
Virchow, Rudolph, 7
Virginia Tech shootings, 24
viruses
    as cancer causing, 45
    hunters, 240
    in obesity, 152–6
    rabies, 202
    volatility of, 242
    West Nile, 8, 234–9, 242
Volvo, 28*n*
vomeronasal organs, 65–6
vomiting, *see* regurgitation
von Mutius, Erika, 208
voodoo curses, 123–4
vulval "winking," 205

wallabies, opium addiction in, 88, 90
Walsh, Seth, 228
warmth, in eating disorder therapy, 189–90
Warren, J. Robin, 155
warthogs, adolescence in, 215
wasting pig syndrome, 188–9
water buffalo, drug-seeking by, 90
Watson, J. B., 96
Watts, Jennifer, 132–3, 145, 150, 156–8
websites, anorexia-nervosa-promoting
    (pro-ana), 190
West Nile virus, 8, 239, 242
    humans affected by, 235–7
    misdiagnosis, of, 234–8
whales
    beluga deaths, 45–7
    cancer in, 38, 41, 42, 43, 45
    R and R in, 191
    weight fluctuation in, 136
Williams, Richard, 113
Wilson, Edward O., 14, 15, 98
Wilson, Gary, 104
Winter (dolphin), 241
Withrow, Stephen, 49
Wolchok, Jedd, 52–4

wolves, reintroduction of, 183–4
"Wondrous Bestiary," 75
worm farms, 185

xenogenic plasmid DNA vaccination,
    52*n*

yaws, 200
Yellowstone National Park, 178–9, 183
yogurt, 192*n*

zebras, capture myopathy in, 121
zeitgebers, 146
Zenyatta (race horse), 106
Zimmer, Carl, 149*n*
Zipes, Doug P., 122
zoobiquity, 238–44
    definition of term, 16–17
    potential sociological application of,
        17–18
    see also multispecies medicine
zoonoses, 8, 239, 241–3
Zoonotic, Vector-Borne, and Enteric
    Diseases department, 239
Zuk, Marlene, 74, 75, 76, 77

A NOTE ABOUT THE AUTHORS

Barbara Natterson-Horowitz, M.D., earned her degrees at Harvard and the University of California, San Francisco. She is a cardiology professor at the David Geffen School of Medicine at UCLA and serves on the medical advisory board of the Los Angeles Zoo as a cardiovascular consultant. Her writing has appeared in many scientific and medical publications.

Kathryn Bowers was a staff editor at *The Atlantic* and a writer and producer at CNN International. She has edited and written popular and academic books and teaches a course at UCLA on medical narrative.

A NOTE ON THE TYPE

This book was set in Minion, a typeface produced by the Adobe Corporation specifically for the Macintosh personal computer, and released in 1990. Designed by Robert Slimbach, Minion combines the classic characteristics of old-style faces with the full complement of weights required for modern typesetting.

Typeset by North Market Street Graphics
Lancaster, Pennsylvania

Printed and bound by Berryville Graphics
Berryville, Virginia

Designed by Maggie Hinders